De Gruyter Studies in Mathematics 60

Vladimir A. Mikhailets
Aleksandr A. Murach

Hörmander Spaces, Interpolation, and Elliptic Problems

De Gruyter

Mathematics Subject Classification 2000: 46E35, 46B70, 35J30, 35J40, 35J45

Authors
Prof. Dr. Vladimir Andreevich Mikhailets
National Academy of Sciences of Ukraine
Institute of Mathematics
Tereshchenkovskaya st.,3
KIEV-4
01601 UKRAINE
mikhailets@imath.kiev.ua

Prof. Dr. Aleksandr Aleksandrovich Murach
National Academy of Sciences of Ukraine
Institute of Mathematics
Tereshchenkovskaya st.,3
KIEV-4
01601 UKRAINE
murach@imath.kiev.ua

Translated by Peter V. Malyshev

ISBN 978-3-11-029685-3
e-ISBN 978-3-11-029689-1
Set-ISBN 978-3-11-029690-7
ISSN 0179-0986

Library of Congress Cataloging-in-Publication Data
A CIP catalog record for this book has been applied for at the Library of Congress.

Bibliographic information published by the Deutsche Nationalbibliothek
The Deutsche Nationalbibliothek lists this publication in the Deutsche Nationalbibliografie; detailed bibliographic data are available in the Internet at http://dnb.dnb.de.

Printing and binding: CPI buch bücher.de GmbH, Birkach
♾ Printed on acid-free paper
Printed in Germany

www.degruyter.com

Preface

The fundamental applications of the Sobolev spaces $W_2^m(G)$ to the investigation of many-dimensional differential equations, in particular of the elliptic type, are well known. Without the theory of spaces of this kind, the investigation of elliptic problems is, in fact, impossible. At the same time, the theory of Hörmander spaces more general than the Sobolev spaces was developed about 40 years ago. At present, there are numerous papers devoted to the applications of Hörmander spaces to differential equations.

However, the applications of Hörmander spaces to boundary-value problems for elliptic equations have been episodic up to now. The main part of the book is devoted to a fairly systematic investigation of the applications of Hörmander spaces to this class of problems. The authors introduce and study Hörmander spaces of the "intermediate" type. The functions from these spaces are characterized by the degree of smoothness intermediate between the smoothness of functions from the spaces $W_2^m(G)$ and $W_2^{m+1}(G)$, where m is an integer. As G, we can take a domain of n-dimensional Euclidean space or a compact manifold of dimension n. The first two chapters of the book are devoted to the detailed introduction and study of these spaces.

In Chapters 3 and 4, the authors consider elliptic equations and homogeneous and inhomogeneous boundary-value problems for these equations. Numerous significant results (similar to the results known for the Sobolev spaces) are obtained for these problems in Hörmander spaces. It is possible to say that the authors managed to transfer the classical "Sobolev" theory of boundary-value problems to the case of Hörmander spaces. It should also be emphasized that some problems posed independently of the notion of Hörmander spaces can be solved with the help of these spaces.

The last fifth chapter of the book is devoted to the transfer of the obtained results to the case of elliptic systems of differential equations.

I think that the book is fairly interesting and useful. It should definitely be translated into English. In this case, the results accumulated there would become accessible for a broader circle of mathematicians. In the case of translation, it would be necessary to include the proofs of various auxiliary facts mentioned in the text, which belong to the other authors. This would significantly increase the circle of possible readers of the book.

Yu. M. Berezansky,
Academician of the Ukrainian National Academy of Sciences

Preface to the English edition

The English translation of the monograph slightly differs from the Russian-language edition.

Thus, in particular, we extended the list of references, corrected the detected misprints, and improved the presentation of some results. In addition, the book is equipped with the index.

V. A. Mikhailets and A. A. Murach

Acknowledgements

The authors are especially grateful to Yu. M. Berezansky for his valuable advice and great influence, which determined, to a significant extent, their scientific interests.

We are also thankful to M. S. Agranovich, B. P. Paneyah, I. V. Skrypnik, and S. D. Eidel'man for stimulating discussions.

The support of M. L. Gorbachuk and A. M. Samoilenko, interest of B. Boyarskii, and kind participation of V. P. Burskii and S. D. Ivasyshen are also highly appreciated.

We also thank all our colleagues for their sincere interest to the new theory and its applications.

Contents

Introduction

In the theory of partial differential equations, the problems of existence, uniqueness, and regularity of solutions are in the focus of investigations. As a rule, the regularity properties of solutions are formulated in terms of belonging of these solutions to the standard classes of function spaces. Moreover, the finer the calibration of the scale of spaces, the more exact and informative the accumulated results.

Unlike the case of ordinary differential equations with smooth coefficients, these problems are fairly complicated. Indeed, some linear partial differential equations with smooth coefficients and right-hand sides are known to have no solutions in the neighborhood of a given point even in the class of distributions [113], [81, Sec. 6.0]. Moreover, some homogeneous equations (specifically, of the elliptic type) with smooth but not analytic coefficients admit nontrivial solutions with compact supports [193], [85, Theorem 13.6.5]. Therefore, the nontrivial null space of this equation cannot be removed by any homogeneous boundary conditions; i.e., the operator corresponding to any boundary-value problem for the analyzed equation is not injective. Finally, the problem of regularity of solutions is also quite complicated. Thus, even for the Laplace operator, it is known that

$$\triangle u = f \in C(\Omega) \not\Rightarrow u \in C^2(\Omega)$$

for any Euclidean domain Ω [64, Chap. 4, Notes].

These problems have been most completely investigated for linear elliptic equations, systems, and boundary-value problems. The fundamental results in this direction were obtained in the 1950s and 1960s by S. Agmon, A. Douglis, and L. Nirenberg [4, 5, 47], M. S. Agranovich and A. S. Dynin [12], Yu. M. Berezansky, S. G. Krein, and Ya. A. Roitberg [22, 21, 202, 203, 209], F. E. Browder [28, 29], L. R. Volevich [267, 268], J.-L. Lions and E. Magenes [121, 126], L. N. Slobodetskii [240, 241], V. A. Solonnikov [245, 246, 247], L. Hörmander [81], M. Schechter [222, 224, 225], and other researchers. In the cited works, the elliptic equations and problems were studied in the classical scales of Hölder spaces (of nonintegral order) and Sobolev spaces (both of positive and negative orders).

As a fundamental result in the theory of elliptic equations, we can mention the fact that they generate bounded Fredholm operators (i.e., operators with finite index) acting between appropriate function spaces. Thus, let $Au = f$

be a linear elliptic differential equation of order m given on a closed smooth manifold Γ. Then

$$A: H^{s+m}(\Gamma) \to H^{s}(\Gamma), \quad \text{with} \quad s \in \mathbb{R},$$

is a bounded Fredholm operator. Moreover, the finite-dimensional spaces formed by the solutions of the homogeneous equations $Au = 0$ and $A^{+}v = 0$ lie in $C^{\infty}(\Gamma)$. Here, A^{+} is the operator formally adjoint to A, whereas $H^{s+m}(\Gamma)$ and $H^{s}(\Gamma)$ are the Sobolev inner product spaces over Γ of the orders $s+m$ and s, respectively. This result implies that each solution u of the elliptic equation $Au = f$ has an important regularity property in the Sobolev scale, namely,

$$(f \in H^{s}(\Gamma) \text{ for some } s \in \mathbb{R}) \Rightarrow u \in H^{s+m}(\Gamma). \tag{1}$$

If a manifold has an edge, then the Fredholm operator is generated by an elliptic boundary-value problem for the equation $Au = f$ (e.g., by the Dirichlet boundary-value problem).

Some of these theorems were extended by H. Triebel [258, 256] and Murach [163, 164] to the scales of Nikol'skii–Besov, Zygmund, and Lizorkin–Triebel function spaces.

The cited results were applied, in various ways, to the theory of differential equations, mathematical physics, and spectral theory of differential operators (see the books by Yu. M. Berezansky [21], Yu. M. Berezansky, G. F. Us, and Z. G. Sheftel [23], O. A. Ladyzhenskaya and N. N. Ural'tseva [111], J.-L. Lions [118, 117], J.-L. Lions and E. Magenes [121], Ya. A. Roitberg [209, 210], I. V. Skrypnik [237], H. Triebel [258], the surveys by M. S. Agranovich [7, 10, 11], and references therein).

From the viewpoint of applications, especially to the spectral theory, the case of Hilbert spaces is of especial importance. Note that, until recently, the scale of Sobolev inner product spaces was the sole scale of Hilbert spaces in which the properties of elliptic operators were systematically studied. However, it was shown that the Sobolev scale is insufficiently fine for various important problems.

We present two typical examples. The first of them deals with the smoothness of the solution of the elliptic equation $Au = f$ on the manifold Γ. According to the Sobolev embedding theorem, we have

$$H^{\sigma}(\Gamma) \subset C^{r}(\Gamma) \quad \Leftrightarrow \quad \sigma > r + n/2, \tag{2}$$

where $r \geq 0$ is an integer and $n := \dim \Gamma$. This fact, together with property (1), allow us to study the classical smoothness of the solution u. Thus, if $f \in H^{s}(\Gamma)$ for some $s > r - m + n/2$, then $u \in H^{s+m}(\Gamma) \subset C^{r}(\Gamma)$. However, this embedding is not true for $s = r - m + n/2$; i.e., the Sobolev scale cannot be used to formulate unimprovable sufficient conditions for the inclusion $u \in C^{r}(\Gamma)$.

A similar situation is also encountered in the theory of elliptic boundary-value problems.

The second typical example corresponds to the spectral theory. We assume that a differential operator A of order $m > 0$ is elliptic on Γ and self-adjoint in the space $L_2(\Gamma)$. Consider the expansion of a function $f \in L_2(\Gamma)$ in the series

$$f = \sum_{j=1}^{\infty} c_j(f)\, h_j, \tag{3}$$

where $(h_j)_{j=1}^{\infty}$ is the complete orthonormal system of eigenfunctions of A and $c_j(f)$ are the Fourier coefficients of the function f in its expansion in h_j. The eigenfunctions are enumerated so that the moduli of the corresponding eigenvalues form a (nonstrictly) increasing sequence. By the Menchoff–Rademacher theorem (valid for general orthogonal series), expansion (3) converges almost everywhere on Γ provided that

$$\sum_{j=1}^{\infty} |c_j(f)|^2 \log^2(j+1) < \infty. \tag{4}$$

This hypotheses cannot be reformulated in equivalent way in terms of the fact that the function f belongs to Sobolev spaces because

$$\|f\|^2_{H^s(\Gamma)} \asymp \sum_{j=1}^{\infty} |c_j(f)|^2\, j^{2s}$$

for any $s > 0$. We can only state that the condition "the fact that $f \in H^s(\Gamma)$ for some $s > 0$" implies the convergence of series (3) almost everywhere on Γ. This condition does not adequately express the hypotheses (4) of the Menchoff–Rademacher theorem.

In 1963, L. Hörmander [81, Sec. 2.2] proposed a significant and useful generalization of Sobolev spaces in the category of Hilbert spaces (see also [85, Sec. 10.1]). He introduced spaces parametrized by a sufficiently general weight function playing the role of an analog of the differentiation order (or smoothness index) used for the Sobolev spaces. In particular, L. Hörmander considered the Hilbert spaces

$$B_{2,\mu}(\mathbb{R}^n) := \{u : \mu\,\mathcal{F}u \in L_2(\mathbb{R}^n)\}, \tag{5}$$

$$\|u\|_{B_{2,\mu}(\mathbb{R}^n)} := \|\mu\,\mathcal{F}u\|_{L_2(\mathbb{R}^n)},$$

where $\mathcal{F}u$ is the Fourier transform of a tempered distribution u given in $\mathbb{R}^n$, and μ is a weight function of n scalar arguments.

In the case where

$$\mu(\xi) = \langle\xi\rangle^s, \quad \langle\xi\rangle := (1 + |\xi|^2)^{1/2}, \quad \xi \in \mathbb{R}^n, \quad s \in \mathbb{R},$$

we get the Sobolev space $B_{2,\mu}(\mathbb{R}^n) = H^s(\mathbb{R}^n)$ of the (differentiation) order s.

In 1965, spaces (5) were independently introduced and studied by L. R. Volevich and B. P. Paneah [269].

The Hörmander spaces occupy an especially important place among the spaces of generalized smoothness characterized by a function parameter instead of a number. These spaces serve as an object of numerous profound investigations, and a good deal of work was performed in the last decades. We refer the reader to the survey by G. A. Kalyabin and P. I. Lizorkin [90], monograph by H. Triebel [259, Sec. 22], and recent papers by A. M. Caetano and H.-G. Leopold [32], W. Farkas, N. Jacob, and R. L. Schilling [55], W. Farkas and H.-G. Leopold [56], P. Gurka, and B. Opic [70], D. D. Haroske and S. D. Moura [74, 75], S. D. Moura [162], B. Opic, and W. Trebels [177], and the references therein. Various classes of spaces of generalized smoothness naturally appear in the embedding theorems for function spaces, in the interpolation theory of function spaces, in the approximation theory, in the theory of differential and pseudodifferential operators, and in the theory of stochastic processes; see the monographs by D. D. Haroske [73], N. Jacob [87], V. G. Maz'ya and T. O. Shaposhnikova [132, Sec. 16], F. Nicola and L. Rodino [175], B. P. Paneah [181], and A. I. Stepanets [248, Chap. 1, § 7], [249, Part I, Chap. 3, Sec. 7.1], the papers by F. Cobos and D. L. Fernandez [35], C. Merucci [136], and M. Schechter [226] devoted to the interpolation of function spaces, and also the papers by D. E. Edmunds and H. Triebel [50, 51] and V. A. Mikhailets and V. Molyboga [140, 141, 142].

As early as in 1963, L. Hörmander [81] used the spaces (5) and more general Banach spaces $B_{p,\mu}(\mathbb{R}^n)$ with $1 \le p \le \infty$ to study the regularity properties of the solutions of partial differential equations with constant coefficients and solutions of some classes of equations with variable coefficients given in Euclidean domains. However, unlike Sobolev spaces, the Hörmander spaces were not widely applied to the general elliptic equations on manifolds and elliptic boundary-value problems. This is explained by the long-term absence of a proper definition of Hörmander spaces on smooth manifolds (this definition should be independent of the choice of local charts covering the manifold) and the absence of analytic tools required for the effective investigation of these spaces.

For the Sobolev spaces, the required tool is available: this is the interpolation of spaces. Thus, every Sobolev space of fractional order can be obtained as a result of the interpolation of a certain pair of Sobolev spaces of integer order. This fact significantly facilitates both the investigation of these spaces and the proofs of various theorems from the theory of elliptic equations because the procedure of interpolation preserves boundedness and the Fredholm property (if the defect is invariant) of linear operators.

Therefore, it seems reasonable to select Hörmander inner product spaces obtained as a result of interpolation (in this case, with a function parameter) between Sobolev inner product spaces. To this end, we introduce a class of isotropic spaces

$$H^{s,\varphi}(\mathbb{R}^n) := B_{2,\mu}(\mathbb{R}^n) \quad \text{for} \quad \mu(\xi) = \langle\xi\rangle^s \varphi(\langle\xi\rangle). \tag{6}$$

Here, $s \in \mathbb{R}$ is a numerical parameter and φ is a positive function parameter slowly varying at infinity in Karamata's sense [26, 235]. (It is possible to assume that φ is constant outside the neighborhood of infinity). Thus, the logarithmic function, its iterations, all their powers, and the products of these functions may play the role of φ.

The class of spaces (6) contains the Hilbert scale of Sobolev spaces $\{H^s\} \equiv \{H^{s,1}\}$ and is attached to it by a number parameter. However, it is calibrated finer than the Sobolev scale. Indeed,

$$H^{s+\varepsilon}(\mathbb{R}^n) \subset H^{s,\varphi}(\mathbb{R}^n) \subset H^{s-\varepsilon}(\mathbb{R}^n) \quad \text{for any} \quad \varepsilon > 0.$$

Therefore, the number parameter s specifies the main (power) smoothness, while the function parameter φ is responsible for an additional (subpower) smoothness in the class of spaces (6). Thus, in particular, if $\varphi(t) \to \infty$ (or $\varphi(t) \to 0$) as $t \to \infty$, then φ specifies an additional positive (or negative) smoothness. In other words, the parameter φ refines the main smoothness s. Hence, it is natural to say that the class of spaces (6) is a refined Sobolev scale or simply the refined scale.

This scale has an important interpolation property, namely, every space $H^{s,\varphi}(\mathbb{R}^n)$ is obtained as a result of the interpolation of a pair of Sobolev spaces $H^{s-\varepsilon}(\mathbb{R}^n)$ and $H^{s+\delta}(\mathbb{R}^n)$, where $\varepsilon,\ \delta > 0$, with an appropriate function parameter. This parameter is a function regularly varying at infinity (in Karamata's sense) of the order $\theta \in (0, 1)$, where $\theta := \varepsilon/(\varepsilon + \delta)$. Moreover, the class of spaces (6) is closed with respect to this interpolation.

Thus, the Hörmander spaces $H^{s,\varphi}(\mathbb{R}^n)$ possess the interpolation property with respect to the Hilbert scale of Sobolev spaces. This means that any linear operator bounded in each space $H^{s-\varepsilon}(\mathbb{R}^n)$ and $H^{s+\delta}(\mathbb{R}^n)$ is also bounded in $H^{s,\varphi}(\mathbb{R}^n)$. In this case, the interpolation property plays a decisive role because it allows us to establish important properties of the refined Sobolev scale guaranteeing the possibility of efficient application of this scale to the theory of elliptic equations. Thus, we can prove with the help of interpolation that the spaces $H^{s,\varphi}(\mathbb{R}^n)$, just as the Sobolev spaces, are invariant under the diffeomorphic transformations of $\mathbb{R}^n$. This enables us to correctly define the space $H^{s,\varphi}(\Gamma)$ over a closed smooth manifold Γ because the set of distributions and the topology in this space are independent of the choice of local charts covering Γ. The spaces $H^{s,\varphi}(\mathbb{R}^n)$ and $H^{s,\varphi}(\Gamma)$ are useful in the theory of elliptic operators on manifolds and in the theory of elliptic boundary-value problems. Moreover, they are implicitly present in various problems of analysis.

We now present some results demonstrating advantages of the refined scale as compared with the Sobolev scale. These results are related to the examples considered above. As earlier, let A be an elliptic differential operator on Γ of order m. Then it specifies bounded Fredholm operators

$$A : H^{s+m,\varphi}(\Gamma) \to H^{s,\varphi}(\Gamma) \ \text{for all} \ s \in \mathbb{R} \ \text{and} \ \varphi \in \mathcal{M}.$$

Here, $\mathcal{M}$ is a class of slowly varying function parameters φ used in (6). Note that the differential operator A leaves the functional parameter φ, which refines the main smoothness s, invariant. In addition, we have the following regularity property of a solution of the equation $Au = f$:

$$(f \in H^{s,\varphi}(\Gamma) \ \text{for some} \ s \in \mathbb{R} \ \text{and} \ \varphi \in \mathcal{M}) \ \Rightarrow \ u \in H^{s+m,\varphi}(\Gamma).$$

An important sharpening of Sobolev's Embedding Theorem is true for the refined scale. Let an integer $r \geq 0$ and a function parameter $\varphi \in \mathcal{M}$ be given. Then the embedding $H^{r+n/2,\varphi}(\Gamma) \subset C^r(\Gamma)$ is equivalent to the inequality

$$\int_1^\infty \frac{dt}{t\,\varphi^2(t)} < \infty. \tag{7}$$

Therefore, if $f \in H^{r-m+n/2,\varphi}(\Gamma)$ for a certain parameter $\varphi \in \mathcal{M}$ satisfying (7), then the solution $u \in C^r(\Gamma)$.

Similar results are also valid for elliptic systems and elliptic boundary-value problems.

We now pass to the analysis of convergence of the spectral expansion (3). It is additionally supposed that the operator A of order $m > 0$ is unbounded and self-adjoint in the space $L_2(\Gamma)$. Condition (4) for the convergence of (3) almost everywhere on Γ is equivalent to the inclusion

$$f \in H^{0,\varphi}(\Gamma) \quad \text{with} \quad \varphi(t) := \max\{1, \log t\}.$$

This condition is much broader than the condition "$f \in H^s(\Gamma)$ for some $s > 0$." In a similar way, we can also represent the conditions of *unconditional* convergence of series (3) almost everywhere or in the Banach space $C^r(\Gamma)$ for integer $r \geq 0$.

These and some other results show that the refined Sobolev scale can be quite useful and convenient. This scale can be also used in the other fields of contemporary analysis (see, e.g., the papers by M. Hegland [78, 79] and P. Mathé, U. Tautenhahn [129]).

The proposed monograph gives the first systematic presentation of the theory of elliptic (scalar and matrix) operators and elliptic boundary-value problems on the refined Sobolev scale. We also dwell upon the related topics concerning the interpolation of Hilbert (abstract and Sobolev) spaces with function parameter. This theory was developed by the authors in the recent papers [143–156, 165–173] (see also the survey [159]). The contents and structure of the monograph are fairly completely reflected in the TOC.

Chapter 1

Interpolation and Hörmander spaces

1.1 Interpolation with function parameter

In the present section, we consider the interpolation of pairs of Hilbert spaces with the help of a function parameter. This is a natural generalization of the classical interpolation method developed by J.-L. Lions and S. G. Krein [121, Chap. 1, Sec. 5], [107, Chap. IV, § 9] to the case where a fairly general function is taken as an interpolation parameter instead of a number. The procedure of interpolation with function parameter is one of the main methods used in the proofs of our results.

1.1.1 Definition of interpolation

We now present the definition of interpolation with the help of a function parameter for pairs of Hilbert spaces and study numerous properties of this interpolation required in what follows. For our purposes, it is sufficient to restrict ourselves to the case of separable spaces.

Definition 1.1. An ordered pair $[X_0, X_1]$ of complex Hilbert spaces X_0 and X_1 is called *admissible*, if the spaces X_0, X_1 are separable and the following dense and continuous embedding $X_1 \hookrightarrow X_0$ is true.

Let $X = [X_0, X_1]$ be a given admissible pair of Hilbert spaces. It is known (see [121, Chap. 1, Sec. 2.1] and [107, Chap. IV, § 9, Sec. 1]) that there exists an isometric isomorphism $J : X_1 \leftrightarrow X_0$ such that J is a self-adjoint positive definite operator with a domain X_1 in the space X_0. The mapping J is called a *generating* operator for the pair X. This operator is uniquely determined by the pair X. Indeed, let J_1 be another generating operator for the pair X. Then the operators J and J_1 are metrically equal:

$$\|Ju\|_{X_0} = \|u\|_{X_1} = \|J_1 u\|_{X_0}$$

for any $u \in X_1$. In addition, these operators are positive definite. This implies that they are equal: $J = J_1$.

By $\mathcal{B}$ we denote the set of all functions $\psi : (0, \infty) \to (0, \infty)$ such that

(a) ψ is Borel measurable on the semiaxis $(0, \infty)$;

(b) the function $1/\psi$ is separated from zero on every set $[r, \infty)$, with $r > 0$;

(c) ψ is bounded on every segment $[a, b]$, with $0 < a < b < \infty$.

Let $\psi \in \mathcal{B}$. The operator $\psi(J)$ is defined in X_0 as a function of J. By $[X_0, X_1]_\psi$ (or, simply, by X_ψ) we denote the domain of $\psi(J)$ equipped with the inner product

$$(u_1, u_2)_{X_\psi} := (\psi(J)u_1, \psi(J)u_2)_{X_0}$$

and the corresponding norm

$$\| u \|_{X_\psi} = (u, u)^{1/2}_{X_\psi}.$$

The space X_ψ is Hilbert and separable and the following dense and continuous embedding is true: $X_\psi \hookrightarrow X_0$. Indeed, $\operatorname{Spec} J \subseteq [r, \infty)$ and $\psi(t) \geq c$ for $t \geq r$ and some positive numbers r and c. (As usual, $\operatorname{Spec} J$ denotes the spectrum of the operator J.) Hence, $\operatorname{Spec} \psi(J) \subseteq [c, \infty)$, which yields the isometric isomorphism $\psi(J) : X_\psi \leftrightarrow X_0$. Thus, we conclude that X_ψ is a complete and separable space (and also that the function $\| \cdot \|_{X_\psi}$ is positive definite and plays the role of norm). Since the operator $\psi^{-1}(J)$ is bounded in the space X_0, the embedding operator

$$I = \psi^{-1}(J)\psi(J) : X_\psi \to X_0$$

is also bounded. This embedding is dense because the domain of $\psi(J)$ is a dense linear manifold in the space X_0.

Remark 1.1. Assume that functions φ, $\psi \in \mathcal{B}$ are such that $\varphi \asymp \psi$ in the vicinity of ∞. Then, by the definition of $\mathcal{B}$, we have $\varphi \asymp \psi$ on $\operatorname{Spec} J$. Hence, $X_\varphi = X_\psi$ up to equivalence of norms. (As usual, the relation $\varphi \asymp \psi$ means that both functions φ/ψ and ψ/φ are bounded on the corresponding set; in this case, φ and ψ are assumed to be positive.)

Definition 1.2. A function $\psi \in \mathcal{B}$ is called an *interpolation parameter* if, for any admissible pairs $X = [X_0, X_1]$ and $Y = [Y_0, Y_1]$ of Hilbert spaces and every linear mapping T given on X_0, the following condition is satisfied: If the restriction of the mapping T to the space X_j is a bounded operator $T : X_j \to Y_j$ for every $j \in \{0, 1\}$, then the restriction of the mapping T to the space X_ψ is also a bounded operator $T : X_\psi \to Y_\psi$.

In other words, the function ψ is an interpolation parameter if and only if the mapping $X \mapsto X_\psi$ is an interpolation functor given on the category of admissible pairs X of Hilbert spaces (see [24, Sec. 2.4] and [258, Sec. 1.2.2]). If ψ is an interpolation parameter, then we say that the space X_ψ is obtained by the *interpolation with function parameter* ψ of an admissible pair X.

In what follows, we study the main properties of the mapping $X \mapsto X_\psi$.

1.1.2 Embeddings of spaces

We now consider some properties of interpolation related to the embeddings of spaces.

Theorem 1.1. *Let $\psi \in \mathcal{B}$ be an interpolation parameter and let $X = [X_0, X_1]$ be an admissible pair of Hilbert spaces. Then the continuous and dense embeddings $X_1 \hookrightarrow X_\psi \hookrightarrow X_0$ are true.*

Proof. By virtue of the results presented above, it remains to show that the continuous and dense embedding $X_1 \hookrightarrow X_\psi$ is true. Consider the bounded embedding operators $I : X_1 \to X_0$ and $I : X_1 \to X_1$. Since ψ is an interpolation parameter, this means that the embedding operator $I : X_1 \to X_\psi$ is bounded, i.e., the embedding $X_1 \hookrightarrow X_\psi$ holds and is continuous.

To prove that this embedding is dense, we choose an arbitrary $u \in X_\psi$. Then $v := \psi(J)u \in X_0$, where J is the generating operator for the pair X. Since X_1 is dense in X_0, there exists a sequence $(v_k) \subset X_1$ such that $v_k \to v$ in X_0 as $k \to \infty$. This yields the convergence

$$u_k := \psi^{-1}(J)v_k \to \psi^{-1}(J)v = u \quad \text{in} \quad X_\psi \quad \text{as} \quad k \to \infty.$$

It remains to note that

$$u_k = \psi^{-1}(J)J^{-1}Jv_k = J^{-1}\psi^{-1}(J)Jv_k \in X_1.$$

Theorem 1.1 is proved.

Theorem 1.2. *Let functions $\psi, \chi \in \mathcal{B}$ be such that the function ψ/χ is bounded in the vicinity of ∞. Then the continuous and dense embedding $X_\chi \hookrightarrow X_\psi$ holds for every admissible pair $X = [X_0, X_1]$ of Hilbert spaces. If the embedding $X_1 \hookrightarrow X_0$ is compact and*

$$\frac{\psi(t)}{\chi(t)} \to 0 \quad \text{as} \quad t \to \infty,$$

then the embedding $X_\chi \hookrightarrow X_\psi$ is also compact.

Proof. Let J be the generating operator for the pair X. Note that $\operatorname{Spec} J \subseteq [r, \infty)$ for some number $r > 0$. By the condition, we have

$$\frac{\psi(t)}{\chi(t)} \leq c$$

for $t \geq r$ and, hence,

$$X_\chi = \operatorname{Dom} \chi(J) \subseteq \operatorname{Dom} \psi(J) = X_\psi, \quad \|\psi(J)\, u\|_{X_0} \leq c\, \|\chi(J)\, u\|_{X_0},$$

whence, by virtue of the definition of X_χ and X_ψ, we conclude that the embedding $X_\chi \hookrightarrow X_\psi$ is continuous.

We now prove that this embedding is dense.

Consider isometric isomorphisms $\psi(J) : X_\psi \leftrightarrow X_0$ and $\chi(J) : X_\chi \leftrightarrow X_0$. Let $u \in X_\psi$. Then $\psi(J)\, u \in X_0$. Since the space X_χ is densely embedded in X_0, there exists a sequence $(v_k) \subset X_\chi$ such that $v_k \to \psi(J)\, u$ in X_0 as $k \to \infty$. Hence,

$$\psi^{-1}(J)\, v_k \to u$$

in X_ψ as $k \to \infty$. Here,

$$\begin{aligned}\psi^{-1}(J)\, v_k &= \psi^{-1}(J)\, \chi^{-1}(J)\, \chi(J)\, v_k \\ &= \chi^{-1}(J)\, \psi^{-1}(J)\, \chi(J)\, v_k \in X_\chi.\end{aligned}$$

This proves that the embedding $X_\chi \hookrightarrow X_\psi$ is dense.

We now assume that the embedding $X_1 \hookrightarrow X_0$ is compact and

$$\psi(t)/\chi(t) \to 0 \quad \text{as} \quad t \to \infty.$$

We prove that the embedding $X_\chi \hookrightarrow X_\psi$ is compact. Let (u_k) be an arbitrary bounded sequence in X_χ. Since the sequence of elements $w_k := J^{-1}\, \chi(J)\, u_k$ is bounded in X_1, we can select a subsequence of elements $w_{k_n} = J^{-1}\, \chi(J)\, u_{k_n}$ fundamental in X_0. We now show that the subsequence (u_{k_n}) is fundamental in X_ψ.

Let E_t with $t \geq r$ be the resolution of identity in X_0 corresponding to the self-adjoint operator J. We can write

$$\begin{aligned}\|u_{k_n} - u_{k_m}\|^2_{X_\psi} &= \|\psi(J)\, (u_{k_n} - u_{k_m})\|^2_{X_0} \\ &= \|\psi(J)\, \chi^{-1}(J)\, J\, (w_{k_n} - w_{k_m})\|^2_{X_0} \\ &= \int\limits_r^\infty \psi^2(t)\, \chi^{-2}(t)\, t^2\, d\, \|E_t(w_{k_n} - w_{k_m})\|^2_{X_0}. \end{aligned} \tag{1.1}$$

Further, we choose an arbitrary $\varepsilon > 0$. There exists a number $\varrho = \varrho(\varepsilon) > r$ such that

$$\frac{\psi(t)}{\chi(t)} \leq (2c_0)^{-1}\varepsilon$$

for $t \geq \varrho$ and

$$c_0 := \sup\{\, \|w_k\|_{X_1} : k \in \mathbb{N} \,\} < \infty.$$

Thus, for any numbers n, m, we can write

$$\int_{\varrho}^{\infty} \psi^2(t)\,\chi^{-2}(t)\,t^2\,d\,\|E_t(w_{k_n} - w_{k_m})\|_{X_0}^2$$

$$\leq (2c_0)^{-2}\,\varepsilon^2 \int_{\varrho}^{\infty} t^2\,d\,\|E_t(w_{k_n} - w_{k_m})\|_{X_0}^2$$

$$\leq (2c_0)^{-2}\,\varepsilon^2\,\|J\,(w_{k_n} - w_{k_m})\|_{X_0}^2$$

$$= (2c_0)^{-2}\,\varepsilon^2\,\|w_{k_n} - w_{k_m}\|_{X_1}^2 \leq \varepsilon^2. \tag{1.2}$$

Moreover, in view of the inequality

$$\frac{\psi(t)}{\chi(t)} \leq c$$

for $t \geq r$, we get

$$\int_{r}^{\varrho} \psi^2(t)\,\chi^{-2}(t)\,t^2\,d\,\|E_t(w_{k_n} - w_{k_m})\|_{X_0}^2$$

$$\leq c^2\varrho^2 \int_{r}^{\varrho} d\,\|E_t(w_{k_n} - w_{k_m})\|_{X_0}^2$$

$$\leq c^2\varrho^2\,\|w_{k_n} - w_{k_m}\|_{X_0}^2 \to 0 \quad \text{for} \quad n, m \to \infty. \tag{1.3}$$

Relations (1.1)–(1.3) now imply the inequality $\|u_{k_n} - u_{k_m}\|_{X_\psi} \leq 2\varepsilon$ for sufficiently large numbers n and m. Hence, the subsequence (u_{k_n}) is fundamental in the space X_ψ, which means that the embedding $X_\chi \hookrightarrow X_\psi$ is compact.

Theorem 1.2 is proved.

1.1.3 Reiteration property

This property can be formulated as follows: The repeated application of the interpolation with function parameter also gives an interpolation with a certain function parameter.

Theorem 1.3. *Let $f, g, \psi \in \mathcal{B}$ and let the function f/g be bounded in the vicinity of ∞. Then $[X_f, X_g]_\psi = X_\omega$ with the equality of norms for every admissible pair X of Hilbert spaces. Here, the function $\omega \in \mathcal{B}$ is defined by the formula*

$$\omega(t) := f(t)\,\psi\Big(\frac{g(t)}{f(t)}\Big)$$

for $t > 0$. If f, g, and ψ are interpolation parameters, then ω is also an interpolation parameter.

Proof. Since the function f/g is bounded in the vicinity of ∞, the pair $[X_f, X_g]$ is admissible by Theorem 1.2 and $\omega \in \mathcal{B}$. Thus, the spaces $[X_f, X_g]_\psi$ and X_ω are defined. We now prove their equality.

Let J be the generating operator for the pair $X = [X_0, X_1]$, where $\operatorname{Spec} J \subseteq [r, \infty)$ for some number $r > 0$. We have the isometric isomorphisms

$$f(J) : X_f \leftrightarrow X_0, \quad g(J) : X_g \leftrightarrow X_0,$$

$$B := f^{-1}(J)\, g(J) : X_g \leftrightarrow X_f.$$

Consider B as a closed operator in the space X_f with the domain X_g. The operator B is the generating operator for the pair $[X_f, X_g]$ because it is positive definite and self-adjoint in X_f. The first property follows from the condition $f(t)/g(t) \leq c$ for $t \geq r$, namely,

$$\begin{aligned}(Bu, u)_{X_f} &= (g(J)\, u, f(J)\, u)_{X_0} \\ &\geq c^{-1}\, (f(J)\, u, f(J)\, u)_{X_0} = c^{-1}\, \|u\|^2_{X_f}, \quad u \in X_f.\end{aligned}$$

The second property now follows from the fact that 0 is a regular point for B.

With the help of the spectral theorem, we can reduce the operator J self-adjoint in the space X_0 to the operator of multiplication by a function. Namely, we can write $J = I^{-1}(\alpha \cdot I)$, where $I : X_0 \leftrightarrow L_2(U, d\mu)$ is an isometric isomorphism, (U, μ) is a space with finite measure, and $\alpha : U \to [r, \infty)$ is a measurable function. The isometric isomorphism $I f(J) : X_f \leftrightarrow L_2(U, d\mu)$ reduces the operator B, self-adjoint in X_f to the operator of multiplication by a function $(g/f) \circ \alpha$. Indeed,

$$\begin{aligned}I f(J)\, B u &= I g(J)\, u = (g \circ \alpha)\, I u \\ &= (g \circ \alpha)\, I f^{-1}(J) f(J)\, u = ((g/f) \circ \alpha)\, I f(J)\, u, \quad u \in X_g.\end{aligned}$$

Hence, for any $u \in X_\psi$, we can write

$$\begin{aligned}\|\psi(B)\, u\|_{X_f} &= \|(\psi \circ (g/f) \circ \alpha) \cdot (I f(J)\, u)\|_{L_2(U,d\mu)} \\ &= |(\omega \circ \alpha) \cdot (I u)\|_{L_2(U,d\mu)} = \|\omega(J)\, u\|_{X_0}.\end{aligned}$$

Note that the function f/ω is bounded in the vicinity of ∞. Hence, $X_\omega \hookrightarrow X_f$ and therefore, $f(J)\, u$ is defined. Thus, the equality $[X_f, X_g]_\psi = X_\omega$ is proved.

We now assume that f, g, and ψ are interpolation parameters and show that ω is an interpolation parameter. Let admissible pairs $X = [X_0, X_1]$, $Y = [Y_0, Y_1]$, and a linear mapping T be the same as in Definition 1.2. Thus, the operators $T : X_f \to Y_f$ and $T : X_g \to Y_g$ are bounded which means that the operator $T : [X_f, X_g]_\psi \to [Y_f, Y_g]_\psi$ is also bounded. As shown above, $[X_f, X_g]_\psi = X_\omega$ and $[Y_f, Y_g]_\psi = Y_\omega$. Hence, the operator $T : X_\omega \to Y_\omega$ is defined and bounded; i.e., ω is an interpolation parameter.

Theorem 1.3 is proved.

1.1.4 Interpolation of dual spaces

Let H be a Hilbert space. As usual, H' denotes the space dual to H; i.e., H' is the normed linear space of all linear continuous functionals $l : H \to \mathbb{C}$. By the Riesz theorem, the mapping $S : v \mapsto (\,\cdot, v)_H$, where $v \in H$, defines an antilinear isometric isomorphism $S : H \leftrightarrow H'$. This means that H' is a Hilbert space; the Hilbert norm in H' is induced by the inner product

$$(l, m)_{H'} := (S^{-1}l, S^{-1}m)_H.$$

Note that we do not identify H and H' with the help of the isomorphism S.

Theorem 1.4. *Let $\psi \in \mathcal{B}$ be a function such that the function $\psi(t)/t$ is bounded in the vicinity of ∞. Then, for any admissible pair $[X_0, X_1]$ of Hilbert spaces, the equality*

$$[X_1', X_0']_\psi = [X_0, X_1]_\chi'$$

holds with equality of the norms. Here, the function $\chi \in \mathcal{B}$ is given by the formula $\chi(t) := t/\psi(t)$ for $t > 0$. If ψ is an interpolation parameter, then χ is also an interpolation parameter.

Proof. Note that the pair $[X_1', X_0']$ is admissible for the natural identification of functionals from X_0' with their restrictions to the space X_1. The condition of the theorem implies that $\chi \in \mathcal{B}$. Hence, the Hilbert spaces $[X_1', X_0']_\psi$ and $[X_0, X_1]_\chi'$ are defined. We now prove that they are equal.

Let $J : X_1 \leftrightarrow X_0$ be the generating operator for the pair $[X_0, X_1]$. Consider the isometric isomorphisms $S_j : X_j \leftrightarrow X_j'$, $j = 0, 1$, appearing in the Riesz theorem. The operator J' adjoint to J satisfies the equality $J' = S_1 J^{-1} S_0^{-1}$. This is a consequence of the following chain of relations:

$$(J'l)u = l(Ju) = (Ju, S_0^{-1}l)_{X_0} = (u, J^{-1}S_0^{-1}l)_{X_1}$$

$$= (S_1 J^{-1} S_0^{-1} l)u \quad \text{for all} \quad l \in X_0',\ u \in X_1.$$

Therefore, the operator J' realizes an isometric isomorphism

$$J' = S_1 J^{-1} S_0^{-1} : X_0' \leftrightarrow X_1'. \tag{1.4}$$

Note that the equalities

$$(u, JS_1^{-1}l)_{X_0} = (J^{-1}u, S_1^{-1}l)_{X_1} = l(J^{-1}u),$$

$$(u, J^{-1}S_0^{-1}l)_{X_0} = (J^{-1}u, S_0^{-1}l)_{X_0} = l(J^{-1}u),$$

valid for any $l \in X_0' \hookrightarrow X_1'$ and $u \in X_0$ yield the following property:

$$JS_1^{-1}l = J^{-1}S_0^{-1}l \in X_1 \quad \text{for all} \quad l \in X_0'. \tag{1.5}$$

We consider J' as a closed operator in the space X_1' with the domain X_0'. The mapping J' is the generating operator for the pair $[X_1', X_0']$ because J' is positive definite and self-adjoint in X_1'. The first property follows from the positive definiteness of the operator J in the space X_0 and property (1.5). Indeed,

$$(J'l, l)_{X_1'} = (S_1 J^{-1}S_0^{-1}l, l)_{X_1'} = (J^{-1}S_0^{-1}l, S_1^{-1}l)_{X_1}$$

$$= (JJ^{-1}S_0^{-1}l, JS_1^{-1}l)_{X_0} = (JJS_1^{-1}l, JS_1^{-1}l)_{X_0}$$

$$\geq c\,\|JS_1^{-1}l\|_{X_0}^2 = c\,\|S_1^{-1}l\|_{X_1}^2 = c\,\|l\|_{X_1'}^2.$$

Here, the number $c > 0$ is independent of $l \in X_0'$. The second property now follows from the fact that 0 is a regular point for the operator J' by virtue of relation (1.4).

We now use the reduction of the operator J to the form $J = I^{-1}(\alpha \cdot I)$ of multiplication by a function. This reduction has already been considered in the proof of Theorem 1.3. The isometric isomorphism

$$IJS_1^{-1} : X_1' \leftrightarrow L_2(U, d\mu) \tag{1.6}$$

reduces the operator J' to the form of multiplication by the same function α. Indeed,

$$(IJS_1^{-1})J'l = IS_0^{-1}l = IJJ^{-1}S_0^{-1}l$$

$$= \alpha \cdot IJ^{-1}S_0^{-1}l = \alpha \cdot IJS_1^{-1}l \quad \text{for any} \quad l \in X_0'$$

[here, we have used relations (1.4) and (1.5)].

By virtue of Theorem 1.2, we get the following continuous and dense embeddings: $X_0' \hookrightarrow [X_1', X_0']_\psi$ and $[X_0, X_1]_\chi \hookrightarrow X_0$. The latter implies that the embedding $X_0' \hookrightarrow [X_0, X_1]_\chi'$ is also continuous and dense. We now show that the norms in the spaces $[X_1', X_0']_\psi$ and $[X_0, X_1]_\chi'$ coincide on the dense subset X_0'. For any $l \in X_0'$ and $u \in [X_0, X_1]_\chi$, we can write

$$l(u) = (u, S_0^{-1}l)_{X_0} = (\chi(J)u, \chi^{-1}(J)S_0^{-1}l)_{X_0}$$

$$= (v, \chi^{-1}(J)S_0^{-1}l)_{X_0},$$

where

$$v := \chi(J)u \in X_0.$$

This yields

$$\begin{aligned}\|l\|_{[X_0,X_1]'_\chi} &= \sup\{\,|l(u)|\,/\,\|u\|_{[X_0,X_1]_\chi} : u \in [X_0,X_1]_\chi,\ u \neq 0\,\}\\ &= \sup\{\,|(v, \chi^{-1}(J)S_0^{-1}l)_{X_0}|\,/\,\|v\|_{X_0} : v \in X_0,\ v \neq 0\,\}\\ &= \|\chi^{-1}(J)S_0^{-1}l\|_{X_0} = \|I\chi^{-1}(J)S_0^{-1}l\|_{L_2(U,d\mu)}\\ &= \|(\chi^{-1}\circ\alpha)\cdot IS_0^{-1}l\|_{L_2(U,d\mu)}.\end{aligned}$$

On the other hand, applying the isomorphisms (1.6) and (1.4), we obtain

$$\begin{aligned}\|l\|_{[X_1',X_0']_\psi} &= \|\psi(J')l\|_{X_1'} = \|\chi^{-1}(J')J'l\|_{X_1'}\\ &= \|(IJS_1^{-1})\chi^{-1}(J')J'l\|_{L_2(U,d\mu)}\\ &= \|(\chi^{-1}\circ\alpha)\cdot(IJS_1^{-1})J'l\|_{L_2(U,d\mu)}\\ &= \|(\chi^{-1}\circ\alpha)\cdot IS_0^{-1}l\|_{L_2(U,d\mu)}.\end{aligned}$$

Thus, the norms in the spaces $[X_1', X_0']_\psi$ and $[X_0, X_1]'_\chi$ coincide on the dense subset X_0'. Therefore, these spaces coincide.

Assume that ψ is an interpolation parameter. It is necessary to show that χ is also an interpolation parameter. Let admissible pairs $X = [X_0, X_1]$, $Y = [Y_0, Y_1]$ and a linear mapping T be the same as in Definition 1.2. Passing to the adjoint operator T', we get bounded operators $T' : Y_j' \to X_j'$ for each $j \in \{0, 1\}$. Since ψ is an interpolation parameter, we obtain a bounded operator

$$T' : [Y_1', Y_0']_\psi \to [X_1', X_0']_\psi.$$

As shown above, $[X_1', X_0']_\psi = [X_0, X_1]'_\chi$ and $[Y_1', Y_0']_\psi = [Y_0, Y_1]'_\chi$ with equality of the norms. Hence, the operator $T' : [Y_0, Y_1]'_\chi \to [X_0, X_1]'_\chi$ is bounded. Passing to the second adjoint operator T'', we get a bounded operator

$$T'' : [X_0, X_1]''_\chi \to [Y_0, Y_1]''_\chi.$$

Identifying the second dual spaces with the original Hilbert spaces, we obtain a bounded operator $T : [X_0, X_1]_\chi \to [Y_0, Y_1]_\chi$. This means that χ is an interpolation parameter.

Theorem 1.4 is proved.

Note that, for a sufficiently broad class of function interpolation parameters, Theorem 1.4 was proved by G. Shlenzak [231, Theorem 2].

In some applications, it is convenient to consider the space of all *anti*linear continuous functionals on H as the dual space H'. For this interpretation of H', Theorem 1.4 is obviously valid. In what follows, it is always clear from the context functionals of what kind (linear or antilinear) are used to form the space H'.

1.1.5 Interpolation of orthogonal sums of spaces

Recall that an orthogonal sum of finitely or countably many separable Hilbert spaces is also a separable Hilbert space.

Theorem 1.5. *Let $X^{(k)} := [X_0^{(k)}, X_1^{(k)}]$, $k \in \omega$, be a given finite (or countable) set of admissible pairs of Hilbert spaces. Assume that the set of norms of the embedding operators $X_1^{(k)} \hookrightarrow X_0^{(k)}$, $k \in \omega$, is bounded. Then, for any function parameter $\psi \in \mathcal{B}$, we have*

$$\Big[\bigoplus_{k\in\omega} X_0^{(k)}, \bigoplus_{k\in\omega} X_1^{(k)}\Big]_\psi = \bigoplus_{k\in\omega} \,[X_0^{(k)}, X_1^{(k)}]_\psi,$$

with equality of the norms.

Proof. Let $\omega = \mathbb{N}$ (the case of a finite set ω is studied similarly and in a simpler way [231, Theorem 4]). Both the spaces

$$X_0 := \bigoplus_{k=1}^{\infty} X_0^{(k)} \quad \text{and} \quad X_1 := \bigoplus_{k=1}^{\infty} X_1^{(k)}$$

are Hilbert and separable. The continuity of the embedding $X_1 \hookrightarrow X_0$ is obvious in view of the condition of the theorem. We now show that this embedding is dense. Let $u := (u_1, u_2, \ldots) \in X_0$. For any numbers n and k, there exists an element $v_{n,k} \in X_1^{(k)}$ such that

$$\|u_k - v_{n,k}\|_{X_0^{(k)}} < 1/n.$$

We now compose a sequence of vectors

$$v^{(n)} := (v_{n,1}, \ldots, v_{n,n}, 0, 0, \ldots) \in X_1.$$

Thus, we get

$$\begin{aligned}\|u - v^{(n)}\|_{X_0}^2 &= \sum_{k=1}^{n} \|u_k - v_{n,k}\|_{X_0}^2 + \sum_{k=n+1}^{\infty} \|u_k\|_{X_0}^2 \\ &\le \frac{n}{n^2} + \sum_{k=n+1}^{\infty} \|u_k\|_{X_0}^2 \to 0 \quad \text{as} \quad n \to \infty.\end{aligned}$$

Hence X_1 is dense in the space X_0 and the pair $X := [X_0, X_1]$ is admissible.

Let J_k be the generating operator for the pair $X^{(k)}$. We can directly show that $J := (J_1, J_2, \ldots)$ is the generating operator for the pair X. Further, we show that $\psi(J) = (\psi(J_1), \psi(J_2), \ldots)$, where $\operatorname{Dom}\psi(J) = \bigoplus_{k=1}^{\infty} X_{\psi}^{(k)}$. The operator J_k is now reduced to the form $I_k J_k = \alpha_k \cdot I_k$ of multiplication by a function. Here, $I_k : X_0^{(k)} \leftrightarrow L_2(V_k, d\mu_k)$ is an isometric isomorphism, V_k is a space with finite measure μ_k, and $\alpha_k : V_k \to (0, \infty)$ is a measurable function. Without loss of generality, we can assume that the sets V_k are pairwise disjoint. We set

$$V := \bigcup_{k=1}^{\infty} V_k.$$

A subset $\Omega \subseteq V$ is called measurable if the set $\Omega \cap V_k$ is μ_k-measurable for every number k. In the σ-algebra of measurable sets $\Omega \subseteq V$, we introduce a σ-finite measure

$$\mu(\Omega) := \sum_{k=1}^{\infty} \mu_k(\Omega \cap V_k).$$

For any vector $u := (u_1, u_2, \ldots) \in X_0$, we define measurable functions Iu and α on the set V by the formulas $(Iu)(\lambda) := (I_k u_k)(\lambda)$ and $\alpha(\lambda) := \alpha_k(\lambda)$ provided that $\lambda \in V_k$. We have an isometric isomorphism $I : X_0 \leftrightarrow L_2(V, d\mu)$. This isomorphism reduces the operator J to the form of multiplication by the function α because

$$(IJu)(\lambda) = (I_k J_k u_k)(\lambda) = \alpha_k(\lambda)(I_k u_k)(\lambda) = \alpha(\lambda)(Iu)(\lambda)$$

for any $u \in X_1$ and $\lambda \in V_k$. This enables us to write

$$\begin{aligned} X_{\psi} &= \operatorname{Dom}\psi(J) = \{\, u \in X_0 : \ (\psi \circ \alpha) \cdot (Iu) \in L_2(V, d\mu) \,\} \\ &= \Big\{ u \in X_0 : \ \sum_{k=1}^{\infty} \|(\psi \circ \alpha_k) \cdot (I_k u_k)\|^2_{L_2(V_k, d\mu_k)} < \infty \Big\} \\ &= \Big\{ u : \ u_k \in \operatorname{Dom}\psi(J_k), \ \sum_{k=1}^{\infty} \|\psi(J_k) u_k\|^2_{X_0^{(k)}} < \infty \Big\} = \bigoplus_{k=1}^{\infty} X_{\psi}^{(k)}. \end{aligned}$$

Moreover, for any $u \in \operatorname{Dom}\psi(J)$, we find

$$\begin{aligned} (I\psi(J)u)(\lambda) &= \psi(\alpha(\lambda))\,(Iu)(\lambda) = \psi(\alpha_k(\lambda))\,(I_k u_k)(\lambda) \\ &= (I_k \psi(J_k) u_k)(\lambda) = \big(I(\psi(J_1)u_1, \psi(J_2)u_2, \ldots)\big)(\lambda) \end{aligned}$$

provided that $\lambda \in V_k$. Hence, $\psi(J)u = (\psi(J_1)u_1, \psi(J_2)u_2, \ldots)$. This yields

$$\|u\|^2_{X_{\psi}} = \|\psi(J)u\|^2_{X_0} = \sum_{k=1}^{\infty} \|\psi(J_k)u_k\|^2_{X_0^{(k)}} = \sum_{k=1}^{\infty} \|u_k\|^2_{X_{\psi}^{(k)}}.$$

Theorem 1.5 is proved.

1.1.6 Interpolation of subspaces and factor spaces

Recall that, by definition, a subspace of a Hilbert space is closed. In what follows, we consider, generally speaking, nonorthogonal projectors onto a subspace.

Theorem 1.6. *Let $X = [X_0, X_1]$ be an admissible pair of Hilbert spaces and let Y_0 be a subspace of X_0. Then $Y_1 := X_1 \cap Y_0$ is a subspace of X_1. Assume that there exists a linear mapping P such that P is a projector of the space X_j onto the subspace Y_j for any $j \in \{0, 1\}$. Then the pairs $[Y_0, Y_1]$ and $[X_0/Y_0, X_1/Y_1]$ are admissible and, for any interpolation parameter $\psi \in \mathcal{B}$, the following spaces are equal (with equivalence of the norms):*

$$[Y_0, Y_1]_\psi = X_\psi \cap Y_0, \quad [X_0/Y_0, X_1/Y_1]_\psi = X_\psi/(X_\psi \cap Y_0). \tag{1.7}$$

Here, $X_\psi \cap Y_0$ is a subspace of X_ψ.

Proof. Since the embeddings $X_1 \hookrightarrow X_0$ and $X_\psi \hookrightarrow X_0$ are continuous, the linear manifolds $Y_1 = X_1 \cap Y_0$ and $X_\psi \cap Y_0$ are closed in the spaces X_1 and X_ψ respectively. The pairs $Y := [Y_0, Y_1]$ and $[X_0/Y_0, X_1/Y_1]$ are admissible because Y_1 is a dense subset of the space Y_0. This fact is proved as follows: For any $u \in Y_0$, there exists a sequence of elements $u_k \in X_1$ such that $u_k \to u$ in X_0 as $k \to \infty$. By the condition of the theorem, this implies that $Pu_k \to Pu = u$ in X_0 as $k \to \infty$, where $Pu_k \in Y_1$. This means that Y_1 is indeed a dense subset of the space Y_0. Thus, the left- and right-hand sides of equalities (1.7) are well defined.

We prove the first equality in (1.7). Suppose that $\psi \in \mathcal{B}$ is an interpolation parameter. Then it follows from the boundedness of the projectors $P : X_0 \to Y_0$ and $P : X_1 \to Y_1$ that the restriction of P to X_ψ is a bounded operator $P : X_\psi \to Y_\psi$. Hence,

$$X_\psi \cap Y_0 = P(X_\psi \cap Y_0) \subseteq Y_\psi.$$

Moreover, since the embedding operators $Y_0 \hookrightarrow X_0$ and $Y_1 \hookrightarrow X_1$ are continuous, we get the continuous embedding $Y_\psi \hookrightarrow X_\psi$. Thus, $Y_\psi = X_\psi \cap Y_0$ and the norms in Y_ψ and $X_\psi \cap Y_0$ are equivalent in view of the Banach theorem on inverse operator.

We now prove the second equality in (1.7). Given any $j \in \{0, 1\}$, we consider bounded linear operators $R : X_j \to X_j/Y_j$ and $T : X_j/Y_j \to X_j$ defined by the formulas $Rx := [x]$ and $T[x] := x - Px$ for any $x \in X_j$, where $[x]$ denotes the coset $\{x + y : y \in Y_j\} \in X_j/Y_j$. Note that $RT = I$ is the identity operator on X_j/Y_j. Since $\psi \in \mathcal{B}$ is an interpolation parameter, we conclude that the corresponding restrictions of R and T specify the bounded operators

$$R : X_\psi \to [X_0/Y_0, X_1/Y_1]_\psi \quad \text{and} \quad T : [X_0/Y_0, X_1/Y_1]_\psi \to X_\psi.$$

Moreover, we have the following bounded operators:

$$R: X_\psi \to X_\psi / Y_\psi \quad \text{and} \quad T: X_\psi / Y_\psi \to X_\psi.$$

Further, we consider the identity mapping I as the product of the second and third operators and arrive at the continuous embedding

$$I = RT: \left[X_0/Y_0, X_1/Y_1\right]_\psi \hookrightarrow X_\psi / Y_\psi.$$

Moreover, multiplying the fourth operator by the first operator, we arrive at a different continuous embedding

$$I = RT: X_\psi / Y_\psi \hookrightarrow \left[X_0/Y_0, X_1/Y_1\right]_\psi.$$

Therefore, $[X_0/Y_0, X_1/Y_1]_\psi = X_\psi/Y_\psi$ up to equivalence of norms with $Y_\psi = X_\psi \cap Y_0$, as shown earlier.

Theorem 1.6 is proved.

1.1.7 Interpolation of Fredholm operators

First, we recall the following definition:

Definition 1.3. A linear bounded operator $T: X \to Y$, where X and Y are Banach spaces, is called a *Fredholm* operator if both its kernel $\ker T$ and cokernel $\operatorname{coker} T := Y/T(X)$ are finite-dimensional. The *index* of a Fredholm operator T is defined by the formula

$$\operatorname{ind} T := \dim \ker T - \dim(Y/T(X)).$$

Note that the domain of a Fredholm operator is closed [86, Lemma 19.1.1].

Theorem 1.7. *Let $X = [X_0, X_1]$ and $Y = [Y_0, Y_1]$ be admissible pairs of Hilbert spaces. Suppose that a linear mapping T is given on X_0 and specifies bounded Fredholm operators $T: X_j \to Y_j$, with $j \in \{0, 1\}$, common kernel N, and the same index $\varkappa$. Then, for any interpolation parameter $\psi \in \mathcal{B}$, the bounded operator $T: X_\psi \to Y_\psi$ is a Fredholm operator with the kernel N, domain $Y_\psi \cap T(X_0)$, and the same index $\varkappa$.*

Proof. By the condition, we have the following bounded operators: $T_j := T: X_j \to Y_j$ and $T_\psi := T: X_\psi \to Y_\psi$. Consider the bounded operators $T_j': Y_j' \to X_j'$ and $T_\psi': Y_\psi' \to X_\psi'$ adjoint to these operators. By Theorem 1.1, we have the following continuous and dense embeddings: $X_1 \hookrightarrow X_\psi \hookrightarrow X_0$ and $Y_1 \hookrightarrow Y_\psi \hookrightarrow Y_0$. Therefore,

$$\ker T_1 \subseteq \ker T_\psi \subseteq \ker T_0 \quad \text{and} \quad \ker T_0' \subseteq \ker T_\psi' \subseteq \ker T_1'.$$

However, by the condition, $\ker T_j = N$ and

$$\dim \ker T_j' = \dim Y_j / T(X_j) = -\varkappa + \dim N.$$

Hence, $\ker T_\psi = N$ and

$$\ker T_\psi' = \ker T_j' =: M,$$

where $\dim M = -\varkappa + \dim N$. Thus, the operator $T : X_\psi \to Y_\psi$ has the finite-dimensional kernel N and the defect subspace M with $\dim N - \dim M = \varkappa$. It remains to verify that $T(X_\psi) = Y_\psi \cap T(X_0)$ because, in this case, $T(X_\psi)$ is closed in the space Y_ψ and $\dim Y_\psi / T(X_\psi) = \dim M$.

For this purpose, we consider the isomorphisms

$$T : X_j/N \leftrightarrow T(X_j) \quad \text{for} \quad j \in \{0, 1\}. \tag{1.8}$$

They are canonically generated by the bounded Fredholm operators T_j. Recall that $T(X_j)$ is a subspace of Y_j. We now apply the interpolation with parameter ψ to (1.8). This gives another isomorphism

$$T : [X_0/N, X_1/N]_\psi \leftrightarrow [T(X_0), T(X_1)]_\psi. \tag{1.9}$$

Note that the pair $[X_0/N, X_1/N]$ is obviously admissible. By virtue of (1.8), this implies that the pair $[T(X_0), T(X_1)]$ is admissible. We now describe the interpolation spaces appearing in (1.9).

Consider the orthogonal sum $X_0 = N \oplus E$. Its restriction to the space X_1 is the direct sum $X_1 = N \dotplus (E \cap X_1)$ of subspaces. By P we denote the orthoprojector of the space X_0 onto N. Then the restriction of P to the space X_1 is the projector of this space onto N corresponding to the second sum. Hence, by virtue of Theorem 1.6 and the equality $X_\psi \cap N = N$, the relation

$$[X_0/N, X_1/N]_\psi = X_\psi/N \tag{1.10}$$

holds with equivalence of the norms.

Further, we consider the orthogonal sum $Y_0 = T(X_0) \oplus Z_0$, where $\dim Z_0 < \infty$. Since Y_1 is dense in the space Y_0, we obtain the decomposition of the space Y_0 into the direct sum $Y_0 = T(X_0) \dotplus Z_1$, where Z_1 is a finite-dimensional subspace of Y_1. Here, we refer the reader to Lemma 2.1 in [65]. The restriction of this sum to the space Y_1 has the form $Y_1 = (T(X_0) \cap Y_1) \dotplus Z_1$. Note that $T(X_0) \cap Y_1 = T(X_1)$. This equality follows from the representation of the closed range of T_j in the form

$$T(X_j) = \{f \in Y_j : l(f) = 0 \text{ for } l \in M\}$$

for $j \in \{0, 1\}$. Thus, we have the direct sums $Y_0 = T(X_0) \dotplus Z_1$ and $Y_1 = T(X_1) \dotplus Z_1$. Moreover, the second sum is a restriction of the first sum. By Q

we denote the projector of the space Y_0 onto the subspace $T(X_0)$ corresponding to the first sum. Then the restriction of Q to Y_1 is the projector of the space Y_1 onto the subspace $T(X_1) = Y_1 \cap T(X_0)$ corresponding to the second sum. Hence, by Theorem 1.4, we can write

$$[T(X_0), T(X_1)]_\psi = Y_\psi \cap T(X_0) \tag{1.11}$$

with equivalence of the norms.

Applying equalities (1.10) and (1.11) to isomorphism (1.9), we obtain the isomorphism $T : X_\psi/N \leftrightarrow Y_\psi \cap T(X_0)$. This yields the equality

$$T(X_\psi) = T(X_\psi/N) = Y_\psi \cap T(X_0),$$

which completes the proof of Theorem 1.7.

Note that an analog of Theorem 1.7 was proved by G. Geymonat [63, p. 281] for any interpolation functor given on the category of all compatible pairs of Banach spaces. The proof presented in the cited work is similar to our proof.

1.1.8 Estimate of the operator norm in interpolation spaces

We now prove the following result:

Theorem 1.8. *Let an interpolation parameter $\psi \in \mathcal{B}$ and a number $m > 0$ be given. Then there exists a number $c = c(\psi, m) > 0$ such that*

$$\|T\|_{X_\psi \to Y_\psi} \le c \max\left\{ \|T\|_{X_j \to Y_j} : j = 0, 1 \right\}. \tag{1.12}$$

Here, $X = [X_0, X_1]$ and $Y = [Y_0, Y_1]$ are arbitrary admissible pairs of Hilbert spaces for which the norms of the embedding operators $X_1 \hookrightarrow X_0$ and $Y_1 \hookrightarrow Y_0$ do not exceed the number m and T is an arbitrary linear mapping given in the space X_0 and specifying the bounded operators $T : X_j \to Y_j$ for each $j \in \{0, 1\}$. The constant c is independent of X, Y, and T.

Proof. Assume that the theorem is not true. Then

$$\|T_k\|_{X^{(k)}_\psi \to Y^{(k)}_\psi} > k\, m_k \quad \text{for every} \quad k \in \mathbb{N}. \tag{1.13}$$

Here, $X^{(k)} := [X^{(k)}_0, X^{(k)}_1]$ and $Y^{(k)} := [Y^{(k)}_0, Y^{(k)}_1]$ are admissible pairs of Hilbert spaces for which the norms of the embedding operators $X^{(k)}_1 \hookrightarrow X^{(k)}_0$ and $Y^{(k)}_1 \hookrightarrow Y^{(k)}_0$ do not exceed the number m and T_k is a linear mapping given in the space $X^{(k)}_0$ and specifying the bounded operators $T_k : X^{(k)}_j \to Y^{(k)}_j$, where $j \in \{0, 1\}$. In this case, we use the notation

$$m_k := \max\left\{ \|T_k\|_{X^{(k)}_0 \to Y^{(k)}_0},\ \|T_k\|_{X^{(k)}_1 \to Y^{(k)}_1} \right\} > 0.$$

Consider bounded operators

$$T : u = (u_1, u_2, \ldots) \mapsto (m_1^{-1} T_1 u_1,\, m_2^{-1} T_2 u_2, \ldots),$$

$$T : \bigoplus_{k=1}^{\infty} X_j^{(k)} \to \bigoplus_{k=1}^{\infty} Y_j^{(k)}, \quad j \in \{0,1\}. \tag{1.14}$$

Their boundedness follows from the inequalities

$$\sum_{k=1}^{\infty} \|m_k^{-1} T_k u_k\|^2_{Y_j^{(k)}} \le \sum_{k=1}^{\infty} m_k^{-2} \|T_k\|^2_{X_j^{(k)} \to Y_j^{(k)}} \|u_k\|^2_{X_j^{(k)}} \le \sum_{k=1}^{\infty} \|u_k\|^2_{X_j^{(k)}}.$$

Since ψ is an interpolation parameter, the boundedness of the operators (1.14) implies that the following operator is defined and bounded:

$$T : \Bigl[\bigoplus_{k=1}^{\infty} X_0^{(k)}, \bigoplus_{k=1}^{\infty} X_1^{(k)}\Bigr]_{\psi} \to \Bigl[\bigoplus_{k=1}^{\infty} Y_0^{(k)}, \bigoplus_{k=1}^{\infty} Y_1^{(k)}\Bigr]_{\psi}.$$

By Theorem 1.5, this yields the boundedness of the operator

$$T : \bigoplus_{k=1}^{\infty} X_{\psi}^{(k)} \to \bigoplus_{k=1}^{\infty} Y_{\psi}^{(k)}.$$

Let c_0 be the norm of this operator. For any number k, we consider a vector $u^{(k)} := (u_1, \ldots, u_k, \ldots)$ such that $u_k \in X_{\psi}^{(k)}$ and $u_j = 0$ for $j \ne k$. We have

$$\|T_k u_k\|_{Y_{\psi}^{(k)}} = m_k \, \|T u^{(k)}\|_{\bigoplus_{j=1}^{\infty} Y_{\psi}^{(j)}}$$

$$\le m_k \, c_0 \, \|u^{(k)}\|_{\bigoplus_{j=1}^{\infty} X_{\psi}^{(j)}} = m_k \, c_0 \, \|u_k\|_{X_{\psi}^{(k)}}$$

for all $u_k \in X_{\psi}^{(k)}$. Hence,

$$\|T_k\|_{X_{\psi}^{(k)} \to Y_{\psi}^{(k)}} \le c_0 \, m_k \quad \text{for any number } k,$$

which contradicts condition (1.13). Thus, our assumption is false. This means that Theorem 1.8 is true.

Note that inequality (1.12), where the constant c is independent of T (but may depend on X and Y) is satisfied for any interpolation functor given on the category of pairs of Hilbert or Banach spaces; see [24, Theorem 2.4.2] or [109, Chap. 1, Lemma 4.3]. Theorem 1.8 sharpens this fact for the interpolation functor $X \mapsto X_{\psi}$.

The admissible pair $[X_0, X_1]$ of Hilbert spaces is called *normal* if $\|u\|_{X_0} \le \|u\|_{X_1}$ for every $u \in X_1$. By Theorem 1.8, the constant c in inequality (1.12) is independent of the admissible pairs X, Y and the operator T whenever these pairs are normal. We note that any admissible pair $[X_0, X_1]$ can be made normal if we change, e.g., the norm $\|u\|_{X_0}$ in the space X_0 by the proportional norm $k\,\|u\|_{X_0}$, where $0 < k \leq m^{-1}$, and m is the norm of the embedding operator $X_1 \hookrightarrow X_0$.

1.1.9 Criterion for a function to be an interpolation parameter

On the basis of the results obtained by J. Peetre [188, 189] (see also [24, Theorem 5.4.4]), we establish the following criterion for a function $\psi \in \mathcal{B}$ to be an interpolation parameter.

Definition 1.4. Let a function $\psi : (0, \infty) \to (0, \infty)$ and a number $r \geq 0$ be given. The function ψ is called *pseudoconcave* on the half line (r, ∞) if there exists a concave function $\psi_1 : (r, \infty) \to (0, \infty)$ such that $\psi(t) \asymp \psi_1(t)$ for $t > r$. The function ψ is called pseudoconcave in the vicinity of ∞ if ψ is pseudoconcave in a half line (r, ∞), where r is a sufficiently large number.

Theorem 1.9. *A function $\psi \in \mathcal{B}$ is an interpolation parameter if and only if ψ is pseudoconcave in the vicinity of ∞.*

Prior to proving this theorem, we establish two lemmas.

Lemma 1.1. *Assume that a function ψ belongs to the set $\mathcal{B}$ and is pseudoconcave in the vicinity of ∞. Then there exists a concave function $\psi_0 : (0, \infty) \to (0, \infty)$ such that, for any number $\varepsilon > 0$, the relation $\psi(t) \asymp \psi_0(t)$ holds for all $t \geq \varepsilon$.*

Proof of Lemma 1.1. By the condition, there exist a number $r \gg 1$ and a concave function $\psi_1 : (r, \infty) \to (0, \infty)$ such that $\psi(t) \asymp \psi_1(t)$ for $t > r$. Since the function ψ_1 is concave and positive on the semiaxis (r, ∞), it is (nonstrictly) increasing on this semiaxis. In addition, for any fixed point $t_0 \in (r, \infty)$, the slope function $(\psi_1(t) - \psi_1(t_0))/(t - t_0)$, $t \in (r, \infty) \setminus \{t_0\}$, (nonstrictly) decreases. Hence, the function ψ_1 has the right tangent at the point $r + 1$, which makes an acute (or zero angle) with the abscissa axis. We now specify a function ψ_2 on the semiaxis $[0, \infty)$ in such a way that its graph coincides with the indicated tangent in the interval $[0, r + 1)$ and with the graph of the function ψ_1 on the semiaxis $[r + 1, \infty)$. The function ψ_2 increases on $[0, \infty)$ and is concave on $(0, \infty)$. The latter follows from the fact that, for every fixed point $t_0 \in (0, \infty)$, the slope function $(\psi_2(t) - \psi_2(t_0))/(t - t_0)$ of $t \in (0, \infty) \setminus \{t_0\}$ decreases. We set

$$\psi_0(t) := \psi_2(t) + |\psi_2(0)| + 1.$$

The function ψ_0 is positive, increases, and is concave on the semiaxis $(0, \infty)$. We choose an arbitrary number $\varepsilon > 0$. Note that $\psi(t) \asymp 1 \asymp \psi_0(t)$ for $t \in [\varepsilon, r+1+\varepsilon]$. Further, since the function ψ_2 increases and is positive on the semiaxis $[r+1, \infty)$, we get

$$|\psi_2(0)| + 1 \leq c\,\psi_2(t)$$

for $t \geq r+1$, where

$$c := \frac{|\psi_2(0)| + 1}{\psi_2(r+1)} > 0.$$

This yields the relation

$$\psi(t) \asymp \psi_1(t) = \psi_2(t) \asymp \psi_0(t)$$

for $t \geq r+1$. Thus, $\psi(t) \asymp \psi_0(t)$ for $t \geq \varepsilon$, Q.E.D.

Lemma 1.2. *Let a function $\psi \in \mathcal{B}$ and a number $r \geq 0$ be given. The function ψ is pseudoconcave on the semiaxis (r, ∞) if and only if there exists a number $c > 0$ such that*

$$\frac{\psi(t)}{\psi(s)} \leq c \max\Big\{1, \frac{t}{s}\Big\} \quad \textit{for any} \quad t, s > r.$$

Proof of Lemma 1.2. For $r = 0$, this lemma was proved by J. Peetre [189] (in this case, the condition $\psi \in \mathcal{B}$ is not necessary; see also [24, Lemma 5.4.3]). For $r > 0$, the sufficiency is proved similarly. The necessity is reduced to the case $r = 0$ with the help of Lemma 1.1. Indeed, assume that ψ is pseudoconcave on (r, ∞) and consider the function ψ_0 from this lemma, where $\varepsilon = r$. Then

$$\frac{\psi(t)}{\psi(s)} \asymp \frac{\psi_0(t)}{\psi_0(s)} \leq c_0 \max\Big\{1, \frac{t}{s}\Big\}$$

for any $t, s > r$. (In fact, $c_0 = 1$ for the concave function ψ_0 [189]). Lemma 1.2 is proved.

Proof of Theorem 1.9. *Sufficiency.* Assume that a function $\psi \in \mathcal{B}$ is pseudoconcave in the vicinity of ∞. We prove that this function is an interpolation parameter.

We arbitrarily choose the same admissible pairs of Hilbert spaces $X = [X_0, X_1]$ and $Y = [Y_0, Y_1]$ and a linear mapping T as in Definition 1.2. Let $J_X : X_1 \leftrightarrow X_0$ and $J_Y : Y_1 \leftrightarrow Y_0$ be the generating operators for the pairs X and Y, respectively. With the help of the spectral theorem, we reduce these operators self-adjoint in X_0 and Y_0 to the form of multiplication by a function; namely,

$$J_X = I_X^{-1}\,(\alpha \cdot I_X) \quad \text{and} \quad J_Y = I_Y^{-1}\,(\beta \cdot I_Y). \tag{1.15}$$

Here, $I_X : X_0 \leftrightarrow L_2(U, d\mu)$ and $I_Y : Y_0 \leftrightarrow L_2(V, d\nu)$ are isometric isomorphisms, (U, μ) and (V, ν) are spaces with finite measures, and $\alpha : U \to (0, \infty)$ and $\beta : V \to (0, \infty)$ are measurable functions. Since both operators $T : X_0 \to Y_0$ and $T : X_1 \to Y_1$ are bounded, we conclude that the following operators are also bounded:

$$I_Y\, T\, I_X^{-1} : L_2(U, d\mu) \to L_2(V, d\nu), \tag{1.16}$$

$$I_Y\, J_Y\, T\, J_X^{-1}\, I_X^{-1} : L_2(U, d\mu) \to L_2(V, d\nu). \tag{1.17}$$

By virtue of (1.15), we find

$$I_Y\, J_Y\, T\, J_X^{-1}\, I_X^{-1} = (\beta \cdot I_Y)\, T\, I_X^{-1}(\alpha^{-1} \cdot).$$

Hence, (1.17) implies that the operator

$$I_Y\, T\, I_X^{-1} = \beta^{-1} \cdot (I_Y\, J_Y\, T\, J_X^{-1}\, I_X^{-1})(\alpha \cdot) : L_2(U, \alpha^2 d\mu) \to L_2(V, \beta^2 d\nu) \tag{1.18}$$

is bounded.

Let $\psi_0 : (0, \infty) \to (0, \infty)$ be the same concave function as in Lemma 1.1. Note that $\psi_0 \in \mathcal{B}$ and

$$X_\psi = X_{\psi_0}, \quad Y_\psi = Y_{\psi_0} \quad \text{with equivalence of the norms} \tag{1.19}$$

(see Remark 1.1). J. Peetre proved [189] that a positive function is pseudoconcave on $(0, \infty)$ if and only if it is an interpolation function of power $p > 0$ (see also [24, Theorem 5.4.4]). For the concave function ψ_0 and the case $p = 2$, his result means that the boundedness of operators (1.16) and (1.18) implies the boundedness of the operator

$$I_Y\, T\, I_X^{-1} : L_2(U, (\psi_0 \circ \alpha^2)\, d\mu) \to L_2(V, (\psi_0 \circ \beta^2)\, d\nu). \tag{1.20}$$

We now pass from operator (1.20) to the operator $T : X_{\psi_0} \to Y_{\psi_0}$ by using the isometric isomorphisms $\psi_0(J_X) : X_{\psi_0} \leftrightarrow X_0$ and $\psi_0(J_Y) : Y_{\psi_0} \leftrightarrow Y_0$. We reduce these isomorphisms to the form of multiplication by a function

$$I_X\, \psi_0(J_X) = (\psi_0 \circ \alpha) \cdot I_X : X_{\psi_0} \leftrightarrow L_2(U, d\mu),$$

$$I_Y\, \psi_0(J_Y) = (\psi_0 \circ \beta) \cdot I_Y : Y_{\psi_0} \leftrightarrow L_2(V, d\nu).$$

As a result, we obtain the isometric isomorphisms

$$I_X = (\psi_0^{-1} \circ \alpha) \cdot (I_X\, \psi_0(J_X)) : X_{\psi_0} \leftrightarrow L_2(U, (\psi^2 \circ \alpha)\, d\mu),$$

$$I_Y = (\psi_0^{-1} \circ \beta) \cdot (I_Y\, \psi_0(J_Y)) : Y_{\psi_0} \leftrightarrow L_2(V, (\psi^2 \circ \beta)\, d\nu).$$

Together with (1.20), they imply the boundedness of the operator

$$T = I_Y^{-1}(I_Y\, T\, I_X^{-1}) I_X : X_{\psi_0} \to Y_{\psi_0}.$$

Thus, in view of equalities (1.19), we find

$$(T : X_j \to Y_j,\ j = 0, 1) \Rightarrow (T : X_{\psi_0} \to Y_{\psi_0}) \Rightarrow (T : X_\psi \to Y_\psi),$$

where the linear operators are bounded. Hence, by Definition 1.2, the function ψ is an interpolation parameter. Sufficiency is proved.

Necessity. Assume that a function $\psi \in \mathcal{B}$ is an interpolation parameter. Let us show that ψ is pseudoconcave in the vicinity of ∞. We proceed by analogy with [189] and [24, Sec. 5.4, p. 117].

Consider a space $L_2(U, d\mu)$, where $U = \{0, 1\}$, $\mu(\{0\}) = \mu(\{1\}) = 1$, and define a linear mapping T on this space by the formulas $(Tu)(0) = 0$ and $(Tu)(1) = u(0)$, where $u \in L_2(U, d\mu)$. We choose arbitrary numbers $s, t > 1$ and set $\omega(0) := s^2$, $\omega(1) := t^2$. Thus, we get an admissible pair of spaces $X := [L_2(U, d\mu), L_2(U, \omega\, d\mu)]$ and bounded operators

$$T : L_2(U, d\mu) \to L_2(U, d\mu) \quad \text{and} \quad T : L_2(U, \omega\, d\mu) \to L_2(U, \omega\, d\mu)$$

whose norms are equal to 1 and t/s respectively. Since ψ is an interpolation parameter, we get a bounded operator $T : X_\psi \to X_\psi$ whose norm satisfies the inequality

$$\|T\|_{X_\psi \to X_\psi} \le c \max\Big\{1, \frac{t}{s}\Big\} \tag{1.21}$$

by virtue of Theorem 1.8, where we take $Y = X$ and $m = 1$. Here, the number $c > 0$ is independent of $t, s > 1$.

One can easily compute the norm in the space X_ψ. Indeed, the operator J of multiplication by the function $\omega^{1/2}$ is the generating operator for the pair X. Thus, since $\psi(J)$ is the operator of multiplication by the function $\psi \circ \omega^{1/2}$, we can write

$$\|u\|^2_{X_\psi} = \|(\psi \circ \omega^{1/2}) \cdot u\|^2_{L_2(U, d\mu)} = \psi^2(s)\, |u(0)|^2 + \psi^2(t)\, |u(1)|^2,$$

$$\|Tu\|^2_{X_\psi} = \psi^2(t)\, |u(0)|^2.$$

This yields

$$\|T\|_{X_\psi \to X_\psi} = \frac{\psi(t)}{\psi(s)}. \tag{1.22}$$

Relations (1.21) and (1.22) now imply the inequality

$$\frac{\psi(t)}{\psi(s)} \le c \max\Big\{1, \frac{t}{s}\Big\} \quad \text{for any} \quad t, s > 1.$$

By Lemma 1.2, this is equivalent to the pseudoconcavity of the function ψ on the semiaxis $(1, \infty)$. Necessity is proved.

Theorem 1.9 is proved.

At the end of this subsection, we give a description of all Hilbert interpolation spaces for a given admissible pair of Hilbert spaces.

Definition 1.5. A Hilbert space H is called an *interpolation* space for an admissible pair of Hilbert spaces $[X_0, X_1]$ if

(i) the continuous embeddings $X_1 \hookrightarrow H \hookrightarrow X_0$ are true;

(ii) any linear operator $T : X_0 \to X_0$ bounded on each space X_0 and X_1 is also a bounded operator on the space H.

The following important result is due to Ovchinnikov [179, p. 511, Theorem 11.4.1].

Proposition 1.1. *Let $X = [X_0, X_1]$ be an arbitrary admissible pair of Hilbert spaces. If a Hilbert space H is an interpolation space for this pair, then there exists a function $\psi \in \mathcal{B}$ pseudoconcave in the vicinity of ∞ such that the spaces H and X_ψ coincide up to equivalence of norms.*

Proposition 1.1 and Theorem 1.9 immediately yield the following assertion:

Corollary 1.1. *Let $X = [X_0, X_1]$ be an arbitrary admissible pair of Hilbert spaces. The class of all Hilbert spaces that are interpolation spaces for X coincides (up to equivalence of norms) with the class of all spaces X_ψ, where $\psi \in \mathcal{B}$ is an arbitrary function pseudoconcave in the vicinity of ∞.*

1.2 Regularly varying functions and their generalization

Regularly varying functions (and the functions weakly equivalent to them) play a fundamental role in our investigation. We use these functions both to parametrize the Hörmander spaces and as interpolation parameters.

1.2.1 Regularly varying functions

The following definition is important for our presentation.

Definition 1.6. A positive function ψ given on a real semiaxis $[b, \infty)$ is said to be *regularly varying* of order $\theta \in \mathbb{R}$ at ∞ if ψ is Borel measurable on $[b_0, \infty)$ for some number $b_0 \geq b$ and satisfies the condition

$$\lim_{t \to \infty} \frac{\psi(\lambda t)}{\psi(t)} = \lambda^\theta \quad \text{for any} \quad \lambda > 0. \tag{1.23}$$

A positive function is called *slowly varying* at ∞ if it is a regularly varying function of order zero at ∞.

The notion of regularly varying function was introduced by J. Karamata [91] (for the class of continuous functions). Regularly varying functions are close to the power functions. They are well studied and have numerous applications mainly due to the special role played by these functions in the Tauberian-type theorems (see the books [26, 128, 199, 235].

By **SV** we denote the set of all functions slowly varying at ∞. It is clear that ψ is a regularly varying function of order θ at ∞ if and only if $\psi(t) = t^{\theta}\varphi(t)$ for $t \gg 1$ and a certain function $\varphi \in \mathbf{SV}$. Therefore, in the study of regularly varying functions, it is sufficient to restrict ourselves to slowly varying functions.

We now present the most known (standard) example [26, Sec. 1.3.3] of a slowly varying function.

Example 1.1. Let $r_1, r_2, \ldots$, and r_k be $k \in \mathbb{N}$ given real numbers. We set

$$\varphi(t) = (\log t)^{r_1} (\log\log t)^{r_2} \ldots (\log \ldots \log t)^{r_k} \quad \text{for} \quad t \gg 1.$$

Then $\varphi \in \mathbf{SV}$.

The functions considered in this example form the so-called *logarithmic multiscale*, which has numerous applications in the theory of function spaces. Some other examples of functions from the class **SV** are presented in what follows.

We now formulate two fundamental properties of slowly varying functions. They were proved by J. Karamata [91, 93] for continuous functions and (somewhat later) by numerous researchers for measurable functions (see the books [26, Sec. 1.2, 1.3], [235, Sec. 1.2], and the references therein).

Proposition 1.2 (Uniform Convergence Theorem). *Let $\varphi \in \mathbf{SV}$. Then, for any fixed compact interval $[a, b]$ with $0 < a < b < \infty$, the ratio $\varphi(\lambda t)/\varphi(t)$ tends to 1 as $t \to \infty$ uniformly in $\lambda \in [a, b]$.*

Proposition 1.3 (Representation Theorem). *Let $\varphi \in \mathbf{SV}$. Then*

$$\varphi(t) = \exp\left(\beta(t) + \int_b^t \frac{\alpha(\tau)}{\tau}\, d\tau\right), \quad \textit{whenever} \quad t \geq b \tag{1.24}$$

for some number $b > 0$, some continuous function $\alpha : [b, \infty) \to \mathbb{R}$ approaching zero at ∞, and some Borel measurable bounded function $\beta : [b, \infty) \to \mathbb{R}$ with finite limit at ∞. The converse assertion is also true: each function of the form (1.24) belongs to the class **SV**.

Remark 1.2. A function of the form (1.24) with $\beta(t) = \text{const}$ is called a *normalized* slowly varying function. For this function,

$$\alpha(\tau) = \frac{\tau\varphi'(\tau)}{\varphi(\tau)}$$

(see [26, Sec. 1.3.2].)

Proposition 1.3 yields the following useful sufficient condition for a function to be slowly varying (see, e.g., [235, Sec. 1.2]):

Proposition 1.4. *Assume that a differentiable function* $\varphi : (b, \infty) \to (0, \infty)$ *satisfies the condition* $t\varphi'(t)/\varphi(t) \to 0$ *as* $t \to \infty$. *Then* $\varphi \in \mathrm{SV}$.

Proposition 1.4 leads to the following three interesting examples of slowly varying functions:

Example 1.2. Let $\psi(t) := \exp\varphi(t)$ for $t \gg 1$, where the function φ is taken from Example 1.1 in which $r_1 < 1$. Then $\psi \in \mathrm{SV}$.

Example 1.3. Let $\alpha, \beta, \gamma \in \mathbb{R}$ with $\beta \neq 0$ and $0 < \gamma < 1$. We set

$$\omega(t) := \alpha + \beta \sin \ln^{\gamma} t \quad \text{and} \quad \varphi(t) := (\ln t)^{\omega(t)}$$

for $t > 1$. Then $\varphi \in \mathrm{SV}$.

Example 1.4. Let $\alpha, \beta, \gamma \in \mathbb{R}$ with $\alpha \neq 0$ and $0 < \gamma < \beta < 1$. We set

$$r(t) := \alpha(\ln t)^{-\beta} \sin \ln^{\gamma} t \quad \text{and} \quad \varphi(t) := t^{r(t)}$$

for $t > 1$. Then $\varphi \in \mathrm{SV}$.

The last two examples show that a function φ slowly varying at ∞ may have infinite oscillation, i.e.,

$$\liminf_{t\to\infty} \varphi(t) = 0 \quad \text{and} \quad \limsup_{t\to\infty} \varphi(t) = \infty.$$

1.2.2 Quasiregularly varying functions

We use regularly varying functions of order $\theta \in (0, 1)$ as interpolation parameters. According to Remark 1.1, the property to be an interpolation parameter is inherited under the transition to a (weakly) equivalent function. Hence, the following generalization of the notion of regularly varying function is useful:

Definition 1.7. A positive function ψ given on the real semiaxis $[b,\infty)$ is called a *quasiregularly varying* function of order $\theta \in \mathbb{R}$ at ∞ if there exist a number $b_1 \geq b$ and a regularly varying function $\psi_1 : [b_1,\infty) \to (0,\infty)$ of order θ at ∞ such that $\psi(t) \asymp \psi_1(t)$ for $t \geq b_1$.

A positive function is called *quasislowly varying* at ∞ if it is a quasiregularly varying function of order zero at ∞.

By QSV we denote the set of all functions quasislowly varying at ∞.

Theorem 1.10. *The class* QSV *consists of all functions of the form* (1.24), *where b is a positive number, $\alpha : [b,\infty) \to \mathbb{R}$ is a continuous function approaching zero at ∞, and $\beta : [b,\infty) \to \mathbb{R}$ is a bounded function.*

Proof. By Definition 1.7, $\varphi \in \mathrm{QSV}$ if and only if $\varphi(t) = \omega(t)\varphi_1(t)$ for $t \gg 1$, where $\varphi_1 \in \mathrm{SV}$, and ω is a positive function such that both ω and $1/\omega$ are bounded in the vicinity of ∞. Therefore, according to Proposition 1.3,

$$\varphi \in \mathrm{QSV} \;\Leftrightarrow\; \varphi(t) = \exp\left(\log\omega(t) + \beta(t) + \int_b^t \frac{\alpha(\tau)}{\tau}\,d\tau\right) \quad \text{for } t \geq b,$$

where the functions α and β satisfy the condition of this proposition and the number $b \gg 1$. This yields Theorem 1.10 because the function $\log\omega + \beta$ is bounded on the semiaxis $[b,\infty)$.

Theorem 1.10 is proved.

The following interpolation property of quasiregularly varying functions plays a fundamental role in our subsequent investigations:

Theorem 1.11. *Assume that $\psi \in \mathcal{B}$ is a quasiregularly varying function of order θ at ∞ with $0 < \theta < 1$. Then ψ is an interpolation parameter.*

Proof. We write $\psi(t) = t^{\theta}\varphi(t)$ for $t > 0$, where $\varphi \in \mathrm{QSV}$. By using Theorem 1.10, we represent the function φ in the form (1.24), where the functions α and β satisfy the condition of this theorem. We set

$$\varepsilon := \min\{\theta, 1-\theta\} > 0$$

and choose a number $b_\varepsilon \geq b$ such that $|\alpha(t)| < \varepsilon$, whenever $t > b_\varepsilon$. For any $t, s > b_\varepsilon$, in view of (1.24), we get

$$\frac{\varphi(t)}{\varphi(s)} = \exp\left(\beta(t) - \beta(s) + \int_s^t \frac{\alpha(\tau)}{\tau}\,d\tau\right)$$

$$\leq c\exp\left|\int_s^t \frac{\varepsilon}{\tau}\,d\tau\right| = c\max\left\{\left(\frac{t}{s}\right)^{\varepsilon}, \left(\frac{s}{t}\right)^{\varepsilon}\right\}.$$

Here, the number $c > 0$ is independent of t and s because the function β is bounded. Further, since $0 \le \theta \pm \varepsilon \le 1$, we obtain

$$\frac{\psi(t)}{\psi(s)} = \frac{t^{\theta}\varphi(t)}{s^{\theta}\varphi(s)} \le c\max\Big\{\Big(\frac{t}{s}\Big)^{\theta+\varepsilon}, \Big(\frac{t}{s}\Big)^{\theta-\varepsilon}\Big\} \le c\max\Big\{1, \frac{t}{s}\Big\} \quad \text{for} \quad t, s > b_{\varepsilon}.$$

Hence, by virtue of Lemma 1.2, the function $\psi \in \mathcal{B}$ is pseudoconcave in the vicinity of ∞. According to Theorem 1.9, this is equivalent to the assertion that ψ is an interpolation parameter. Theorem 1.11 is proved.

Remark 1.3. The direct proof of Theorem 1.11 (without using Theorem 1.9) can be found in [144, Sec. 2].

Remark 1.4. Theorem 1.11 is not true in the limiting cases $\theta = 0$ and $\theta = 1$ even if we additionally assume that $\psi(t) \to \infty$ as $t \to \infty$ for $\theta = 0$ or $\psi(t)/t \to 0$ as $t \to \infty$ for $\theta = 1$.

We now present the corresponding examples.

Example 1.5. Case $\theta = 0$. Let

$$h(t) := (\ln t)^{-1/2} \sin \ln^{1/4} t$$

for $t > 1$. We define the function ψ as follows: $\psi(t) := t^{h(t)} + \ln t$ for $t \ge 3$, and $\psi(t) := 1$ for $0 < t < 3$. Note that $\psi \in \mathcal{B}$ and $\psi(t) \to \infty$ as $t \to \infty$. We immediately establish the convergence $t\psi'(t)/\psi(t) \to 0$ as $t \to \infty$. By virtue of Proposition 1.4, this means that the function ψ is slowly varying at ∞. Let us show that ψ is not pseudoconcave in the vicinity of ∞ and, hence, by Theorem 1.9, it is not an interpolation parameter. Consider sequences of numbers

$$t_k := \exp((2\pi k + \pi/2)^4) \quad \text{and} \quad s_k := \exp((2\pi k + \pi)^4),$$

where $k \in \mathbb{N}$. Further, we compute $h(t_k) = (2\pi k + \pi/2)^{-2}$ and $h(s_k) = 0$. This yields

$$\ln \psi(t_k) \ge h(t_k) \ln t_k = (2\pi k + \pi/2)^2,$$

$$\psi(s_k) = 1 + (2\pi k + \pi)^4.$$

Thus,

$$\frac{\psi(t_k)}{\psi(s_k)} \ge \frac{\exp((2\pi k + \pi/2)^2)}{(1 + (2\pi k + \pi)^4)} \to \infty \quad \text{as} \quad k \to \infty.$$

However $t_k < s_k$ and, hence, by Lemma 1.2, we conclude that the function ψ cannot be pseudoconcave in the vicinity of ∞.

Example 1.6. Case $\theta = 1$. By using the function ψ from Example 1.5, we set $\psi_1(t) := t/\psi(t)$ for $t > 0$. By the definition, ψ_1 is a regularly varying function of order $\theta = 1$ at ∞. Note that $\psi_1 \in \mathcal{B}$ and $\psi_1(t)/t = 1/\psi(t) \to 0$ as $t \to \infty$. By Theorem 1.4, the function ψ_1 is not an interpolation parameter because, otherwise, the function $\psi(t) = t/\psi_1(t)$ must be an interpolation parameter. However, as shown in the previous example, this is impossible.

Further, we use the following properties of the class QSV.

Theorem 1.12. *Let $\varphi, \chi \in$ QSV. The following assertions are true:*

(i) *there exists a positive function $\varphi_1 \in C^\infty((0;\infty)) \cap$ SV such that $\varphi(t) \asymp \varphi_1(t)$ for $t \gg 1$;*

(ii) *for any number $\theta > 0$, $t^{-\theta}\varphi(t) \to 0$ and $t^{\theta}\varphi(t) \to \infty$ as $t \to \infty$;*

(iii) *the functions $\varphi + \chi$, $\varphi\chi$, φ/χ, and φ^σ with $\sigma \in \mathbb{R}$ belong to the class* QSV;

(iv) *let $\theta \geq 0$ be an arbitrary number and, in addition, $\varphi(t) \to \infty$ as $t \to \infty$ for $\theta = 0$; then the composite function $\chi(t^\theta\varphi(t))$ of argument t belongs to the class* QSV.

Proof. If we additionally assume that $\varphi, \chi \in$ SV, then we get the well-known [235, Sec. 1.5] properties of slowly varying functions. (Under this assumption, we get even the relation of strong equivalence $\varphi(t) \sim \varphi_1(t)$ as $t \to \infty$ in assertion (i).) This immediately yields assertions (i), (ii), and (iii) for the functions $\varphi, \chi \in$ QSV.

We now prove assertion (iv). Let $\lambda > 0$ be an arbitrary number. Since $\varphi \in$ QSV, the quantities $\varphi(\lambda t)/\varphi(t)$ and $\varphi(t)/\varphi(\lambda t)$ are both bounded as functions of $t \gg 1$. Hence, for any positive function $\chi_1 \in$ SV satisfying the condition $\chi_1(\tau) \asymp \chi(\tau)$ whenever $\tau \gg 1$, by virtue of Proposition 1.2, we get

$$\frac{\chi_1((\lambda t)^\theta \varphi(\lambda t))}{\chi_1(t^\theta\varphi(t))} = \frac{\chi_1\left(\frac{\lambda^\theta\varphi(\lambda t)}{\varphi(t)}\, t^\theta\varphi(t)\right)}{\chi_1(t^\theta\varphi(t))} \to 1 \quad \text{as} \quad t \to \infty.$$

Here, we have used the fact that $t^\theta\varphi(t) \to \infty$ as $t \to \infty$. Hence, the function $\chi_1(t^\theta\varphi(t))$ slowly varies at ∞. At the same time, $\chi(t^\theta\varphi(t)) \asymp \chi_1(t^\theta\varphi(t))$ whenever $t \gg 1$. Thus, the function $\chi(t^\theta\varphi(t))$ of t belongs to the class QSV. Assertion (iv) and, therefore, Theorem 1.12 are proved.

In what follows, we also need the following generalization of assertion (iv) in Theorem 1.12 to the case where $\theta = 0$ and $\varphi(t) \not\to \infty$ as $t \to \infty$.

Theorem 1.13. *Suppose that $\varphi : [b, \infty) \to (0, \infty)$ and $\chi : (0, \infty) \to (0, \infty)$ are functions from the class* QSV. *Assume that the function $1/\varphi$ is bounded on the semiaxis $[b, \infty)$ and that the functions χ and $1/\chi$ are bounded on every segment*

$[a_1, a_2]$, *where* $0 < a_1 < a_2 < \infty$. *Then the composite function* $\chi(\varphi(t))$ *of the argument* t *belongs to the class* **QSV**.

Proof. By the condition, $\varphi(t) \geq r$ for all $t \geq b$ and some $r > 0$. According to Theorem 1.12(i), there exist a number $b_1 \geq \max\{b, r\}$ and continuous functions $\varphi_1, \chi_1 : [b_1, \infty) \to (0, \infty)$ slowly varying at ∞ and such that $\varphi(t) \asymp \varphi_1(t)$ and $\chi(t) \asymp \chi_1(t)$ whenever $t \geq b_1$. In this case, it is possible to assume that $\varphi_1(t) \geq r$ for $t \geq b_1$. In addition, let $\varphi_1(t) := \varphi_1(b_1)$ and $\chi_1(t) := \chi_1(b_1)$ for $0 < t < b_1$. Then $\varphi_1, \chi_1 \in C((0, \infty))$ and $\varphi_1(t) \geq r$ for $t > 0$ and, moreover, $\chi(t) \asymp \chi_1(t)$ for $t \geq r$ in view of the condition of this theorem. We now show that $\chi(\varphi(t)) \asymp \chi_1(\varphi_1(t))$ whenever $t \geq b_1$ and that $\chi_1(\varphi_1(t))$ is a function slowly varying at ∞. This would imply the required property $\chi \circ \varphi \in$ **QSV**.

Let $t \geq b_1$. We have

$$c^{-1} \leq \varphi_1(t)/\varphi(t) \leq c$$

for some number $c \geq 1$. Thus, by Proposition 1.2, the function $\chi_1 \in$ **SV** possesses the property

$$\frac{\chi_1(\varphi_1(t))}{\chi_1(\varphi(t))} = \frac{\chi_1\left(\frac{\varphi_1(t)}{\varphi(t)}\,\varphi(t)\right)}{\chi_1(\varphi(t))} \to 1 \quad \text{as} \quad \varphi(t) \to \infty.$$

Hence, there exists a number $\varrho \geq r$ such that $\chi_1(\varphi(t)) \asymp \chi_1(\varphi_1(t))$ whenever $\varphi(t) \geq \varrho$. Moreover,

$$\chi_1(\varphi(t)) \asymp \chi(\varphi(t)) \asymp 1 \asymp \chi(\varphi_1(t)) \asymp \chi_1(\varphi_1(t))$$

provided that $r \leq \varphi(t) \leq \varrho$. Hence,

$$\chi(\varphi(t)) \asymp \chi_1(\varphi(t)) \asymp \chi_1(\varphi_1(t)) \quad \text{whenever} \quad t \geq b_1.$$

We fix a number $\lambda > 0$ and prove that

$$\frac{\chi_1(\varphi_1(\lambda t))}{\chi_1(\varphi_1(t))} \to 1 \quad \text{as} \quad t \to \infty.$$

We choose an arbitrary number $\varepsilon > 0$. Since $\varphi_1 \in$ **SV**, the function $\beta_\lambda(t) := \varphi_1(\lambda t)/\varphi_1(t) \to 1$ as $t \to \infty$. In particular, $1/2 \leq \beta_\lambda(t) \leq 2$ for $t \geq t_\lambda > 0$. Therefore, by virtue of Proposition 1.2, for the function $\chi_1 \in$ **SV**, there exists a number $k = k(\varepsilon) > r$ such that

$$\left|\frac{\chi_1(\varphi_1(\lambda t))}{\chi_1(\varphi_1(t))} - 1\right| = \left|\frac{\chi_1(\beta_\lambda(t)\varphi_1(t))}{\chi_1(\varphi_1(t))} - 1\right| < \varepsilon \qquad (1.25)$$

whenever $t \geq t_\lambda$ and $\varphi_1(t) > k$. In addition, since the function $\chi_1 > 0$ is uniformly continuous on the segment $[r, k+1]$, there exists a number $m =$

$m(\varepsilon) > 0$ such that

$$\begin{aligned}|\chi_1(\varphi_1(\lambda t)) - \chi_1(\varphi_1(t))| &= |\chi_1(\beta_\lambda(t)\varphi_1(t)) - \chi_1(\varphi_1(t))| \\ &< \varepsilon \min\{\chi_1(\tau) : r \le \tau \le k\}\end{aligned}$$

whenever $t \ge m$ and $r \le \varphi_1(t) \le k$. Hence,

$$\left|\frac{\chi_1(\varphi_1(\lambda t))}{\chi_1(\varphi_1(t))} - 1\right| = \frac{|\chi_1(\varphi_1(\lambda t)) - \chi_1(\varphi_1(t))|}{\chi_1(\varphi_1(t))} < \varepsilon \tag{1.26}$$

for $t \ge m$ and $r \le \varphi_1(t) \le k$. Relations (1.25) and (1.26) now yield the inequality

$$\left|\frac{\chi_1(\varphi_1(\lambda t))}{\chi_1(\varphi_1(t))} - 1\right| < \varepsilon \quad \text{for} \quad t \ge \max\{t_\lambda, m\}.$$

Since the positive number ε is chosen arbitrarily, this inequality means that

$$\frac{\chi_1(\varphi_1(\lambda t))}{\chi_1(\varphi_1(t))} \to 1 \quad \text{as} \quad t \to \infty.$$

Hence, the function $\chi_1(\varphi_1(t))$ is slowly varying at ∞.

Theorem 1.13 is proved.

1.2.3 Auxiliary results

In the present subsection, we prove two auxiliary assertions about the properties of the class QSV. They will be used in what follows.

Lemma 1.3. *Suppose that a function $\varphi \in$ QSV is positive on the semiaxis $[1, \infty)$ and bounded, together with the function $1/\varphi$, on every segment $[1, b]$, where $1 < b < \infty$. Then, for any number $\varepsilon > 0$, there exists a number $c(\varepsilon) > 0$ such that*

$$\frac{\varphi(t)}{\varphi(s)} \le c(\varepsilon)\,(1 + |t - s|)^\varepsilon \quad \textit{for any} \quad t \ge 1 \quad \textit{and} \quad s \ge 1. \tag{1.27}$$

Proof. Without loss of generality, we can assume that $0 < \varepsilon < 1$. According to Theorem 1.10, we represent φ in the form (1.24). Since, in this representation, we have $\alpha(\tau) \to 0$ as $\tau \to \infty$, there exists a number $b_\varepsilon \ge 1$ such that $|\alpha(\tau)| \le \varepsilon$ for $\tau \ge b_\varepsilon$. We choose arbitrary numbers $t \ge 1$ and $s \ge 1$. In our proof, c_1, c_2, and c_3 are finite positive constants independent of t and s. We prove (1.27) separately for four possible cases of location of the numbers t and s relative to b_ε.

In the first case, $t \geq b_\varepsilon$ and $s \geq b_\varepsilon$. By virtue of (1.24), we find

$$\frac{\varphi(t)}{\varphi(s)} = \exp \int_s^t \frac{\alpha(\tau)}{\tau}\, d\tau \leq \exp\left| \int_s^t \frac{\varepsilon}{\tau}\, d\tau \right| = \max\left\{ \left(\frac{t}{s}\right)^\varepsilon, \left(\frac{s}{t}\right)^\varepsilon \right\}$$

$$= \max\left\{ \left(1 + \frac{t-s}{s}\right)^\varepsilon, \left(1 + \frac{s-t}{t}\right)^\varepsilon \right\} \leq (1 + |t-s|)^\varepsilon.$$

In the second case, $t \geq b_\varepsilon$ and $1 \leq s \leq b_\varepsilon$. By virtue of Theorem 1.12(ii) and the condition of the lemma, we get $\varphi(t) \leq c_1 t^\varepsilon$ and $1/\varphi(s) \leq c_1$. This yields

$$\frac{\varphi(t)}{\varphi(s)} \leq c_1^2 t^\varepsilon = c_1^2 (s + (t-s))^\varepsilon$$

$$\leq c_1^2 (b_\varepsilon + |t-s|)^\varepsilon \leq c_1^2 b_\varepsilon (1 + |t-s|)^\varepsilon.$$

In the third case, $1 \leq t \leq b_\varepsilon$ and $s \geq b_\varepsilon$. By analogy with the previous case, we obtain $1/\varphi(s) \leq c_2 s^\varepsilon$, $\varphi(t) \leq c_2$, and hence,

$$\frac{\varphi(t)}{\varphi(s)} \leq c_2^2 s^\varepsilon = c_2^2 (t + (s-t))^\varepsilon$$

$$\leq c_2^2 (b_\varepsilon + |s-t|)^\varepsilon \leq c_2^2 b_\varepsilon (1 + |t-s|)^\varepsilon.$$

In the fourth case, $1 \leq t \leq b_\varepsilon$ and $1 \leq s \leq b_\varepsilon$. This case is trivial. Indeed,

$$\frac{\varphi(t)}{\varphi(s)} \leq c_3 \leq c_3 (1 + |t-s|)^\varepsilon.$$

Thus, inequality (1.27) holds for any $t \geq 1$ and $s \geq 1$.
Lemma 1.3 is proved.

Lemma 1.4. *Let $\psi_1 \in$ QSV be a function positive and continuous on the semi-axis $[1, \infty)$ and satisfying the condition*

$$I_1 := \int_1^\infty \frac{dt}{t\, \psi_1(t)} < \infty. \tag{1.28}$$

Then there exists a positive continuous function $\psi_0 \in$ SV given on $[1, \infty)$ such that $\psi_0(t)/\psi_1(t) \to 0$ as $t \to \infty$ and

$$\int_1^\infty \frac{dt}{t\, \psi_0(t)} < \infty. \tag{1.29}$$

Proof. In view of Theorem 1.12(i), without loss of generality, we can assume that $\psi_1 \in C((0,\infty)) \cap \mathrm{SV}$. We set

$$\varphi(t) := \int\limits_t^\infty \frac{dt}{t\,\psi_1(t)} \quad \text{and} \quad \psi_0(t) := \psi_1(t)\,\sqrt{\varphi(t)} \quad \text{for } t \geq 1. \tag{1.30}$$

By the condition, φ is a finite positive function on $[1,\infty)$ such that $\varphi(t) \to 0$ as $t \to \infty$. Moreover, φ has the continuous derivative $\varphi'(t) = -(t\,\psi_1(t))^{-1}$, whenever $t \geq 1$. Thus, in view of the inclusion $\psi_1 \in \mathrm{SV}$, we get, by using the L'Hospital rule,

$$\lim_{t\to\infty} \frac{\varphi(\lambda t)}{\varphi(t)} = \lim_{t\to\infty} \frac{\lambda\,(\lambda\,t\,\psi_1(\lambda\,t))^{-1}}{(t\,\psi_1(t))^{-1}} = \lim_{t\to\infty} \frac{\psi_1(t)}{\psi_1(\lambda\,t)} = 1$$

for any $\lambda > 0$. Hence, $\varphi \in \mathrm{SV}$. We now consider the function ψ_0. It is positive and continuous on $[1,\infty)$. Since $\psi_1, \varphi \in \mathrm{SV}$, the function $\psi_0 \in \mathrm{SV}$ is slowly varying at ∞ (by definition). Furthermore, $\psi_0(t)/\psi_1(t) = \sqrt{\varphi(t)} \to 0$ as $t \to \infty$ and

$$\int\limits_1^\infty \frac{dt}{t\,\psi_0(t)} = \int\limits_1^\infty \frac{dt}{t\,\psi_1(t)\,\sqrt{\varphi(t)}} = -\int\limits_1^\infty \frac{d\varphi(t)}{\sqrt{\varphi(t)}} = -\int\limits_{I_1}^0 \frac{d\tau}{\sqrt{\tau}} < \infty.$$

Thus, ψ_0 satisfies all conditions of the lemma.

Lemma 1.4 is proved.

1.3 Hörmander spaces and the refined Sobolev scale

In the present section, we consider an important class of Hörmander inner product spaces parametrized with the help of regularly varying functions. This is the main class of function spaces in which we study elliptic operators. It is called the refined (Sobolev) scale.

1.3.1 Preliminary information and notation

We use the following generally accepted notation for complex linear topological spaces of test functions and distributions given in the Euclidean space $\mathbb{R}^n$:

- $C_0^\infty(\mathbb{R}^n)$ and $\mathcal{D}(\mathbb{R}^n)$ denote the space of all infinitely differentiable functions $u : \mathbb{R}^n \to \mathbb{C}$ with compact support;
- $\mathcal{S}(\mathbb{R}^n)$ is the Schwartz space of all infinitely differentiable functions $u : \mathbb{R}^n \to \mathbb{C}$ such that

$$\max\bigl\{(1+|x|)^m\,|\partial_x^\alpha u(x)| : x \in \mathbb{R}^n,\ |\alpha| \leq m\bigr\} < \infty$$

for any integer $m \geq 0$; here and in what follows $\alpha = (\alpha_1, \ldots, \alpha_n)$ is a multiindex (a vector with nonnegative integer coordinates), $|\alpha| := \alpha_1 + \ldots + \alpha_n$, and

$$\partial_x^\alpha u(x) := \frac{\partial^{|\alpha|} u(x)}{\partial x_1^{\alpha_1} \ldots \partial x_n^{\alpha_n}}$$

is the partial derivative corresponding to the multiindex α;

- $\mathcal{S}'(\mathbb{R}^n)$ is the space of all tempered distributions given in $\mathbb{R}^n$; this space is dual to $\mathcal{S}(\mathbb{R}^n)$;
- $\mathcal{D}'(\mathbb{R}^n)$ is the space of all distributions given in $\mathbb{R}^n$; this space is dual to $\mathcal{D}(\mathbb{R}^n)$.

From the viewpoint of applications to differential operators, it is convenient to interpret distributions as antilinear functionals. Thus, we consider antilinear continuous functionals on $\mathcal{S}(\mathbb{R}^n)$ or $\mathcal{D}(\mathbb{R}^n)$ as elements of the dual spaces $\mathcal{S}'(\mathbb{R}^n)$ or $\mathcal{D}'(\mathbb{R}^n)$. The mutual duality of the spaces of test functions and distributions in $\mathbb{R}^n$ is considered with respect to the expansion by continuity of a sesquilinear form

$$(u, v)_{\mathbb{R}^n} := \int_{\mathbb{R}^n} u(x)\, \overline{v(x)}\, dx.$$

This expansion is also denoted by $(u, v)_{\mathbb{R}^n}$. It is equal to the value of the distribution u on the test function v.

Every locally Lebesgue integrable function $u : \mathbb{R}^n \to \mathbb{C}$ is identified with a distribution (antilinear functional) $v \mapsto (u, v)_{\mathbb{R}^n}$ given on the test functions $v \in \mathcal{D}(\mathbb{R}^n)$. This distribution is called regular. In this sense, we have the following dense and continuous embeddings:

$$\mathcal{D}(\mathbb{R}^n) \hookrightarrow \mathcal{S}(\mathbb{R}^n) \hookrightarrow \mathcal{S}'(\mathbb{R}^n) \hookrightarrow \mathcal{D}'(\mathbb{R}^n).$$

The Fourier transform of an arbitrary distribution $u \in \mathcal{S}'(\mathbb{R}^n)$ is denoted by $\mathcal{F}u$ or simply by $\widehat{u}$. If $u \in \mathcal{S}(\mathbb{R}^n)$, then we use the following formula for the Fourier transform:

$$(\mathcal{F}u)(\xi) = \widehat{u}(\xi) := \int_{\mathbb{R}^n} e^{ix\cdot\xi}\, u(x)\, dx, \quad \xi \in \mathbb{R}^n.$$

Here, as usual, i is the imaginary unit and $x \cdot \xi := x_1\xi_1 + \ldots + x_n\xi_n$ is the inner product of vectors $x, \xi \in \mathbb{R}^n$. The Fourier transform is an isomorphism of the space $\mathcal{S}(\mathbb{R}^n)$ onto itself. The inverse Fourier transform is given by the formula

$$u(x) = (\mathcal{F}^{-1}\widehat{u})(x) = (2\pi)^{-n} \int_{\mathbb{R}^n} e^{-ix\cdot\xi}\, \widehat{u}(\xi)\, d\xi, \quad x \in \mathbb{R}^n.$$

The Fourier transform is uniquely continued to an isomorphism of the space $\mathcal{S}'(\mathbb{R}^n)$ onto itself. In this case, the Parseval equality is preserved, namely,

$$(\widehat{u},\widehat{v})_{\mathbb{R}^n}=(2\pi)^n(u,v)_{\mathbb{R}^n}\quad\text{for all }\ u\in\mathcal{S}'(\mathbb{R}^n),\ \ v\in\mathcal{S}(\mathbb{R}^n).$$

This equality can be used as a definition of the Fourier transform of an arbitrary distribution $u\in\mathcal{S}'(\mathbb{R}^n)$.

We now recall some additional standard notation for function spaces. The space $L_p(\mathbb{R}^n)$ consists of all Lebesgue measurable functions $u:\mathbb{R}^n\to\mathbb{C}$ such that

$$\|u\|^p_{L_p(\mathbb{R}^n)}:=\int\limits_{\mathbb{R}^n}|u(x)|^p\,dx<\infty\quad\text{for}\quad 1<p<\infty,$$

$$\|u\|_{L_\infty(\mathbb{R}^n)}:=\operatorname{ess\,sup}\{\,|u(x)|:\,x\in\mathbb{R}^n\}<\infty\quad\text{for}\quad p=\infty.$$

The space $L_p(\mathbb{R}^n)$ with $1\le p\le\infty$ is a Banach space with respect to the norm $\|u\|_{L_p(\mathbb{R}^n)}$. Naturally, we identify the functions that coincide almost everywhere on $\mathbb{R}^n$ as elements of $L_p(\mathbb{R}^n)$. In the case $p=2$, this space turns into a Hilbert space. The norm in this space is generated by the inner product $(u,v)_{\mathbb{R}^n}$. The Fourier transform multiplied by $(2\pi)^{-n/2}$ is an isometric isomorphism of the space $L_2(\mathbb{R}^n)$ onto itself.

As usual, $C^k(\mathbb{R}^n)$, where $k\ge 0$ is an integer, denotes the space of all functions $u:\mathbb{R}^n\to\mathbb{C}$ with continuous partial derivatives of any order $\le k$. The subspace $C^k_{\mathrm{b}}(\mathbb{R}^n)$ is formed by the functions $u\in C^k(\mathbb{R}^n)$ for which all partial derivatives of the orders $\le k$ are bounded on $\mathbb{R}^n$. The space $C^k_{\mathrm{b}}(\mathbb{R}^n)$ is Banach with respect to the norm

$$\sup\{\,|\partial^\alpha_x u(x)|:\,x\in\mathbb{R}^n,\ |\alpha|\le k\}.$$

We also use the spaces

$$C^\infty(\mathbb{R}^n):=\bigcap_{k\ge 0}C^k(\mathbb{R}^n)\quad\text{and}\quad C^\infty_{\mathrm{b}}(\mathbb{R}^n):=\bigcap_{k\ge 0}C^k_{\mathrm{b}}(\mathbb{R}^n).$$

1.3.2 Hörmander spaces

We now present the definition of function spaces introduced and investigated by L. Hörmander in [81, Sec. 2.2] (see also [85, Sec. 10.1]). These spaces are formed by the distributions in $\mathbb{R}^n$, where $n\in\mathbb{N}$, They are denoted by $B_{p,\mu}(\mathbb{R}^n)$. In the present subsection, the scalar index p satisfies the inequality $1\le p\le\infty$. Moreover, the index μ is a continuous positive function $\mu=\mu(\xi)$ of $\xi\in\mathbb{R}^n$ playing the role of a weight function in the following sense:

Definition 1.8. A function $\mu:\mathbb{R}^n\to(0,\infty)$ is called a *weight* function if there exist numbers $c\ge 1$ and $l>0$ such that

$$\frac{\mu(\xi)}{\mu(\eta)}\le c\,(1+|\xi-\eta|)^l\quad\text{for any}\quad \xi,\eta\in\mathbb{R}^n. \tag{1.31}$$

The following definition is basic for our presentation.

Definition 1.9. A *Hörmander space* $B_{p,\mu}(\mathbb{R}^n)$ is defined as a linear space of all tempered distributions $u \in \mathcal{S}'(\mathbb{R}^n)$ such that their Fourier transform $\widehat{u}$ is locally Lebesgue integrable on $\mathbb{R}^n$ and satisfies the inclusion $\mu\,\widehat{u} \in L_p(\mathbb{R}^n)$. The norm in the linear space $B_{p,\mu}(\mathbb{R}^n)$ is introduced by the formula

$$\|u\|_{B_{p,\mu}(\mathbb{R}^n)} := \|\mu\,\widehat{u}\|_{L_p(\mathbb{R}^n)}.$$

The space $B_{p,\mu}(\mathbb{R}^n)$ is complete with respect to this norm and is continuously embedded in $\mathcal{S}'(\mathbb{R}^n)$. If $1 \leq p < \infty$, then the space $B_{p,\mu}(\mathbb{R}^n)$ is separable and the set $C_0^\infty(\mathbb{R}^n)$ is dense in this space [81, Theorem 2.2.1]. The case $p = 2$ in which $B_{p,\mu}(\mathbb{R}^n)$ becomes a Hilbert space is of especial interest .

Remark 1.5. L. Hörmander [81, Definition 2.1.1] first supposed that the function μ must satisfy a stronger condition than (1.31); namely, that there exist positive numbers c and l such that

$$\frac{\mu(\xi)}{\mu(\eta)} \leq (1 + c\,|\xi - \eta|)^l \quad \text{for each} \quad \xi, \eta \in \mathbb{R}^n. \tag{1.32}$$

However, later he understood [81, see Remark at the end of Sec. 2.1] that the sets of functions satisfying conditions (1.31) or (1.32) lead to the same class of spaces $B_{p,\mu}(\mathbb{R}^n)$.

Remark 1.6. L. R. Volevich and B. P. Paneah [269] introduced and studied spaces $H_p^\mu(\mathbb{R}^n)$, $1 < p < \infty$, closely related to the Hörmander spaces, namely,

$$H_p^\mu(\mathbb{R}^n) := \bigl\{u \in \mathcal{S}'(\mathbb{R}^n) : \mathcal{F}^{-1}(\mu\,\widehat{u}) \in L_p(\mathbb{R}^n)\bigr\}.$$

Thus, for $p \neq 2$, some additional conditions were imposed on the weight function μ. In the Hilbert case $p = 2$, the spaces $B_{p,\mu}(\mathbb{R}^n)$ and $H_p^\mu(\mathbb{R}^n)$ coincide.

Among the properties of Hörmander spaces, we mention the following important embedding theorem [81, Theorem 2.2.7].:

Proposition 1.5. *Let $p, q \in [1, \infty]$, $1/p + 1/q = 1$, and let $\mu : \mathbb{R}^n \to (0, \infty)$ be a continuous weight function. Assume that $k \geq 0$ is an arbitrary integer. Then the condition*

$$(1 + |\xi|)^k\,\mu^{-1}(\xi) \in L_q(\mathbb{R}^n_\xi) \tag{1.33}$$

implies the continuous embedding $B_{p,\mu}(\mathbb{R}^n) \hookrightarrow C_{\mathrm{b}}^k(\mathbb{R}^n)$. Conversely, if

$$\{u \in B_{p,\mu}(\mathbb{R}^n) : \operatorname{supp} u \subset V\} \subset C^k(\mathbb{R}^n)$$

for some open nonempty set $V \subseteq \mathbb{R}^n$, then condition (1.33) is satisfied.

1.3.3 Refined Sobolev scale

From the viewpoint of applications to elliptic operators, it is reasonable to restrict ourselves to the isotropic Hörmander inner product spaces $B_{2,\mu}(\mathbb{R}^n)$, where $\mu(\xi) = \langle\xi\rangle^s \varphi(\langle\xi\rangle)$, $s \in \mathbb{R}$, and $\varphi \in \mathrm{QSV}$. Here and in what follows,

$$\langle\xi\rangle := (1 + \xi_1^2 + \ldots + \xi_n^2)^{1/2}$$

is the smoothed modulus of a vector $\xi = (\xi_1, \ldots, \xi_n) \in \mathbb{R}^n$. This space is denoted by $H^{s,\varphi}(\mathbb{R}^n)$. Making the choice of a function parameter φ somewhat more specific, we give the definition of the space $H^{s,\varphi}(\mathbb{R}^n)$:

By $\mathcal{M}$ we denote the set of all functions $\varphi : [1;\infty) \to (0;\infty)$ such that

(i) φ is Borel measurable on the semiaxis $[1;\infty)$;

(ii) the functions φ and $1/\varphi$ are bounded on every compact interval $[1;b]$, where $1 < b < \infty$;

(iii) $\varphi \in \mathrm{QSV}$.

Theorem 1.10 immediately yields the following description of the class $\mathcal{M}$:

$$\varphi \in \mathcal{M} \Leftrightarrow \varphi(t) = \exp\left(\beta(t) + \int_1^t \frac{\alpha(\tau)}{\tau}\, d\tau\right) \quad \text{for } t \geq 1.$$

Here, α is a continuous function such that $\alpha(\tau) \to 0$ as $\tau \to \infty$. Moreover, β is a Borel measurable function bounded on the semiaxis $[1,\infty)$.

Let $s \in \mathbb{R}$ and let $\varphi \in \mathcal{M}$.

Definition 1.10. The linear space $H^{s,\varphi}(\mathbb{R}^n)$ consists of all tempered distributions $u \in \mathcal{S}'(\mathbb{R}^n)$ such that their Fourier transforms $\widehat{u}$ are locally Lebesgue summable on $\mathbb{R}^n$ and satisfy the condition

$$\int_{\mathbb{R}^n} \langle\xi\rangle^{2s} \varphi^2(\langle\xi\rangle)\, |\widehat{u}(\xi)|^2\, d\xi < \infty.$$

The inner product in the space $H^{s,\varphi}(\mathbb{R}^n)$ is defined by the formula

$$(u_1, u_2)_{H^{s,\varphi}(\mathbb{R}^n)} := \int_{\mathbb{R}^n} \langle\xi\rangle^{2s} \varphi^2(\langle\xi\rangle)\, \widehat{u_1}(\xi)\, \overline{\widehat{u_2}(\xi)}\, d\xi$$

and induces the norm in this space in a standard way.

By Lemma 1.3, $\mu(\xi) = \langle\xi\rangle^s \varphi(\langle\xi\rangle)$ is a weight function. Indeed, for any $\xi, \eta \in \mathbb{R}^n$, we get

$$\frac{\mu(\xi)}{\mu(\eta)} = \left(\frac{\langle\xi\rangle}{\langle\eta\rangle}\right)^s \frac{\varphi(\langle\xi\rangle)}{\varphi(\langle\eta\rangle)} \leq c\,(1 + |\langle\xi\rangle - \langle\eta\rangle|)^{|s|+1} \leq c\,(1 + |\xi - \eta|)^{|s|+1},$$

where $c > 0$ is a constant independent of ξ and η. If the function φ is continuous, then the space $H^{s,\varphi}(\mathbb{R}^n) = B_{2,\mu}(\mathbb{R}^n)$ is a special case of the Hörmander space. Note that the replacement of the continuity condition by a weaker condition of Borel measurability realized in the definition of the set $\mathcal{M}$ of function parameters φ does not lead to new spaces. This follows from Theorem 1.12(i). Indeed, for any $\varphi \in \mathcal{M}$, there exists a function $\varphi_1 \in C^\infty([1,\infty)) \cap \mathcal{M}$ such that $\varphi \asymp \varphi_1$ on the semiaxis $[1,\infty)$. Hence, the spaces $H^{s,\varphi}(\mathbb{R}^n)$ and $H^{s,\varphi_1}(\mathbb{R}^n)$ coincide up to equivalence of norms.

In a special case where $\varphi \equiv 1$, the space $H^{s,\varphi}(\mathbb{R}^n)$ is equal to the Sobolev inner product space $H^s(\mathbb{R}^n)$ of order s. In the general case, we get the following lemma:

Lemma 1.5. *Let $s \in \mathbb{R}$ and let $\varphi \in \mathcal{M}$. Then the following continuous embeddings are true:*

$$H^{s+\varepsilon}(\mathbb{R}^n) \hookrightarrow H^{s,\varphi}(\mathbb{R}^n) \hookrightarrow H^{s-\varepsilon}(\mathbb{R}^n) \quad \textit{for each} \quad \varepsilon > 0. \tag{1.34}$$

Proof. Let $\varepsilon > 0$. Since $\varphi \in \mathcal{M} \subset \mathrm{QSV}$, it follows from Theorem 1.12(ii) that $t^{-\varepsilon} \le \varphi(t) \le t^{\varepsilon}$ for $t \gg 1$. In view of the definition of $\mathcal{M}$, this means that there exists a number $c > 1$ such that

$$c^{-1} t^{-\varepsilon} \le \varphi(t) \le c\, t^{\varepsilon}$$

for all $t \ge 1$. Therefore,

$$c^{-1}\langle\xi\rangle^{s-\varepsilon} \le \langle\xi\rangle^{s}\varphi(\langle\xi\rangle) \le c\langle\xi\rangle^{s+\varepsilon} \quad \text{for any} \quad \xi \in \mathbb{R}^n.$$

This relation immediately yields the continuous embeddings(1.34).

Lemma 1.5 is proved.

It is useful to represent embeddings (1.34) in the form

$$\bigcup_{\varepsilon>0} H^{s+\varepsilon}(\mathbb{R}^n) =: H^{s+}(\mathbb{R}^n) \subset H^{s,\varphi}(\mathbb{R}^n) \subset H^{s-}(\mathbb{R}^n) := \bigcap_{\varepsilon>0} H^{s-\varepsilon}(\mathbb{R}^n). \tag{1.35}$$

It is easy to see that the number parameter s specifies the main (power) smoothness in the class of spaces

$$\{H^{s,\varphi}(\mathbb{R}^n) : s \in \mathbb{R}, \varphi \in \mathcal{M}\}, \tag{1.36}$$

whereas the function parameter φ specifies an additional smoothness subordinated to the main smoothness. Depending on the convergence $\varphi(t) \to \infty$ or $\varphi(t) \to 0$ as $t \to \infty$, the parameter φ specifies either the positive or negative additional smoothness. In other words, the parameter φ refines the main s-smoothness. Therefore, it is natural to give the following definition:

Definition 1.11. The class of function spaces (1.36) is called the *refined Sobolev scale* or simply the refined scale given over $\mathbb{R}^n$.

1.3.4 Properties of the refined scale

The relationship between the refined scale and the Sobolev scale is not exhausted by embeddings (1.35). It turns out that every space on the refined scale can be obtained by the interpolation of a pair of Sobolev inner product spaces with an appropriate function parameter.

Theorem 1.14. *Let a function $\varphi \in \mathcal{M}$ and positive numbers ε and δ be given. Also let*

$$\psi(t) := \begin{cases} t^{\varepsilon/(\varepsilon+\delta)}\, \varphi(t^{1/(\varepsilon+\delta)}) & \text{for } t \geq 1, \\ \varphi(1) & \text{for } 0 < t < 1. \end{cases}$$

Then

(i) *the function ψ belongs to the set $\mathcal{B}$ and is an interpolation parameter;*

(ii) *for any $s \in \mathbb{R}$,*

$$\left[H^{s-\varepsilon}(\mathbb{R}^n), H^{s+\delta}(\mathbb{R}^n)\right]_\psi = H^{s,\varphi}(\mathbb{R}^n)$$

with equality of the norms.

Proof. (i) By virtue of Theorem 1.12(ii) and (iv), the function ψ belongs to $\mathcal{B}$ and is quasiregularly varying at ∞ of the order $\theta = \varepsilon/(\varepsilon+\delta) \in (0, 1)$. Hence, in view of Theorem 1.11, ψ is an interpolation parameter. Assertion (i) is proved.

(ii) Let $s \in \mathbb{R}$. The pair of the Sobolev spaces $H^{s-\varepsilon}(\mathbb{R}^n)$ and $H^{s+\delta}(\mathbb{R}^n)$ is admissible and the pseudodifferential operator with symbol $\langle\xi\rangle^{\varepsilon+\delta}$ is the generating operator J for this pair. By using the Fourier transform

$$\mathcal{F} : H^{s-\varepsilon}(\mathbb{R}^n) \leftrightarrow L_2\big(\mathbb{R}^n, \langle\xi\rangle^{2(s-\varepsilon)} d\xi\big),$$

the operator is reduced J to the form of multiplication by the function $\langle\xi\rangle^{\varepsilon+\delta}$ of the argument $\xi \in \mathbb{R}^n$. Hence, the operator $\psi(J)$ is reduced to the form of multiplication by the function

$$\psi(\langle\xi\rangle^{\varepsilon+\delta}) = \langle\xi\rangle^{\varepsilon}\varphi(\langle\xi\rangle).$$

Thus, in view of (1.35), we obtain

$$\begin{aligned} &\left[H^{s-\varepsilon}(\mathbb{R}^n), H^{s+\delta}(\mathbb{R}^n)\right]_\psi \\ &\qquad = \Big\{u \in H^{s-\varepsilon}(\mathbb{R}^n) : \langle\xi\rangle^{\varepsilon}\varphi(\langle\xi\rangle)\, \widehat{u}(\xi) \in L_2\big(\mathbb{R}^n, \langle\xi\rangle^{2(s-\varepsilon)} d\xi\big)\Big\} \end{aligned}$$

$$= \left\{ u \in H^{s-\varepsilon}(\mathbb{R}^n) : \int\limits_{\mathbb{R}^n} \langle\xi\rangle^{2s} \varphi^2(\langle\xi\rangle) \, |\widehat{u}(\xi)|^2 \, d\xi < \infty \right\}$$

$$= H^{s-\varepsilon}(\mathbb{R}^n) \cap H^{s,\varphi}(\mathbb{R}^n) = H^{s,\varphi}(\mathbb{R}^n).$$

Moreover, the norm in the space $\left[H^{s-\varepsilon}(\mathbb{R}^n), H^{s+\delta}(\mathbb{R}^n)\right]_\psi$ is equal to

$$\|\psi(J)u\|_{H^{s-\varepsilon}(\mathbb{R}^n)} = \left(\int\limits_{\mathbb{R}^n} |\langle\xi\rangle^{\varepsilon} \varphi(\langle\xi\rangle) \, \widehat{u}(\xi)|^2 \, \langle\xi\rangle^{2(s-\varepsilon)} \, d\xi \right)^{1/2} = \|u\|_{H^{s,\varphi}(\mathbb{R}^n)}.$$

Assertion (ii) is proved.

Theorem 1.14 is proved.

The properties of the refined scale over $\mathbb{R}^n$ required for our subsequent presentation are collected in the following theorem:

Theorem 1.15. *Let $s \in \mathbb{R}$ and let $\varphi, \varphi_1 \in \mathcal{M}$. The following assertions hold:*

(i) *For any $\varepsilon > 0$, the following continuous and dense embedding is true:*

$$H^{s+\varepsilon,\varphi_1}(\mathbb{R}^n) \hookrightarrow H^{s,\varphi}(\mathbb{R}^n)$$

(ii) *The function $\varphi(t)/\varphi_1(t)$ is bounded in the vicinity of infinity if and only if $H^{s,\varphi_1}(\mathbb{R}^n) \hookrightarrow H^{s,\varphi}(\mathbb{R}^n)$. This embedding is continuous and dense.*

(iii) *For a given integer $k \geq 0$, the condition*

$$\int\limits_1^\infty \frac{dt}{t\,\varphi^2(t)} < \infty \tag{1.37}$$

is equivalent to the embedding

$$H^{k+n/2,\varphi}(\mathbb{R}^n) \hookrightarrow C^k_{\mathrm{b}}(\mathbb{R}^n). \tag{1.38}$$

This embedding is continuous.

(iv) *The spaces $H^{s,\varphi}(\mathbb{R}^n)$ and $H^{-s,1/\varphi}(\mathbb{R}^n)$ are mutually dual with respect to the expansion by continuity of the inner product in $L_2(\mathbb{R}^n)$.*

Proof. (i) By virtue of Lemma 1.5, we have the following continuous embeddings:

$$H^{s+\varepsilon,\varphi_1}(\mathbb{R}^n) \hookrightarrow H^{s+\varepsilon/2}(\mathbb{R}^n) \hookrightarrow H^{s,\varphi}(\mathbb{R}^n) \quad \text{for any } \varepsilon > 0.$$

They are dense because the set $C_0^\infty(\mathbb{R}^n)$ is dense in each of these spaces. Assertion (i) is proved.

(ii) Since $\varphi, \varphi_1 \in \mathcal{M}$, the function φ/φ_1 is bounded in the vicinity of infinity if and only if $\mu(\xi) \leq c\mu_1(\xi)$ for any $\xi \in \mathbb{R}^n$. Here,

$$\mu(\xi) := \langle\xi\rangle^s \varphi(\langle\xi\rangle), \quad \mu_1(\xi) := \langle\xi\rangle^s \varphi_1(\langle\xi\rangle),$$

and the constant $c > 0$ is independent of ξ. Hörmander [81, Theorem 2.2.2] proved that the inequality $\mu(\xi) \leq c\mu_1(\xi)$ is equivalent to the continuous embedding

$$H^{s,\varphi_1}(\mathbb{R}^n) = B_{2,\mu_1}(\mathbb{R}^n) \hookrightarrow B_{2,\mu}(\mathbb{R}^n) = H^{s,\varphi}(\mathbb{R}^n).$$

This embedding is dense because the set $C_0^\infty(\mathbb{R}^n)$ is dense in these spaces. Assertion (ii) is proved.

(iii) For a given integer $k \geq 0$, by virtue of Proposition 1.5 in which we set

$$\mu(\xi) := \langle\xi\rangle^{k+n/2}\, \varphi(\langle\xi\rangle)$$

and $p = 2$, it is possible to conclude that

$$\int\limits_{\mathbb{R}^n} \frac{d\xi}{\langle\xi\rangle^n\, \varphi^2(\langle\xi\rangle)} < \infty \;\Leftrightarrow\; H^{k+n/2,\varphi}(\mathbb{R}^n) \hookrightarrow C_b^k(\mathbb{R}^n). \tag{1.39}$$

In this case, the embedding is continuous. Passing to the spherical coordinates in the integral, after the change of variables $t = (1+r^2)^{1/2}$, we find

$$\begin{aligned} \int\limits_{\mathbb{R}^n} \frac{d\xi}{\langle\xi\rangle^n\, \varphi^2(\langle\xi\rangle)} &= c \int\limits_1^\infty \frac{r^{n-1}\, dr}{(1+r^2)^{n/2}\, \varphi^2((1+r^2)^{1/2})} \\ &= c \int\limits_1^\infty \frac{(t^2-1)^{(n-1)/2}}{t^n\, \varphi^2(t)} \frac{t\, dt}{(t^2-1)^{1/2}} = c \int\limits_1^\infty \frac{\omega_n(t)\, dt}{t\, \varphi^2(t)}. \end{aligned} \tag{1.40}$$

Here, $c > 0$ is a number and

$$\omega_n(t) := (\sqrt{t^2-1}/t)^{n-2}.$$

Note that the functions ω_n and $1/\omega_n$ are bounded on the semiaxis $[2, \infty)$. Moreover, since $n \geq 1$, we get

$$\int\limits_1^2 \omega_n(t) dt < \infty.$$

In view of the definition of the set $\mathcal{M}$, the function $\varphi^{-2}(t)$ is bounded on the segment $[1, 2]$. Hence,

$$\int_1^\infty \frac{dt}{t\varphi^2(t)} < \infty \Leftrightarrow \int_2^\infty \frac{dt}{t\varphi^2(t)} < \infty$$

$$\Leftrightarrow \int_2^\infty \frac{\omega_n(t)dt}{t\varphi^2(t)} < \infty \Leftrightarrow \int_1^\infty \frac{\omega_n(t)dt}{t\varphi^2(t)} < \infty.$$

By virtue of (1.40), this yields

$$\int_1^\infty \frac{dt}{t\varphi^2(t)} < \infty \Leftrightarrow \int_{\mathbb{R}^n} \frac{d\xi}{\langle\xi\rangle^n \varphi^2(\langle\xi\rangle)} < \infty. \tag{1.41}$$

Relations (1.39) and (1.41) now imply assertion (iii).

(iv) Assertion (iv) is a special case of Hörmander's result [81, Theorem 2.2.9]. Note that $\varphi \in \mathcal{M} \Leftrightarrow 1/\varphi \in \mathcal{M}$. Hence, the space $H^{-s,1/\varphi}(\mathbb{R}^n)$ belongs to the refined scale.

Theorem 1.15 is proved.

1.4 Uniformly elliptic operators on the refined scale

In the present section, we study uniformly elliptic pseudodifferential operators given on the Euclidean space $\mathbb{R}^n$. We obtain an *a priori* estimate for the solutions of an elliptic equation considered on the refined Sobolev scale and study their internal smoothness.

1.4.1 Pseudodifferential operators

The detailed presentation of the theory of pseudodifferential operators (PsDOs) can be found, e.g., in the monographs [86, 232, 253, 254] and in the survey [10]. For the sake of convenience, we recall the definition of pseudodifferential operators on $\mathbb{R}^n$ and discuss some related notions necessary for what follows. We mainly use the terminology and notation from the survey by M. S. Agranovich [10, § 1, 3].

Let $m \in \mathbb{R}$. By $S^m(\mathbb{R}^{2n})$ we denote the set of all functions $a \in C^\infty(\mathbb{R}^{2n})$ satisfying the condition: For any multiindices α and β, there exists a number $c_{\alpha,\beta} > 0$ such that

$$|\, \partial_x^\alpha \partial_\xi^\beta a(x,\xi)\,| \le c_{\alpha,\beta}\, \langle\xi\rangle^{m-|\beta|} \quad \text{for any} \quad x, \xi \in \mathbb{R}^n.$$

Definition 1.12. A *pseudodifferential operator* (PsDO) A on $\mathbb{R}^n$ with *symbol* $a \in S^m(\mathbb{R}^{2n})$ is defined by the formula

$$(Au)(x) := (2\pi)^{-n} \int_{\mathbb{R}^n} e^{-ix\cdot\xi}\, a(x,\xi)\, \widehat{u}(\xi)\, d\xi, \quad x \in \mathbb{R}^n;$$

here, u is an arbitrary function from the space $\mathcal{S}(\mathbb{R}^n)$.

By $\Psi^m(\mathbb{R}^n)$ we denote the class of all PsDOs on $\mathbb{R}^n$ with symbols from $S^m(\mathbb{R}^{2n})$.

An important example of PsDO in $\Psi^m(\mathbb{R}^n)$ is given by a linear differential operator

$$A(x,D) := \sum_{|\mu|\le m} a_\mu(x)\, D^\mu \tag{1.42}$$

of order m with coefficients $a_\mu \in C_{\mathrm{b}}(\mathbb{R}^n)$. Its symbol is

$$a(x,\xi) = \sum_{|\mu|\le m} a_\mu(x)\, \xi^\mu, \quad x,\xi \in \mathbb{R}^n.$$

Here, $\mu = (\mu_1, \ldots, \mu_n)$ is a multiindex,

$$D^\mu := i^{|\mu|}\, \partial_x^\mu$$

and, as usual, $\xi^\mu := \xi_1^{\mu_1} \ldots \xi_n^{\mu_n}$ for a vector $\xi = (\xi_1, \ldots, \xi_n)$. (The Fourier transformation transforms the differentiation operator D^μ into the operator of multiplication by ξ^μ.)

Note that, as the parameter m increases, the class $\Psi^m(\mathbb{R}^n)$ is extended. We set

$$\Psi^{-\infty}(\mathbb{R}^n) := \bigcap_{m\in\mathbb{R}} \Psi^m(\mathbb{R}^n), \quad \Psi^{\infty}(\mathbb{R}^n) := \bigcup_{m\in\mathbb{R}} \Psi^m(\mathbb{R}^n).$$

Every PsDO $A \in \Psi^\infty(\mathbb{R}^n)$ is continuous in $\mathcal{S}(\mathbb{R}^n)$ and can be uniquely extended to a linear continuous operator on $\mathcal{S}'(\mathbb{R}^n)$. This operator is also denoted by A.

We now consider a narrower family of polyhomogeneous (or classical) PsDOs most important for applications. First, we introduce a set of homogeneous symbols of order $m \in \mathbb{R}$.

By $S_{\mathrm{h}}^m(\mathbb{R}^{2n})$ we denote the set of all infinitely differentiable complex-valued functions $b(x,\xi)$ of $x,\xi \in \mathbb{R}^n$ with $\xi \ne 0$ satisfying the following conditions:

(i) $b(x,\lambda\xi) = \lambda^m b(x,\xi)$ for any $\lambda > 0$;

(ii) for any multiindices α and β, there exists a number $c_{\alpha,\beta} > 0$ such that

$$|\, \partial_x^\alpha\, \partial_\xi^\beta\, b(x,\xi)\,| \le c_{\alpha,\beta}\, |\xi|^{m-|\beta|} \quad \text{for any } x,\xi \in \mathbb{R}^n \text{ with } \xi \ne 0.$$

Definition 1.13. Let A be a PsDO in $\Psi^m(\mathbb{R}^n)$ with symbol $a(x,\xi)$. The operator A is called a *polyhomogeneous* (or *classical*) PsDO of order m if there exist a (strictly) decreasing sequence of real numbers $(m_j)_{j=0}^{\infty}$ with $m_0 = m$ approaching $-\infty$ and a sequence of homogeneous symbols $(a_j)_{j=0}^{\infty} \subset S_{\mathrm{h}}^{m_j}(\mathbb{R}^{2n})$ such that

$$a(x,\xi) - \theta(\xi) \sum_{j=0}^{k} a_j(x,\xi) \in S^{m_{k+1}}(\mathbb{R}^{2n}_{(x,\xi)})$$

for every integer $k \geq 0$. In this case, the function $a_0(x,\xi)$ is assumed to be different from the zero function. Moreover, the function $\theta \in C_{\mathrm{b}}^{\infty}(\mathbb{R}^n)$ is set equal to 1 outside a certain neighborhood of the origin and equal to 0 in its somewhat smaller neighborhood. The function $a_0(x,\xi)$ is called the *principal symbol* of the polyhomogeneous PsDO A.

By $\Psi_{\mathrm{ph}}^m(\mathbb{R}^n)$ we denote the class of all polyhomogeneous PsDOs of order m on $\mathbb{R}^n$. This class is independent of the indicated choice of θ. Since the principal symbol of a polyhomogeneous PsDO is assumed to be not identically equal to zero, we note that $\Psi_{\mathrm{ph}}^m(\mathbb{R}^n) \cap \Psi_{\mathrm{ph}}^r(\mathbb{R}^n) = \varnothing$ for $m \neq r$.

The differential operator (1.42) belongs to the class $\Psi_{\mathrm{ph}}^m(\mathbb{R}^n)$ if at least one coefficient $a_\mu(x)$ with $|\mu| = m$ is not identically zero. The principal symbol of this operator has the form

$$a_0(x,\xi) = \sum_{|\mu|=m} a_\mu(x)\,\xi^\mu \quad \text{for all} \quad x,\xi \in \mathbb{R}^n.$$

We are interested in polyhomogeneous PsDOs uniformly elliptic on $\mathbb{R}^n$. We recall their definition.

Definition 1.14. A PsDO $A \in \Psi_{\mathrm{ph}}^m(\mathbb{R}^n)$ and its principal symbol $a_0(x,\xi)$ are called *uniformly elliptic* on $\mathbb{R}^n$ if there exists a number $c > 0$ such that $|a_0(x,\xi)| \geq c$ for any $x,\xi \in \mathbb{R}^n$ with $|\xi| = 1$.

It is worth noting that every uniformly elliptic PsDO A has a parametrix B; i.e., the following proposition is true [10, Theorem 1.8.3]:

Proposition 1.6. *Let a PsDO $A \in \Psi_{\mathrm{ph}}^m(\mathbb{R}^n)$ be uniformly elliptic on $\mathbb{R}^n$. Then there exists a PsDO $B \in \Psi_{\mathrm{ph}}^{-m}(\mathbb{R}^n)$ uniformly elliptic on $\mathbb{R}^n$ and such that*

$$BA = I + T_1, \quad AB = I + T_2, \tag{1.43}$$

where T_1 and T_2 are some PsDOs from $\Psi^{-\infty}(\mathbb{R}^n)$ and I is the identity operator on $\mathcal{S}'(\mathbb{R}^n)$.

1.4.2 *A priori* estimate of the solutions

Consider a PsDO $A \in \Psi^m_{\mathrm{ph}}(\mathbb{R}^n)$ with $m \in \mathbb{R}$. In this and next subsections, we assume that the PsDO A is uniformly elliptic on $\mathbb{R}^n$. Consider the equation $Au = f$ in $\mathbb{R}^n$. For its solution u, we establish an *a priori* estimate in the refined Sobolev scale.

Prior to do this, we prove the following lemma on the boundedness of PsDOs in the refined scale.

Lemma 1.6. *Let G be a PsDO in $\Psi^r(\mathbb{R}^n)$ with $r \in \mathbb{R}$. Then the restriction of the mapping $u \mapsto Gu$, $u \in \mathcal{S}'(\mathbb{R}^n)$, is a bounded linear operator*

$$G : H^{\sigma,\varphi}(\mathbb{R}^n) \to H^{\sigma-r,\varphi}(\mathbb{R}^n) \tag{1.44}$$

for arbitrary parameters $\sigma \in \mathbb{R}$ and $\varphi \in \mathcal{M}$.

Proof. In the Sobolev case with $\varphi \equiv 1$, this lemma is known (see, e.g., [10, Theorem 1.1.2] or [86, Theorem 18.1.13]). We choose arbitrary $\sigma \in \mathbb{R}$ and $\varphi \in \mathcal{M}$. Consider bounded linear operators

$$G : H^{\sigma\mp1}(\mathbb{R}^n) \to H^{\sigma\mp1-r}(\mathbb{R}^n).$$

We apply the interpolation with the function parameter ψ from Theorem 1.14, where we set $\varepsilon = \delta = 1$. According to assertion (i) of this theorem, we get the bounded operator

$$G : \left[H^{\sigma-1}(\mathbb{R}^n), H^{\sigma+1}(\mathbb{R}^n)\right]_{\psi} \to \left[H^{\sigma-r-1}(\mathbb{R}^n), H^{\sigma-r+1}(\mathbb{R}^n)\right]_{\psi}.$$

By virtue of assertion (ii), this implies that the PsDO G defines the bounded operator (1.44).

Lemma 1.6 is proved.

It is easy to see that each PsDO in $\Psi^r(\mathbb{R}^n)$ decreases the main smoothness σ by r and preserves invariant the additional smoothness φ in the space $H^{\sigma,\varphi}(\mathbb{R}^n)$.

By virtue of Lemma 1.6, the operator

$$A : H^{s+m,\varphi}(\mathbb{R}^n) \to H^{s,\varphi}(\mathbb{R}^n)$$

is bounded for all $s \in \mathbb{R}$ and $\varphi \in \mathcal{M}$.

Theorem 1.16. *Let $s \in \mathbb{R}$, $\sigma > 0$, and $\varphi \in \mathcal{M}$. There exists a number $c = c(s, \sigma, \varphi) > 0$ such that, for any distribution $u \in H^{s+m,\varphi}(\mathbb{R}^n)$, the following a priori estimate is true:*

$$\|u\|_{H^{s+m,\varphi}(\mathbb{R}^n)} \leq c\left(\|Au\|_{H^{s,\varphi}(\mathbb{R}^n)} + \|u\|_{H^{s+m-\sigma,\varphi}(\mathbb{R}^n)}\right). \tag{1.45}$$

Proof. We use Proposition 1.6. By virtue of the first equality in (1.43), we can write $u = BAu - T_1 u$. This yields estimate (1.45). Indeed,

$$\begin{aligned}\|u\|_{H^{s+m,\varphi}(\mathbb{R}^n)} &= \|BAu - T_1 u\|_{H^{s+m,\varphi}(\mathbb{R}^n)}\\ &\leq \|BAu\|_{H^{s+m,\varphi}(\mathbb{R}^n)} + \|T_1 u\|_{H^{s+m,\varphi}(\mathbb{R}^n)}\\ &\leq c\,\|Au\|_{H^{s,\varphi}(\mathbb{R}^n)} + c\,\|u\|_{H^{s+m-\sigma,\varphi}(\mathbb{R}^n)}.\end{aligned}$$

Here, c is the maximum of norms of the operators

$$B : H^{s,\varphi}(\mathbb{R}^n) \to H^{s+m,\varphi}(\mathbb{R}^n), \tag{1.46}$$

$$T_1 : H^{s+m-\sigma,\varphi}(\mathbb{R}^n) \to H^{s+m,\varphi}(\mathbb{R}^n). \tag{1.47}$$

These operators are bounded by Lemma 1.6 and Proposition 1.6.

Theorem 1.16 is proved.

This theorem improves the well-known *a priori* estimate for the solutions of uniformly elliptic equations considered on the Sobolev scale (the case $\varphi \equiv 1$) [10, Theorem 1.8.4].

1.4.3 Smoothness of the solutions

Assume that the right-hand side of the equation $Au = f$ is characterized by a certain internal smoothness on a given open nonempty set $V \subseteq \mathbb{R}^n$ with respect to the refined Sobolev scale. We study the internal smoothness of the solution u on this set. First, we consider the case where $V = \mathbb{R}^n$. By $H^{-\infty}(\mathbb{R}^n)$ we denote the union of all spaces $H^{s,\varphi}(\mathbb{R}^n)$ with $s \in \mathbb{R}$ and $\varphi \in \mathcal{M}$. The linear space $H^{-\infty}(\mathbb{R}^n)$ is endowed with the topology of inductive limit of spaces [23, Chap. 14, Sec. 2.3].

Theorem 1.17. *Assume that $u \in H^{-\infty}(\mathbb{R}^n)$ is a solution of the equation $Au = f$ on $\mathbb{R}^n$, where $f \in H^{s,\varphi}(\mathbb{R}^n)$ for certain parameters $s \in \mathbb{R}$ and $\varphi \in \mathcal{M}$. Then $u \in H^{s+m,\varphi}(\mathbb{R}^n)$.*

Proof. By Theorem 1.15(i), there exists a number $\sigma > 0$ such that

$$u \in H^{s+m-\sigma,\varphi}(\mathbb{R}^n). \tag{1.48}$$

Thus, the required property follows from this inclusion, the condition of the theorem, and relations (1.43), (1.46), and (1.47):

$$u = BAu - T_1 u = Bf - T_1 u \in H^{s+m,\varphi}(\mathbb{R}^n).$$

Theorem 1.17 is proved.

We now consider the general case where V is an arbitrary open nonempty subset of $\mathbb{R}^n$. We set

$$H^{\sigma,\varphi}_{\mathrm{int}}(V) := \big\{w \in H^{-\infty}(\mathbb{R}^n) : \chi\, w \in H^{\sigma,\varphi}(\mathbb{R}^n)$$

$$\text{for all } \chi \in C^{\infty}_{\mathrm{b}}(\mathbb{R}^n),\ \operatorname{supp}\chi \subset V,\ \operatorname{dist}(\operatorname{supp}\chi, \partial V) > 0\big\},$$

where $\sigma \in \mathbb{R}$ and $\varphi \in \mathcal{M}$. The topology in the space $H^{\sigma,\varphi}_{\mathrm{int}}(V)$ is given by the seminorms $w \mapsto \|\chi\, w\|_{H^{\sigma,\varphi}(\mathbb{R}^n)}$, where the functions χ are the same as in the definition of this space.

Theorem 1.18. *Assume that $u \in H^{-\infty}(\mathbb{R}^n)$ is a solution of the equation $Au = f$ on the set V, where $f \in H^{s,\varphi}_{\mathrm{int}}(V)$ for some parameters $s \in \mathbb{R}$ and $\varphi \in \mathcal{M}$. Then $u \in H^{s+m,\varphi}_{\mathrm{int}}(V)$.*

Proof. It is necessary to show that the condition $f \in H^{s,\varphi}_{\mathrm{int}}(V)$ yields the following property of increase in the internal smoothness of the solutions of equation $Au = f$: for every number $r \geq 1$,

$$u \in H^{s-r+m,\varphi}_{\mathrm{int}}(V) \ \Rightarrow\ u \in H^{s-r+1+m,\varphi}_{\mathrm{int}}(V). \tag{1.49}$$

We arbitrarily choose a function $\chi \in C^{\infty}_{\mathrm{b}}(\mathbb{R}^n)$ such that

$$\operatorname{supp}\chi \subset V \quad \text{and} \quad \operatorname{dist}(\operatorname{supp}\chi, \partial V) > 0. \tag{1.50}$$

For this function, there exists a function $\eta \in C^{\infty}_{\mathrm{b}}(\mathbb{R}^n)$ such that

$$\begin{gathered}\operatorname{supp}\eta \subset V,\ \operatorname{dist}(\operatorname{supp}\eta, \partial V) > 0,\\ \text{and } \eta = 1 \text{ in the vicinity of } \operatorname{supp}\chi.\end{gathered} \tag{1.51}$$

Indeed, we can define this function with the help of the operation of convolution, namely, $\eta := \chi_{2\varepsilon} * \omega_\varepsilon$, where $\varepsilon := \operatorname{dist}(\operatorname{supp}\chi, \partial V)/4$, $\chi_{2\varepsilon}$ is the indicator function for the 2ε-vicinity of $\operatorname{supp}\chi$, and $\omega_\varepsilon \in C^{\infty}(\mathbb{R}^n)$ is a function satisfying the conditions

$$\omega_\varepsilon \geq 0,\ \operatorname{supp}\omega_\varepsilon \subseteq \{x \in \mathbb{R}^n : \|x\| \leq \varepsilon\},\ \text{and} \int\limits_{\mathbb{R}^n} \omega_\varepsilon(x)\, dx = 1.$$

We can directly verify that the function η belongs to $C^{\infty}_{\mathrm{b}}(\mathbb{R}^n)$ and has the following properties: $\eta \equiv 1$ in the ε-vicinity of the set $\operatorname{supp}\chi$, and $\eta \equiv 0$ outside the 3ε-vicinity of $\operatorname{supp}\chi$. Hence, η satisfies conditions (1.51).

As a result of permutation of the PsDO A and the operator of multiplication by the function χ, we can write

$$\begin{aligned} A\chi u &= A\chi\eta u = \chi\, A\eta u + A'\eta u \\ &= \chi\, Au + \chi\, A(\eta - 1)u + A'\eta u \\ &= \chi f + \chi\, A(\eta - 1)u + A'\eta u \quad \text{on} \quad \mathbb{R}^n. \end{aligned} \tag{1.52}$$

Here, the PsDO $A' \in \Psi^{m-1}(\mathbb{R}^n)$ is the commutator of A and the operator of multiplication by χ. By Lemma 1.6, we get the bounded operator

$$A' : H^{s-r+m,\varphi}(\mathbb{R}^n) \to H^{s-r+1,\varphi}(\mathbb{R}^n).$$

Hence,

$$u \in H_{\text{int}}^{s-r+m,\varphi}(V) \;\Rightarrow\; A'\eta u \in H^{s-r+1,\varphi}(\mathbb{R}^n). \tag{1.53}$$

Further, in view of the condition $f \in H_{\text{int}}^{s,\varphi}(V)$ and the inequality $r \geq 1$, we find

$$\chi f \in H^{s,\varphi}(\mathbb{R}^n) \hookrightarrow H^{s-r+1,\varphi}(\mathbb{R}^n). \tag{1.54}$$

In addition, since the supports of the functions χ and $\eta - 1$ are disjoint, the PsDO $\chi A(\eta - 1)$ belongs to $\Psi^{-\infty}(\mathbb{R}^n)$. This immediately follows from the formula for the symbol of composition of the following two PsDOs: χA and the operator of multiplication by the function $\eta - 1$ (see [10, Theorem 1.2.4]). Since relation (1.48) holds for $u \in H^{-\infty}(\mathbb{R}^n)$ and some $\sigma > 0$, by virtue of Lemma 1.6, we arrive at the inclusion

$$\chi A(\eta - 1)u \in H^{s-r+1,\varphi}(\mathbb{R}^n) \tag{1.55}$$

In view of relations (1.52)–(1.55) and Theorem 1.17, we conclude that

$$u \in H_{\text{int}}^{s-r+m,\varphi}(V) \;\Rightarrow\; A\chi u \in H^{s-r+1,\varphi}(\mathbb{R}^n) \;\Rightarrow \chi u \in H^{s-r+1+m,\varphi}(\mathbb{R}^n).$$

This proves implication (1.49) in view of the arbitrariness of the choice of the function $\chi \in C_{\text{b}}^{\infty}(\mathbb{R}^n)$ satisfying (1.50).

Further, by using implication (1.49), one can readily deduce the inclusion $u \in H_{\text{int}}^{s,\varphi}(V)$. It is possible to assume that the number $\sigma > 0$ in (1.48) is integer. Hence, $u \in H_{\text{int}}^{s-\sigma+m,\varphi}(V)$. We now successively apply implication (1.49) for $r = \sigma$, $r = \sigma - 1$,..., and $r = 1$. This enables us to deduce the required inclusion, namely,

$$u \in H_{\text{int}}^{s-\sigma+m,\varphi}(V) \;\Rightarrow\; u \in H_{\text{int}}^{s-\sigma+1+m,\varphi}(V) \;\Rightarrow \ldots \Rightarrow\; u \in H_{\text{int}}^{s+m,\varphi}(V).$$

Theorem 1.18 is proved.

As applied to the spaces $H^{s,\varphi}(\mathbb{R}^n)$, Theorem 1.18 refines the well-known assertions on the increase in internal smoothness of the solutions of linear elliptic equations considered in the Sobolev scale (see, e.g., [21, Chap. III, Theorem 4.3], [52, Chap. II, Lemma 3.2], and [253, Chap. III, Corollary 1.5]). It is easy to see that the refined smoothness φ of the right-hand side of an elliptic equation is inherited by its solutions. If the operator A is differential and the set V is bounded, then Theorem 1.18 is contained in Hörmander's theorem [81, Theorem 7.4.1] on the regularity of solutions of hypoelliptic equations.

Remark 1.7. It is necessary to distinguish *internal* smoothness from the *local* smoothness in an open set $V \subset \mathbb{R}^n$. The space of distributions with given local smoothness on this set is defined as follows:

$$H^{\sigma,\varphi}_{\mathrm{loc}}(V) := \{w \in H^{-\infty}(\mathbb{R}^n) :$$

$$\chi w \in H^{\sigma,\varphi}(\mathbb{R}^n) \text{ for all } \chi \in C^\infty_0(\mathbb{R}^n),\ \operatorname{supp}\chi \subset V\}.$$

In the case where the set V is bounded, the spaces $H^{\sigma,\varphi}_{\mathrm{int}}(V)$ and $H^{\sigma,\varphi}_{\mathrm{loc}}(V)$ coincide. At the same time, if V is unbounded, then we may get the strict inclusion $H^{\sigma,\varphi}_{\mathrm{int}}(V) \subset H^{\sigma,\varphi}_{\mathrm{loc}}(V)$. An analog of Theorem 1.18 is true for the local refined smoothness. One must only replace int with loc in the notation of the spaces. This analog readily follows from Theorem 1.18.

By using Theorems 1.18 and 1.15(iii), we can establish the existence of continuous generalized partial derivatives of the solutions of equation $Au = f$.

Theorem 1.19. *Let $r \geq 0$ be a given integer and let $\varphi \in \mathcal{M}$ be a function satisfying condition* (1.37). *Assume that a distribution $u \in H^{-\infty}(\mathbb{R}^n)$ is a solution of the equation $Au = f$ on the open set $V \subseteq \mathbb{R}^n$ and that*

$$f \in H^{r-m+n/2,\varphi}_{\mathrm{int}}(V). \tag{1.56}$$

Then the solution u has continuous partial derivatives up to the order r, inclusively, on the set V and these derivatives are bounded on each set $V_0 \subset V$ such that $\operatorname{dist}(V_0, \partial V) > 0$. *In particular, if $V = \mathbb{R}^n$, then $u \in C^r_{\mathrm{b}}(\mathbb{R}^n)$.*

Proof. The inclusion $u \in H^{r+n/2,\varphi}_{\mathrm{int}}(V)$ holds by virtue of Theorem 1.18 in which we set $s := r-m+n/2$. Let a function $\eta \in C^\infty_{\mathrm{b}}(\mathbb{R}^n)$ satisfy the conditions

$$\operatorname{supp}\eta \subset V,\ \ \operatorname{dist}(\operatorname{supp}\eta,\, \partial V) > 0, \ \text{ and } \ \eta = 1 \text{ in the vicinity of } V_0.$$

This function can be constructed in exactly the same way as in the proof of Theorem 1.18 if we replace the set $\operatorname{supp}\chi$ by V_0. According to Theorem 1.15(iii), the distribution ηu satisfies the inclusion

$$\eta u \in H^{r+n/2,\varphi}(\mathbb{R}^n) \hookrightarrow C^r_{\mathrm{b}}(\mathbb{R}^n).$$

This implies that all generalized partial derivatives of the function u up to the order r, inclusively, are continuous and bounded in a certain neighborhood of the set V_0. Hence, these derivatives are also continuous on the set V because we can take $V_0 := \{x_0\}$ for any point $x_0 \in V$.

Theorem 1.19 is proved.

Remark 1.8. If we use an analog of Theorem 1.19 for the Sobolev scale, then, instead of (1.56), it is necessary to demand that

$$f \in H^{r-m+n/2+\varepsilon,1}_{\text{int}}(V) \quad \text{for some} \quad \varepsilon > 0.$$

This condition is much stronger than (1.56).

Remark 1.9. Condition (1.37) is not only sufficient in Theorem 1.19 but also necessary in the class of all solutions of the equation $Au = f$. Namely, (1.37) is equivalent to the implication

$$\big(u \in H^{-\infty}(\mathbb{R}^n) \text{ and } f := Au \in H^{r-m+n/2,\varphi}_{\text{int}}(V)\big) \Rightarrow u \in C^r(V). \tag{1.57}$$

Indeed, if $u \in H^{r+n/2,\varphi}_{\text{int}}(V)$, then

$$f = Au \in H^{r-m+n/2,\varphi}_{\text{int}}(V),$$

whence it follows that $u \in C^r(\Omega)$ provided that (1.57) is true. Therefore, (1.57) implies (1.37) in view of Proposition 1.5 and relation (1.41).

1.5 Remarks and comments

Section 1.1. The first method of interpolation of spaces was independently proposed by J.-L. Lions [114] and S. G. Krein [105]. This was the interpolation of pairs of Hilbert spaces with power parameter in which the exponent was used as a parameter of interpolation. This method was extended to pairs of normed spaces by J.-L. Lions and J. Peetre [115, 122, 187] (real interpolation) and by S. G. Krein [105, 106], J.-L. Lions [116], A. P. Calderon [33], and M. Schechter [226] (complex or holomorphic interpolation). Generally speaking, the complex and real methods of interpolation lead to different interpolation spaces. The principles of construction of general interpolation methods were developed by E. Gagliardo [59].

At present, we have various real and complex interpolation methods for normed and more general topological spaces with finite collections of numbers playing the role of interpolation parameters; see the monographs by K. Bennet and R. Sharpley [20], J. Bergh and J. Löfström [24], Yu. A. Brudnyi and N. Ya. Krugljak [30], S. G. Krein, Yu. I. Petunin, and E. M. Semenov [109], J.-L. Lions and E. Magenes [121], V. I. Ovchinnikov [179], L. Tartar [258], and H. Triebel [258] and the great number of references therein.

A method of interpolation of normed spaces with function parameter was first introduced by C. Foiaş and J.-L. Lions in [57], where the Hilbert case was considered separately. The interpolation of Hilbert spaces with function parameter was investigated by W. F. Donoghue [46], G. Shlenzak [231],

E. I. Pustyl'nik [195], V. I. Ovchinnikov [179], and the authors [144, 146, 153]. Ovchinnikov [179] described (in terms of interpolation with a function parameter) all Hilbert spaces obtained as interpolation spaces with respect to a given pair of Hilbert spaces. Note that, in some applications, this interpolation is called the method of variable Hilbert scales; see, e.g., the works by M. Hegland [78, 79], and P. Mathé and U. Tautenhahn [129].

Various methods of interpolation of normed spaces with general function parameters were introduced and studied by T. F. Kalugina [89], J. Gustavsson [71], S. Janson [88], C. Merucci [135], L.-E. Persson [190], N. Ya. Krugljak [110], and V. I. Ovchinnikov [180] for the case of real interpolation and by M. J. Carro and J. Cerdà [34] for the case of complex interpolation. Note that the interpolation methods proposed by S. Janson [88] involve a fairly broad class of interpolation parameters, namely, arbitrary positive pseudoconcave functions.

All theorems presented in Section 1.1 were proved by the authors. Thus, Theorems 1.1–1.5, 1.8, and 1.9 were proved in [153, Sec. 2] and Theorems 1.6 and 1.7 in [146, Sec. 3]. The proof of Theorem 1.6 is similar to the proof presented in [258, Sec. 1.17] and the proof of Theorem 1.7 is close to the proof proposed by Geymonat [63, p. 133]. Triebel and Geymonat studied interpolation functors on the category of all compatible pairs of Banach spaces.

Section 1.2. The notion of regularly varying function was introduced by J. Karamata [91] (in the case of continuous functions). He also established [92, 93] the main properties of regularly varying functions. The theory of these functions and its various applications can be found in the monographs by N. H. Bingham, C. M. Goldie, and J. L. Teugels [26], J. L. Geluk and L. de Haan [62], L. de Haan [72], V. Maric [128], S. I. Reshnick [199], and E. Seneta [235].

The notion of quasiregularly varying function was introduced in [146, p. 15] and [153, p. 90]. This notion is convenient in the theory of interpolation of spaces. All theorems of Section 1.2 (except Theorem 1.13) were proved in [153, Sec. 3.1]. The proof of Theorem 1.13 has never been presented earlier. The direct proof of the important interpolation theorem 1.11 was given in [144, Sec. 2]. This proof is based on a modification [231, p. 49] of the interpolation method of traces proposed in [121, Chap. 1, § 3 and 5]. The auxiliary lemmas 1.3 and 1.4 were established in [144, Sec. 1].

Section 1.3. The theory of distributions can be traced back to S. L. Sobolev and L. Schwartz. Sobolev introduced important Banach spaces of distributions, which are named after him and play a key role in the modern theory of partial differential equations. The theory of distributions and Sobolev spaces is presented, e.g., in the monographs by S. L. Sobolev [242, 243], L. Schwartz [228, 229], R. A. Adams [1], I. M. Gel'fand and G. E. Shilov [60], S. G. Mikhlin [161], L. Tartar [252], and V. S. Vladimirov [266]. Various applications of Sobolev spaces stimulated their profound investigation, which led to the construction of new important classes of the spaces of distributions, such as the Nikol'skii spaces, the Besov spaces, the scale of Lizorkin–Triebel spaces, and their various weight and anisotropic analogs; see the monographs by O. V. Besov, V. P. Il'in, and S. M. Nikol'skii [25], S. M. Nikol'skii [176], H. Triebel [256, 257, 258, 260] and the references therein. These spaces are parametrized by finite collections of numbers.

The subsequent generalization of the Sobolev spaces was obtained as a result of the transition from numerical to function parameters. The function parameters enable one to give a finer description of the regularity properties of distributions contained in these spaces as compared with the number parameters. This generalization was realized by B. Malgrange [127] and, systematically, by L. Hörmander in his monograph [81, Chap. II] and L. R. Volevich and B. P. Paneah in [269] who introduced and investigated various spaces parametrized with the help of fairly general function parameters. The Hörmander spaces coincide with the Volevich–Paneah spaces in the Hilbert case. Some applications of these spaces to the theory of partial differential equations can be found in the monographs by L. Hörmander [81, 85] and B. P. Paneah [181].

The Hörmander and Volevich–Paneah spaces occupy a central place in the collection of spaces of generalized smoothness. In the last decades, these spaces serve as the object of numerous profound investigations; see the surveys by G. A. Kalyabin and P. I. Lizorkin [90], the monographs by N. Jacob [87], F. Nicola and L. Rodino [175], and H. Triebel [259, Chapt. III], recent papers by V. I. Burenkov [31], A. M. Caetano and H.-G. Leopold [32], D. E. Edmunds, P. Gurka, and B. Opic [48], D. E. Edmunds and D. D. Haroske [49], W. Farkas, N. Jacob, and R. L. Schilling [54, 55], W. Farkas and H.-G. Leopold [56], P. Gurka and B. Opic [70], D. D. Haroske and S. D. Moura [74, 75], H.-G. Leopold [112], S. D. Moura [162], and B. Opic and W. Trebels [177], and the references therein. Various analogs of the Nikol'skii-Besov and Lizorkin–Triebel spaces parametrized with the help of function parameters were constructed. The exact embedding theorems, extension theorems, theorems on traces, and other results were established for these spaces. The interpolation properties of the spaces of generalized smoothness were studied by M. Schechter [226], C. Merucci [136], and F. Cobos and D. L. Fernandez [35].

The definition of the refined Sobolev scale on $\mathbb{R}^n$ was given in [143]; its properties (Theorems 1.14 and 1.15) were established in [145, Sec. 3].

The term "refined scale" was earlier used by G. Shlenzak [231] for a different class of Hörmander inner product spaces. This class has no constructive description and is not attached to the Sobolev scale. The corresponding class of interpolation parameters is quite narrow and obeys a redundant condition imposed on the behavior of the parameters near the origin.

Triebel [259, Chapt. III] and Haroske and Moura [74] introduced and studied some analogs of the Nikol'skii–Besov and Lizorkin–Triebel normed spaces in which the regularity of distributions is characterized, as for the refined Sobolev scale, by two parameters, namely, by the main numerical parameter and by an additional function parameter. In his works, Triebel used logarithmic function parameters, whereas Haroske and Moura applied more general slowly varying function parameters.

Section 1.4. The algebra of pseudodifferential operators (PsDOs) was constructed and investigated mainly by J. J. Kohn, L. Nirenberg [96], and L. Hörmander [82, 84]. Elliptic operators form a very important class of PsDOs and have various applications to the theory of elliptic boundary-value problems for differential equations, spectral theory of differential operators, theory of function spaces, *etc.* The detailed presentation of the theory of PsDOs can be found, e.g., in the monographs by L. Hörmander [86], M. A. Shubin [232], M. Taylor [253], and F. Treves [254] and in the surveys by M. S. Agranovich [7, 10]. In our presentation, we mainly follow the notation and terminology used in [10].

For elliptic differential equations, we know internal *a priori* estimates of the solutions in suitable pairs of Hölder spaces (of fractional positive order) or Sobolev spaces (of arbitrary real order) and theorems on the local regularity (or smoothness) of the solutions. The presentation of these results and the corresponding references can be found in Yu. M. Berezansky's book [21, Chap. III, § 4]. The regularity of solutions of hypoelliptic differential equations in Hörmander spaces was investigated by Hörmander in [81, Theorems 4.1.5 and 7.4.1] and [85, Theorems 11.1.8 and 13.4.1]. For elliptic pseudodifferential equations on the Sobolev scale, the internal *a priori* estimates of the solutions and theorems on the local regularity of solutions are presented in the cited papers and books devoted to the theory of PsDOs. If a PsDO is uniformly elliptic on $\mathbb{R}^n$, then the *a priori* estimate holds in the entire $\mathbb{R}^n$. This is also true for the assertion concerning the increase in the regularity of solutions; see, e.g., the survey by Agranovich [10, Sec. 1.8].

All theorems in Section 1.4 were proved in [172] for more general matrix PsDOs uniformly elliptic in Petrovskii's sense.

Chapter 2

Hörmander spaces on closed manifolds and their applications

2.1 Hörmander spaces on closed manifolds

In the present section, we consider Hörmander spaces from the refined Sobolev scale on a smooth closed manifold and present the equivalent definitions of these spaces similar to the definitions used for the Sobolev spaces.

2.1.1 Equivalent definitions

Throughout the chapter, Γ denotes a closed (i.e., compact and without boundary) infinitely smooth oriented manifold of dimension $n \geq 1$. Assume that a C^∞-density dx is given on Γ. Recall that $\mathcal{D}'(\Gamma)$ is the topological linear space of all distributions on Γ. This space is dual to the space $C^\infty(\Gamma)$ with respect to the extension by continuity of the inner product in the space $L_2(\Gamma, dx) =: L_2(\Gamma)$ of square integrable functions $f : \Gamma \to \mathbb{C}$. We denote this extension by $(f, w)_\Gamma$, where $f \in \mathcal{D}'(\Gamma)$ and $w \in C^\infty(\Gamma)$.

Let $s \in \mathbb{R}$ and $\varphi \in \mathcal{M}$. We now give three equivalent definitions of the Hörmander space $H^{s,\varphi}(\Gamma)$.

The first definition characterizes $H^{s,\varphi}(\Gamma)$ in terms of the local properties of distributions $f \in \mathcal{D}'(\Gamma)$. We choose an arbitrary finite atlas from the C^∞-structure on Γ. Assume that this atlas is formed by local charts $\alpha_j : \mathbb{R}^n \leftrightarrow \Gamma_j$, where $j = 1, \ldots, r$. Here, the open sets Γ_j form a finite covering of the manifold Γ. We also choose an arbitrary finite collection of functions $\chi_j \in C^\infty(\Gamma)$, $j = 1, \ldots, r$, such that $0 \leq \chi_j(x) \leq 1$ and $\sum_{j=1}^{r} \chi_j(x) = 1$ for each $x \in \Gamma$ and, in addition, $\operatorname{supp} \chi_j \subset \Gamma_j$. These functions form a decomposition of unit on Γ.

Definition 2.1 (local). By definition, the linear space $H^{s,\varphi}(\Gamma)$ consists of all distributions $f \in \mathcal{D}'(\Gamma)$ such that

$$(\chi_j f) \circ \alpha_j \in H^{s,\varphi}(\mathbb{R}^n)$$

for any $j \in \{1, \ldots, r\}$. Here, $(\chi_j f) \circ \alpha_j$ is the representation of the distribution $\chi_j f$ in the local chart α_j. The inner product in $H^{s,\varphi}(\Gamma)$ is defined by the formula

$$(f_1, f_2)_{H^{s,\varphi}(\Gamma)} := \sum_{j=1}^{r} ((\chi_j f_1) \circ \alpha_j, (\chi_j f_2) \circ \alpha_j)_{H^{s,\varphi}(\mathbb{R}^n)}$$

for all $f_1, f_2 \in H^{s,\varphi}(\Gamma)$. This inner product induces the Hilbert norm in a standard way.

In the important special case $\varphi \equiv 1$, the space $H^{s,\varphi}(\Gamma)$ coincides with the Sobolev space $H^s(\Gamma)$ of order s. The latter is complete and, up to equivalence of norms, is independent of the indicated choice of an atlas and the decomposition of unit on Γ (see, e.g., [81, Sec. 2.6] or [232, Sec. 7.5]).

The second definition connects the space $H^{s,\varphi}(\Gamma)$ with the Sobolev spaces by means of interpolation and shows that $H^{s,\varphi}(\Gamma)$ is also complete and independent of the choice of an atlas and the decomposition of unit mentioned above.

Definition 2.2 (via interpolation). Let k_0 and k_1 be two integers such that $k_0 < s < k_1$. By definition,

$$H^{s,\varphi}(\Gamma) := \left[H^{k_0}(\Gamma), H^{k_1}(\Gamma)\right]_{\psi},$$

where the interpolation parameter ψ is given by the formula

$$\psi(t) = \begin{cases} t^{(s-k_0)/(k_1-k_0)}\,\varphi(t^{1/(k_1-k_0)}) & \text{for } \ t \geq 1, \\ \varphi(1) & \text{for } \ 0 < t < 1. \end{cases}$$

Remark 2.1. The function ψ in Definition 2.2 is an interpolation parameter by virtue of Theorem 1.11 because ψ is a regularly varying function at infinity of the order $\theta = (s - k_0)/(k_1 - k_0) \in (0, 1)$.

The third definition of the space $H^{s,\varphi}(\Gamma)$ is useful for the spectral theory of differential operators. This definition connects the norm in $H^{s,\varphi}(\Gamma)$ with a function of $1 - \Delta_\Gamma$, where Δ_Γ is the Beltrami–Laplace operator on Γ. (In this case, the Riemannian metric is introduced on the manifold Γ; see, e.g., [232, Sec. 22.1] or [255, Chap. XII, § 1].)

Definition 2.3 (via the operator). The Hilbert space $H^{s,\varphi}(\Gamma)$ is defined as the completion of $C^\infty(\Gamma)$ with respect to the norm

$$f \mapsto \|(1 - \Delta_\Gamma)^{s/2}\varphi((1 - \Delta_\Gamma)^{1/2})\, f\|_{L_2(\Gamma)}, \quad f \in C^\infty(\Gamma).$$

Theorem 2.1. *Definitions* 2.1, 2.2, *and* 2.3 *are equivalent; namely, they specify the same Hilbert space $H^{s,\varphi}(\Gamma)$ up to equivalence of norms.*

We prove Theorem 2.1 in Subsections 2.1.2 and 2.1.3. In connection with this theorem, it is reasonable to give the following definition.

Definition 2.4. A family of Hilbert spaces

$$\{H^{s,\varphi}(\Gamma) : \ s \in \mathbb{R}, \ \varphi \in \mathcal{M}\}$$

is called the refined Sobolev scale or simply the refined scale over the closed manifold Γ.

2.1.2 Interpolation properties

In this subsection, we consider the interpolation properties of the refined Sobolev scale on Γ and obtain, as a consequence, the proof of the equivalence of Definitions 2.1 and 2.2. We use the local definition 2.1 as the original definition of the space $H^{s,\varphi}(\Gamma)$.

Theorem 2.2. *Let $\varphi \in \mathcal{M}$ be an arbitrary function and let ε and δ be arbitrary positive numbers. Then, for any $s \in \mathbb{R}$,*

$$\big[H^{s-\varepsilon}(\Gamma),\ H^{s+\delta}(\Gamma)\big]_{\psi} = H^{s,\varphi}(\Gamma) \tag{2.1}$$

with the equivalence of norms. Here, ψ is the interpolation parameter in Theorem 1.14.

Proof. The pair of Sobolev spaces appearing on the left-hand side of equality (2.1) is admissible (see, e.g., [232, Proposition 7.4 and Theorem 7.4]). We deduce this equality from Theorem 1.14 with the help of the well-known procedures of "flattening" and "sewing" of the manifold Γ. By Definition 2.1, the linear "flattening" mapping

$$T : f \mapsto \big(\,(\chi_1 f)\circ\alpha_1, \ldots, (\chi_r f)\circ\alpha_r\,\big), \quad f \in \mathcal{D}'(\Gamma),$$

specifies the isometric operators

$$T:\ H^{\sigma}(\Gamma) \to (H^{\sigma}(\mathbb{R}^n))^r, \quad \sigma \in \mathbb{R}, \tag{2.2}$$

$$T:\ H^{s,\varphi}(\Gamma) \to (H^{s,\varphi}(\mathbb{R}^n))^r. \tag{2.3}$$

Since the function ψ is an interpolation parameter, the boundedness of operators (2.2), where $\sigma \in \{s-\varepsilon, s+\delta\}$, implies the boundedness of the operator

$$T:\ \big[H^{s-\varepsilon}(\Gamma), H^{s+\delta}(\Gamma)\big]_{\psi} \to \big[(H^{s-\varepsilon}(\mathbb{R}^n))^r, (H^{s+\delta}(\mathbb{R}^n))^r\big]_{\psi}. \tag{2.4}$$

By virtue of Theorems 1.5 and 1.14, we get the following equalities for the spaces and norms in these spaces:

$$\begin{aligned}\big[\,(H^{s-\varepsilon}(\mathbb{R}^n))^r, (H^{s+\delta}(\mathbb{R}^n))^r\,\big]_{\psi} &= \big(\,\big[H^{s-\varepsilon}(\mathbb{R}^n), H^{s+\delta}(\mathbb{R}^n)\big]_{\psi}\,\big)^r \\ &= (H^{s,\varphi}(\mathbb{R}^n))^r.\end{aligned} \tag{2.5}$$

Hence, the boundedness of operator (2.4) implies the boundedness of the operator

$$T:\ \big[H^{s-\varepsilon}(\Gamma), H^{s+\delta}(\Gamma)\big]_{\psi} \to (H^{s,\varphi}(\mathbb{R}^n))^r. \tag{2.6}$$

Further, we construct the left inverse “sewing” operator K for T. For any $j \in \{1,\dots,r\}$, we choose a function $\eta_j \in C_0^\infty(\mathbb{R}^n)$ such that $\eta_j = 1$ on the set $\alpha_j^{-1}(\operatorname{supp}\chi_j)$. Consider a linear mapping

$$K:\,(h_1,\dots,h_r)\mapsto\sum_{j=1}^{r}\Theta_j\big((\eta_j h_j)\circ\alpha_j^{-1}\big),\quad h_1,\dots,h_r\in\mathcal{S}'(\mathbb{R}^n).$$

Here, $(\eta_j h_j)\circ\alpha_j^{-1}$ is a distribution defined in the open set $\Gamma_j\subseteq\Gamma$ and satisfying the following condition: the representative of this distribution in the local chart α_j has the form $\eta_j h_j$. In addition, Θ_j denotes the operator of extension of a function/distribution by zero from Γ_j onto the entire manifold Γ. The operator Θ_j is well defined on the distributions whose support lies in Γ_j. By the choice of χ_j and η_j, we can write

$$\begin{aligned}KTf&=\sum_{j=1}^{r}\Theta_j\left(\left(\eta_j\left((\chi_j f)\circ\alpha_j\right)\right)\circ\alpha_j^{-1}\right)\\&=\sum_{j=1}^{r}\Theta_j\left((\chi_j f)\circ\alpha_j\circ\alpha_j^{-1}\right)=\sum_{j=1}^{r}\chi_j f=f.\end{aligned}$$

Thus,

$$KTf=f\quad\text{for any}\quad f\in\mathcal{D}'(\Gamma).\tag{2.7}$$

We now show that the linear mapping K defines a bounded operator

$$K:\,(H^{s,\varphi}(\mathbb{R}^n))^r\to H^{s,\varphi}(\Gamma).\tag{2.8}$$

For every vector $h=(h_1,\dots,h_r)\in(H^{s,\varphi}(\mathbb{R}^n))^r$, we can write

$$\begin{aligned}\|Kh\|^2_{H^{s,\varphi}(\Gamma)}&=\sum_{l=1}^{r}\|(\chi_l\,Kh)\circ\alpha_l\|^2_{H^{s,\varphi}(\mathbb{R}^n)}\\&=\sum_{l=1}^{r}\Big\|\Big(\chi_l\sum_{j=1}^{r}\Theta_j\big((\eta_j h_j)\circ\alpha_j^{-1}\big)\Big)\circ\alpha_l\Big\|^2_{H^{s,\varphi}(\mathbb{R}^n)}\\&=\sum_{l=1}^{r}\Big\|\sum_{j=1}^{r}(\eta_{j,l}\,h_j)\circ\beta_{j,l}\Big\|^2_{H^{s,\varphi}(\mathbb{R}^n)}\\&\le\sum_{l=1}^{r}\Big(\sum_{j=1}^{r}\|(\eta_{j,l}\,h_j)\circ\beta_{j,l}\|_{H^{s,\varphi}(\mathbb{R}^n)}\Big)^2.\end{aligned}\tag{2.9}$$

Here,

$$\eta_{j,l} := (\chi_l \circ \alpha_j)\, \eta_j \in C_0^\infty(\mathbb{R}^n)$$

and $\beta_{j,l} : \mathbb{R}^n \leftrightarrow \mathbb{R}^n$ is a C^∞-diffeomorphism such that

$$\beta_{j,l} = \alpha_j^{-1} \circ \alpha_l$$

in the vicinity of the set $\operatorname{supp} \eta_{j,l}$ and $\beta_{j,l}(x) = x$ for all $x \in \mathbb{R}^n$ sufficiently large in the absolute value. It is well known that the operator of multiplication by a function in $C_0^\infty(\mathbb{R}^n)$ and the operator of the change of variables $u \mapsto u \circ \beta_{j,l}$ are bounded in each space $H^\sigma(\mathbb{R}^n)$ with $\sigma \in \mathbb{R}$ (see, e.g., [86, Theorems B.1.7 and B.1.8]). Therefore, the linear operator $v \mapsto (\eta_{j,l}\, v) \circ \beta_{j,l}$ is bounded in the space $H^\sigma(\mathbb{R}^n)$. By Theorem 1.14, this implies its boundedness in the space $H^{s,\varphi}(\mathbb{R}^n)$. Hence, relations (2.9) yield the estimate

$$\|Kh\|^2_{H^{s,\varphi}(\Gamma)} \leq c \sum_{j=1}^{r} \|h_j\|^2_{H^{s,\varphi}(\mathbb{R}^n)},$$

where the number $c > 0$ is independent of $h = (h_1, \ldots, h_r)$. Thus, operator (2.8) is bounded for all $s \in \mathbb{R}$ and $\varphi \in \mathcal{M}$.

Specifically, the operators $K : (H^\sigma(\mathbb{R}^n))^r \to H^\sigma(\Gamma)$ with $\sigma \in \mathbb{R}$ are bounded. Taking the values $\sigma \in \{s - \varepsilon,\, s + \delta\}$ and using the interpolation with a parameter ψ, in view of relation (2.5), we conclude that the operator

$$K : (H^{s,\varphi}(\mathbb{R}^n))^r \to \left[H^{s-\varepsilon}(\Gamma),\, H^{s+\delta}(\Gamma)\right]_\psi \tag{2.10}$$

is bounded. Relations (2.3), (2.10), and (2.7) now imply that the identity operator KT realizes a continuous embedding of the space $H^{s,\varphi}(\Gamma)$ in the interpolation space $[H^{s-\varepsilon}(\Gamma),\, H^{s+\delta}(\Gamma)]_\psi$. Moreover, it follows from relations (2.6) and (2.8) that the same operator KT realizes the inverse continuous embedding.

Theorem 2.2 is proved.

In the case where the numbers $k_0 := s - \varepsilon$ and $k_1 := s + \delta$ in Theorem 2.2 are integer, we obtain the following important result:

Corollary 2.1. *Definitions* 2.1 *and* 2.2 *are equivalent for any parameters* $s \in \mathbb{R}$ *and* $\varphi \in \mathcal{M}$.

Applying Theorem 2.2 and the properties of interpolation, we get the following properties of the refined Sobolev scale over the manifold Γ :

Theorem 2.3. *Let* $s \in \mathbb{R}$ *and let* $\varphi, \varphi_1 \in \mathcal{M}$. *The following assertions are true:*

(i) *The space* $H^{s,\varphi}(\Gamma)$ *is complete (Hilbert) and, up to equivalence of norms, is independent of the choice of an atlas of the manifold* Γ *and the decomposition of unit used in Definition* 2.1.

(ii) *The set* $C^\infty(\Gamma)$ *is dense in the space* $H^{s,\varphi}(\Gamma)$.

(iii) *The compact and dense embedding* $H^{s+\varepsilon,\varphi_1}(\Gamma) \hookrightarrow H^{s,\varphi}(\Gamma)$ *is true for any number* $\varepsilon > 0$.

(iv) *The function* φ/φ_1 *is bounded in the vicinity of* ∞ *if and only if* $H^{s,\varphi_1}(\Gamma) \hookrightarrow H^{s,\varphi}(\Gamma)$. *This embedding is dense and continuous. It is compact if and only if* $\varphi(t)/\varphi_1(t) \to 0$ *as* $t \to \infty$.

(v) *The spaces* $H^{s,\varphi}(\Gamma)$ *and* $H^{-s,1/\varphi}(\Gamma)$ *are mutually dual up to equivalence of norms with respect to the extension by continuity of the inner product in* $L_2(\Gamma)$.

Proof. (i) By Theorem 2.2, the space $H^{s,\varphi}(\Gamma)$ is complete because it is obtained as a result of the interpolation of Sobolev inner product spaces. We arbitrarily choose two pairs $\mathcal{A}_1$ and $\mathcal{A}_2$ each of which consists of a finite atlas of Γ and a relevant decomposition of unit on Γ used in Definition 2.1. By $H^{s,\varphi}(\Gamma, \mathcal{A}_j)$ we denote the space $H^{s,\varphi}(\Gamma)$ corresponding to the pair $\mathcal{A}_j$ with $j \in \{1, 2\}$. Similarly, $H^\sigma(\Gamma, \mathcal{A}_j)$ stands for the Sobolev space $H^\sigma(\Gamma)$ corresponding to this pair. For the Sobolev spaces, the identity mapping defines an isomorphism

$$I : H^\sigma(\Gamma, \mathcal{A}_1) \leftrightarrow H^\sigma(\Gamma, \mathcal{A}_2)$$

for every $\sigma \in \mathbb{R}$. We now take $\sigma := s \mp 1$ and apply the interpolation with the same parameter ψ as in Theorem 1.14. According to Theorem 2.2, we get the isomorphism

$$I : H^{s,\varphi}(\Gamma, \mathcal{A}_1) \leftrightarrow H^{s,\varphi}(\Gamma, \mathcal{A}_2).$$

This means that the space $H^{s,\varphi}(\Gamma)$ is independent of the indicated choice of the atlas and the decomposition of unity.

Assertion (i) is proved.

(ii) By virtue of Theorems 1.1 and 2.2, the embedding $H^{s+\delta}(\Gamma) \hookrightarrow H^{s,\varphi}(\Gamma)$ is continuous and dense. It is known that the set $C^\infty(\Gamma)$ is dense in the Sobolev space $H^{s+\delta}(\Gamma)$ (see, e.g., [232, Proposition 7.4]). Hence, this set is dense in $H^{s,\varphi}(\Gamma)$.

Assertion (ii) is proved.

(iii) Let $\varepsilon > 0$. According to Theorem 2.2 there exist interpolation parameters $\chi, \eta \in \mathcal{B}$ such that

$$\big[H^{s+\varepsilon/2}(\Gamma), H^{s+2\varepsilon}(\Gamma)\big]_\chi = H^{s+\varepsilon,\varphi_1}(\Gamma),$$

$$\big[H^{s-\varepsilon}(\Gamma), H^{s+\varepsilon/3}(\Gamma)\big]_\eta = H^{s,\varphi}(\Gamma)$$

up to equivalence of norms. Hence, by Theorem 1.1, we get a chain of continuous embeddings

$$H^{s+\varepsilon,\varphi_1}(\Gamma) \hookrightarrow H^{s+\varepsilon/2}(\Gamma) \hookrightarrow H^{s+\varepsilon/3}(\Gamma) \hookrightarrow H^{s,\varphi}(\Gamma),$$

where the central embedding of Sobolev spaces is compact (see, e.g., [232, Theorem 7.4]). Thus, the embedding $H^{s+\varepsilon,\varphi_1}(\Gamma) \hookrightarrow H^{s,\varphi}(\Gamma)$ is also compact. It is dense in view of assertion (ii).

Assertion (iii) is proved.

(iv) Assume that the function φ/φ_1 is bounded in the vicinity of ∞. According to Theorem 2.2, we arrive at the following equalities of spaces up to equivalence of norms:

$$\big[H^{s-1}(\Gamma), H^{s+1}(\Gamma)\big]_{\psi} = H^{s,\varphi}(\Gamma),$$

$$\big[H^{s-1}(\Gamma), H^{s+1}(\Gamma)\big]_{\psi_1} = H^{s,\varphi_1}(\Gamma).$$

Here, the interpolation parameters $\psi, \psi_1 \in \mathcal{B}$ satisfy the condition

$$\frac{\psi(t)}{\psi_1(t)} = \frac{\varphi(t^{1/2})}{\varphi_1(t^{1/2})}$$

for $t \geq 1$. Hence, the function ψ/ψ_1 is bounded in the vicinity of ∞ and the embedding $H^{s,\varphi_1}(\Gamma) \hookrightarrow H^{s,\varphi}(\Gamma)$ is continuous and dense by virtue of Theorem 1.2. Further, if $\varphi(t)/\varphi_1(t) \to 0$ as $t \to \infty$, then $\psi(t)/\psi_1(t) \to 0$ as $t \to \infty$. Therefore, it follows from the compactness of the embedding of Sobolev spaces $H^{s+1}(\Gamma) \hookrightarrow H^{s-1}(\Gamma)$ that the embedding $H^{s,\varphi_1}(\Gamma) \hookrightarrow H^{s,\varphi}(\Gamma)$ is compact by Theorem 1.2 provided that $\varphi(t)/\varphi_1(t) \to 0$ as $t \to \infty$.

We now show that the embedding $H^{s,\varphi_1}(\Gamma) \hookrightarrow H^{s,\varphi}(\Gamma)$ implies that the function φ/φ_1 is bounded on the semiaxis $[1, \infty)$. Assume that this embedding is true. We use the local definition 2.1. It is possible to assume that the conditions $U \subset \Gamma_1$ and $U \cap \Gamma_j = \varnothing$ with $j \neq 1$ are satisfied for some open nonempty set $U \subset \Gamma$. For any distribution $w \in H^{s,\varphi_1}(\mathbb{R}^n)$ such that $\operatorname{supp} w \subset \alpha_1^{-1}(U)$, we find

$$\Theta(w \circ \alpha_1^{-1}) \in H^{s,\varphi_1}(\Gamma) \hookrightarrow H^{s,\varphi}(\Gamma).$$

Here, Θ is the operator of extension of a distribution by zero from the set U onto the entire Γ. Thus,

$$w = \big(\chi_1(\Theta(w \circ \alpha_1^{-1}))\big) \circ \alpha_1 \in H^{s,\varphi}(\mathbb{R}^n).$$

Hence, according to Hörmander's result [81, Theorem 2.2.2], the function

$$\varphi(\langle\xi\rangle)/\varphi_1(\langle\xi\rangle)$$

of $\xi \in \mathbb{R}^n$ is bounded. Therefore, the function φ/φ_1 is bounded on the semi-axis $[1, \infty)$.

Finally, we show that the compactness of the embedding $H^{s,\varphi_1}(\Gamma) \hookrightarrow H^{s,\varphi}(\Gamma)$ implies the convergence $\varphi(t)/\varphi_1(t) \to 0$ as $t \to \infty$. If this embedding is compact, then the embedding operator

$$\{w \in H^{s,\varphi_1}(\mathbb{R}^n) : \operatorname{supp} w \subset \alpha_1^{-1}(U)\} \hookrightarrow H^{s,\varphi}(\mathbb{R}^n)$$

is also compact. Thus, by virtue of the above-mentioned result [81, Theorem 2.2.3], we conclude that $\varphi(\langle\xi\rangle)/\varphi_1(\langle\xi\rangle) \to 0$ as $|\xi| \to \infty$. Hence, $\varphi(t)/\varphi_1(t) \to 0$ as $t \to \infty$.

Assertion (iv) is proved.

(v) Assertion (v) is well known in the Sobolev case $\varphi \equiv 1$ (see, e.g., [232, Theorem 7.7]). Thus, the Sobolev spaces $H^{s\pm 1}(\Gamma)$ and $H^{-s\mp 1}(\Gamma)$ are mutually dual with respect to the extension of the inner product in $L_2(\Gamma)$ by continuity. This means that the linear mapping $Q : w \mapsto (w, \cdot)_\Gamma$, where $w \in C^\infty(\Gamma)$, can be extended by continuity to the isomorphisms

$$Q : H^{s\mp 1}(\Gamma) \leftrightarrow (H^{-s\pm 1}(\Gamma))'.$$

Applying the interpolation with the same parameter ψ as in Theorem 2.2 for $\varepsilon = \delta = 1$, we get one more isomorphism

$$Q : \left[H^{s-1}(\Gamma), H^{s+1}(\Gamma)\right]_\psi \leftrightarrow \left[(H^{-s+1}(\Gamma))', (H^{-s-1}(\Gamma))'\right]_\psi. \tag{2.11}$$

Here, the interpolation space on the left-hand side is $H^{s,\varphi}(\Gamma)$, whereas the interpolation space on the right-hand side can be represented (in view of Theorem 1.4) in the form

$$\left[(H^{-s+1}(\Gamma))', (H^{-s-1}(\Gamma))'\right]_\psi = \left[H^{-s-1}(\Gamma), H^{-s+1}(\Gamma)\right]'_\chi = (H^{-s,1/\varphi}(\Gamma))'.$$

Note that the last equality is true because

$$\chi(t) := \frac{t}{\psi(t)} = \frac{t^{1/2}}{\varphi(t^{1/2})} \quad \text{for} \quad t \geq 1.$$

Thus, relation (2.11) implies the isomorphism

$$Q : H^{s,\varphi}(\Gamma) \leftrightarrow (H^{-s,1/\varphi}(\Gamma))',$$

i.e., the indicated mutual duality of the spaces $H^{s,\varphi}(\Gamma)$ and $H^{-s,1/\varphi}(\Gamma)$.

Assertion (v) is proved.

Theorem 2.3 is proved.

Theorem 2.2 and the reiteration property (Theorem 1.3) now imply the following assertion showing that the refined Sobolev scale over Γ is closed with respect to the interpolation with function parameters quasiregularly varying at ∞.

Theorem 2.4. *Let $s_0, s_1 \in \mathbb{R}$, $s_0 \leq s_1$, and let $\varphi_0, \varphi_1 \in \mathcal{M}$. Assume that the function φ_0/φ_1 is bounded in the vicinity of ∞ for $s_0 = s_1$. Suppose that $\psi \in \mathcal{B}$ is a function quasiregularly varying at ∞ of order θ, where $0 < \theta < 1$. By Theorem 1.11, this function is an interpolation parameter. It can be represented in the form $\psi(t) = t^\theta \chi(t)$ for some $\chi \in$ QSV. Also let $s := (1-\theta)s_0 + \theta s_1$ and let*

$$\varphi(t) := \varphi_0^{1-\theta}(t)\, \varphi_1^{\theta}(t)\, \chi\left(t^{s_1-s_0}\varphi_1(t)/\varphi_0(t)\right) \quad \text{for} \quad t \geq 1. \tag{2.12}$$

Then $\varphi \in \mathcal{M}$ and

$$[\, H^{s_0,\varphi_0}(\Gamma) H^{s_1,\varphi_1}(\Gamma)\,]_\psi = H^{s,\varphi}(\Gamma) \tag{2.13}$$

with equivalence of the norms.

Proof. The positive function φ is Borel measurable on the set $[1,\infty)$ and bounded, together with the function $1/\varphi$, on every segment $[1,b]$, $1 < b < \infty$, because the functions φ_0, φ_1, and χ have similar properties. Moreover, the condition $\varphi_0, \varphi_1, \chi \in$ QSV implies the inclusion $\varphi \in$ QSV by virtue of Theorem 1.12(iii) and (iv) for $s_0 < s_1$ or Theorem 1.13 for $s_0 = s_1$. Thus, $\varphi \in \mathcal{M}$. Further, according to Theorem 2.3(iii) and (iv), the pair of spaces

$$[H^{s_0,\varphi_0}(\Gamma), H^{s_1,\varphi_1}(\Gamma)]$$

is admissible.

We now prove equality (2.13). We set

$$\varrho := s_1 - s_0 + 1 \quad \varepsilon_j := s_j - s + \varrho, \quad \text{and} \quad \delta_j := s - s_j + \varrho$$

for any $j \in \{0,1\}$. The numbers ϱ, ε_j, and δ_j are positive because $s_0 \leq s \leq s_1$. Note that they have the following properties:

$$\varepsilon_j + \delta_j = 2\varrho, \quad \varepsilon_1 - \varepsilon_0 = s_1 - s_0, \quad (1-\theta)\varepsilon_0 + \theta\varepsilon_1 = \varrho. \tag{2.14}$$

By virtue of Theorem 2.2, we get the equalities

$$\left[H^{s_j-\varepsilon_j}(\Gamma),\ H^{s_j+\delta_j}(\Gamma)\right]_{\psi_j} = H^{s_j,\varphi_j}(\Gamma) \quad \text{for each} \quad j \in \{0,1\}$$

with equivalence of the norms. Here, the interpolation parameter ψ_j is given by the formula

$$\psi_j(t) := \begin{cases} t^{\varepsilon_j/(\varepsilon_j+\delta_j)}\, \varphi_j(t^{1/(\varepsilon_j+\delta_j)}) & \text{for } t \geq 1, \\ \varphi_j(1) & \text{for } 0 < t < 1. \end{cases} \tag{2.15}$$

Since ψ is an interpolation parameter, $s_j - \varepsilon_j = s - \varrho$, and $s_j + \delta_j = s + \varrho$, this yields

$$\begin{aligned} &[\, H^{s_0,\varphi_0}(\Gamma), H^{s_1,\varphi_1}(\Gamma)\,]_\psi \\ &\quad = \Big[\left[H^{s-\varrho}(\Gamma), H^{s+\varrho}(\Gamma)\right]_{\psi_0},\ \left[H^{s-\varrho}(\Gamma), H^{s+\varrho}(\Gamma)\right]_{\psi_1} \Big]_\psi \end{aligned} \tag{2.16}$$

up to equivalence of norms. Note (see (2.14)) that the function

$$\frac{\psi_0(t)}{\psi_1(t)} = \frac{t^{(s_0-s_1)/(2\varrho)}\varphi_0(t^{1/(2\varrho)})}{\varphi_1(t^{1/(2\varrho)})} \quad \text{of} \quad t \geq 1$$

is bounded in the vicinity of ∞ by virtue of Theorem 1.12(ii) in the case where $s_0 < s_1$ and by virtue of the condition for $s_0 = s_1$. Applying the reiteration theorem 1.3 to (2.16), we get

$$\left[\, H^{s_0,\varphi_0}(\Gamma), H^{s_1,\varphi_1}(\Gamma)\, \right]_{\psi} = \left[\, H^{s-\varrho}(\Gamma), H^{s+\varrho}(\Gamma)\, \right]_{\omega} \tag{2.17}$$

up to equivalence of norms. Here, the interpolation parameter ω is given by the formula

$$\omega(t) := \psi_0(t)\, \psi(\psi_1(t)/\psi_0(t)) \quad \text{for} \ \ t > 0.$$

By virtue of relations (2.14), (2.15), and (2.12), after elementary calculations, we obtain the equalities:

$$\omega(t) = t^{1/2}\varphi(t^{1/(2\varrho)})$$

for $t \geq 1$ and $\omega(t) = \varphi(1)$ for $0 < t < 1$. Thus, according to Theorem 2.2, we find

$$\left[\, H^{s-\varrho}(\Gamma), H^{s+\varrho}(\Gamma)\, \right]_{\omega} = H^{s,\varphi}(\Gamma) \tag{2.18}$$

up to equivalence of norms. Relations (2.17) and (2.18) now yield (2.13).

Theorem 2.4 is proved.

Remark 2.2. Theorem 2.4 is true in the limiting cases $\theta = 0$ or $\theta = 1$ if we additionally assume that ψ is a pseudoconcave function in the vicinity of ∞. Hence, by Theorem 1.9, the function ψ is an interpolation parameter, and the proof presented above remains valid. Thus, Theorem 2.4 remains true for each function

$$\psi(t) := \ln^r t \quad \text{and} \quad \psi(t) := t/\ln^r t,$$

where $t \gg 1$ and $r > 0$.

Remark 2.3. Theorem 2.4 remains true if we replace Γ in its formulation by $\mathbb{R}^n$. The proof is similar to the proof presented above. Thus, the refined Sobolev scale over $\mathbb{R}^n$ (just as its analog over Γ) is closed with respect to the interpolation with function parameters quasiregularly varying at ∞.

2.1.3 Equivalent norms

In this subsection, we construct equivalent norms in the space $H^{s,\varphi}(\Gamma)$ with the help of elliptic positive-definite PsDOs. As a consequence, we prove that Definitions 2.1 and 2.3 are equivalent.

Let A be an elliptic polyhomogeneous PsDO of order $m > 0$ on the manifold Γ. (The definition of PsDO on closed manifolds and all related notions are presented in Subsection 2.2.1.) Suppose that the operator $A : C^\infty(\Gamma) \to C^\infty(\Gamma)$ is positive definite in the space $L_2(\Gamma)$, i.e., there exists a number $\varkappa > 0$ such that

$$(Au, u)_\Gamma \geq \varkappa\,(u, u)_\Gamma \quad \text{for any} \quad u \in C^\infty(\Gamma). \tag{2.19}$$

By A_0 we denote the closure of the operator $A : C^\infty(\Gamma) \to C^\infty(\Gamma)$ in the space $L_2(\Gamma)$. Since PsDO A is elliptic on Γ, this closure exists and is defined on $H^m(\Gamma)$. Condition (2.19) implies that the PsDO A is formally self-adjoint. Therefore, A_0 is an unbounded self-adjoint operator in the space $L_2(\Gamma)$ with $\operatorname{Spec} A_0 \subseteq [\varkappa, \infty)$. (See, e.g., [10, Theorems 2.3.5 and 2.3.7] or [232, Proposition 8.4 and Theorem 8.3]).

Let $s \in \mathbb{R}$ and $\varphi \in \mathcal{M}$. We set

$$\varphi_{s,m}(t) := \begin{cases} t^{s/m}\varphi(t^{1/m}) & \text{for } t \geq 1, \\ \varphi(1) & \text{for } 0 < t < 1. \end{cases} \tag{2.20}$$

Since the function $\varphi_{s,m}$ is positive and Borel measurable on the semiaxis $(0, \infty)$, the unbounded self-adjoint operator $\varphi_{s,m}(A_0)$ is defined as a function of A_0 on a certain linear manifold in $L_2(\Gamma)$.

Lemma 2.1. *The following assertions are true:*

(i) *The domain of the operator $\varphi_{s,m}(A_0)$ contains $C^\infty(\Gamma)$;*

(ii) *The mapping*

$$f \mapsto \|\varphi_{s,m}(A_0)f\|_{L_2(\Gamma)}, \quad f \in C^\infty(\Gamma), \tag{2.21}$$

is a norm in the space $C^\infty(\Gamma)$.

Proof. (i) We choose an integer $k > s/m$. Since $\varphi \in \mathcal{M}$, the function $\varphi_{s,m}$ is bounded on every compact subset of the semiaxis $(0, \infty)$, and, moreover, $t^{-k}\varphi_{s,m}(t) \to 0$ as $t \to \infty$ by Theorem 1.12(ii) and (iv). Hence, there exists a number $c > 0$ such that $\varphi_{s,m}(t) \leq c\,t^k$ for all $t \geq \varkappa$. Consider an unbounded operator A_0^k acting in the space $L_2(\Gamma)$. Since $A : C^\infty(\Gamma) \to C^\infty(\Gamma)$, we can write

$$C^\infty(\Gamma) \subset \operatorname{Dom} A_0^k \subset \operatorname{Dom} \varphi_{s,m}(A_0).$$

Assertion (i) is proved.

(ii) According to assertion (i), mapping (2.21) is well defined. For this mapping, all properties of the norm are obvious except the property of positive definiteness.

We now establish this property. In view of the spectral theorem, for any function $f \in C^\infty(\Gamma)$, we find

$$\|\varphi_{s,m}(A_0)f\|^2_{L_2(\Gamma)} = \int\limits_{\varkappa}^{\infty} \varphi^2_{s,m}(t)\, d(E_t f, f)_\Gamma, \tag{2.22}$$

$$\|f\|^2_{L_2(\Gamma)} = \int\limits_{\varkappa}^{\infty} d(E_t f, f)_\Gamma. \tag{2.23}$$

Here, E_t with $t \geq \varkappa$, is the resolution of the identity in $L_2(\Gamma)$ corresponding to the self-adjoint operator A_0. Further, if

$$\|\varphi_{s,m}(A_0)f\|^2_{L_2(\Gamma)} = 0,$$

then it follows from relation (2.22) and the positiveness of the function $\varphi_{s,m}$ that the measure $(E(\cdot)f, f)_\Gamma$ of the set $[\varkappa, \infty)$ is equal to zero. Hence, by virtue of (2.23), we arrive at the equality $f = 0$ on Γ.

Assertion (ii) is proved.

Lemma 2.1 is proved.

Theorem 2.5. *For all $s \in \mathbb{R}$ and $\varphi \in \mathcal{M}$, the norm* (2.21) *and the norm in the space $H^{s,\varphi}(\Gamma)$ are equivalent on $C^\infty(\Gamma)$. Thus, the space $H^{s,\varphi}(\Gamma)$ coincides (up to equivalence of norms) with the completion of the linear space $C^\infty(\Gamma)$ with respect to the norm* (2.21).

Proof. First, we assume that $s > 0$. We choose an integer $k \geq 1$ such that $k\,m > s$. Since the operator A_0^k is closed and positive definite on $L_2(\Gamma)$, its domain $\operatorname{Dom} A_0^k$ is a Hilbert space with respect to the inner product $(A_0^k f, A_0^k g)_\Gamma$ of functions f, g. In this case, the pair of spaces $[L_2(\Gamma), \operatorname{Dom} A_0^k]$ is admissible and A_0^k is a generating operator for this pair. In addition, since A_0^k is the closure of the elliptic PsDO A^k acting in $L_2(\Gamma)$, the spaces $\operatorname{Dom} A_0^k$ and $H^{km}(\Gamma)$ are equal up to equivalence of norms. Let ψ be the interpolation function parameter used in Theorems 1.14 and 2.2, where we set $\varepsilon = s$ and $\delta = k\,m - s$. Then

$$\psi(t^k) = \varphi_{s,m}(t)$$

for all $t > 0$ and, by virtue of Theorem 2.2, we get

$$\|f\|_{H^{s,\varphi}(\Gamma)} \asymp \|f\|_{[H^0(\Gamma),\, H^{km}(\Gamma)]_\psi} \asymp \|f\|_{[L_2(\Gamma),\, \operatorname{Dom} A_0^k]_\psi}$$

$$= \|\psi(A_0^k)f\|_{L_2(\Gamma)} = \|\varphi_{s,m}(A_0)f\|_{L_2(\Gamma)},$$

where $f \in C^\infty(\Gamma)$.

Now let the real number s be arbitrary. We choose an integer $k \geq 1$ such that $s + k\,m > 0$. In the previous paragraph, we have proved that

$$\|g\|_{H^{s+km,\varphi}(\Gamma)} \asymp \|\varphi_{s+km,m}(A_0)\, g\|_{L_2(\Gamma)}, \quad g \in C^\infty(\Gamma). \tag{2.24}$$

The PsDO A^k realizes the isomorphisms

$$A^k : H^{\sigma+km}(\Gamma) \leftrightarrow H^\sigma(\Gamma), \quad \sigma \in \mathbb{R}, \tag{2.25}$$

$$A^k : H^{s+km,\varphi}(\Gamma) \leftrightarrow H^{s,\varphi}(\Gamma). \tag{2.26}$$

This is proved in the next paragraph. By A^{-k} we denote the operator inverse to A^k. In view of (2.25), we find

$$g := A^{-k} f \in \bigcap_{\sigma\in\mathbb{R}} H^{\sigma+km}(\Gamma) = C^\infty(\Gamma) \quad \text{and} \quad A_0^k A^{-k} f = f$$

for any function $f \in C^\infty(\Gamma)$. By virtue of (2.26) and (2.24), this yields the required equivalence of norms:

$$\|f\|_{H^{s,\varphi}(\Gamma)} \asymp \|A^{-k} f\|_{H^{s+km,\varphi}(\Gamma)} \asymp \|\varphi_{s+km,m}(A_0) A^{-k} f\|_{L_2(\Gamma)}$$

$$= \|\varphi_{s,m}(A_0) A_0^k A^{-k} f\|_{L_2(\Gamma)} = \|\varphi_{s,m}(A_0) f\|_{L_2(\Gamma)}, \quad f \in C^\infty(\Gamma).$$

It remains to show that A^k realizes isomorphisms (2.25) and (2.26). Since the PsDO A is elliptic on Γ, it specifies a bounded Fredholm operator

$$A : H^{\sigma+m}(\Gamma) \to H^\sigma(\Gamma)$$

for any $\sigma \in \mathbb{R}$. Both the kernel and the index of this operator are independent of σ (see, e.g., [86, Theorem 19.2.1] or [232, Theorem 8.1]). Since $0 \notin \operatorname{Spec} A_0$, the self-adjoint operator A_0 isomorphically maps its own domain $H^m(\Gamma)$ onto the space $H^0(\Gamma) = L_2(\Gamma)$. Hence, the kernel is trivial and the index is equal to zero. Thus, the PsDO A establishes the isomorphisms

$$A : H^{\sigma+m}(\Gamma) \leftrightarrow H^\sigma(\Gamma), \quad \sigma \in \mathbb{R},$$

whence, as a result of k iterations, we arrive at isomorphism (2.25). Finally, we set $\sigma = s \mp 1$ in (2.25) and apply the interpolation Theorem 2.2. This yields isomorphism (2.26).

Theorem 2.5 is proved.

For $A_0 = 1 - \Delta_\Gamma$, Theorem 2.5 gives the following important result:

Corollary 2.2. *Definitions* 2.1 *and* 2.3 *are equivalent for all parameters $s \in \mathbb{R}$ and $\varphi \in \mathcal{M}$.*

It is worth to distinguish the case where the space $H^{s,\varphi}(\Gamma)$ coincides with the domain of the operator $\varphi_{s,m}(A_0)$.

Theorem 2.6. *Let $s \geq 0$ and let $\varphi \in \mathcal{M}$. In the case where $s = 0$, we additionally assume that the function $1/\varphi$ is bounded in the vicinity of ∞. Then the space $H^{s,\varphi}(\Gamma)$ coincides with the domain of the operator $\varphi_{s,m}(A_0)$ and the norm in the space $H^{s,\varphi}(\Gamma)$ is equivalent to the graph norm of the unbounded operator $\varphi_{s,m}(A_0)$ acting in $L_2(\Gamma)$.*

Proof. The domain $\operatorname{Dom}\varphi_{s,m}(A_0)$ of the closed operator $\varphi_{s,m}(A_0)$ is a Hilbert space with respect to the inner product of the graph of this operator. We prove that the graph norm of the operator $\varphi_{s,m}(A_0)$ is equivalent to norm (2.21) on the linear manifold $C^\infty(\Gamma)$ and that this manifold is dense in the space $\operatorname{Dom}\varphi_s(A_0)$. By virtue of Theorem 2.5, this yields Theorem 2.6.

By the condition and Theorem 1.12(ii), there exists a number $c > 0$ such that $\varphi_{s,m}(t) \geq c$ for any $t > 0$. Therefore,

$$\|\varphi_{s,m}(A_0)f\|_{L_2(\Gamma)} \geq c\,\|f\|_{L_2(\Gamma)} \quad \text{for any} \quad f \in C^\infty(\Gamma).$$

This yields the indicated equivalence of norms. It remains to show that the set $C^\infty(\Gamma)$ is dense in the space $\operatorname{Dom}\varphi_{s,m}(A_0)$.

Let $f \in \operatorname{Dom}\varphi_{s,m}(A_0)$. Since $\varphi_{s,m}(A_0)f \in L_2(\Gamma)$, there exists a sequence of functions $h_j \in C^\infty(\Gamma)$ such that $h_j \to \varphi_{s,m}(A_0)f$ in $L_2(\Gamma)$ as $j \to \infty$. Since $1/\varphi_{s,m}(t) \leq 1/c$ for every $t > 0$, the operator $\varphi_{s,m}^{-1}(A_0)$ is bounded in the space $L_2(\Gamma)$. Hence,

$$f_j := \varphi_{s,m}^{-1}(A_0)h_j \to f,$$

$$\varphi_{s,m}(A_0)f_j = h_j \to \varphi_{s,m}(A_0)f \quad \text{in} \quad L_2(\Gamma) \quad \text{as} \quad j \to \infty.$$

In other words, $f_j \to f$ with respect to the graph norm of the operator $\varphi_{s,m}(A_0)$.

In addition, since $h_j \in C^\infty(\Gamma)$, the relation

$$f_j = A_0^{-k}\varphi_{s,m}^{-1}(A_0)A_0^k\, h_j \in H^{km}(\Gamma)$$

holds for any $k \in \mathbb{N}$. Thus, $f_j \in C^\infty(\Gamma)$. Hence, we have shown that the set $C^\infty(\Gamma)$ is dense in the Hilbert space $\operatorname{Dom}\varphi_{s,m}(A_0)$.

Theorem 2.6 is proved.

At the end of this subsection, we endow the space $H^{s,\varphi}(\Gamma)$ with an equivalent norm expressed in terms of sequences.

Since the PsDO A is elliptic on Γ and positive definite on $L_2(\Gamma)$, it has the following spectral properties (see, e.g., [10, Sec. 6.1], [253, Chap. 2, § 2], and [232, Sec. 8.3 and 15.2]). The space $L_2(\Gamma)$ has an orthonormal basis $(h_j)_{j=1}^\infty$

formed by the eigenfunctions $h_j \in C^\infty(\Gamma)$ of the operator A_0. In this case, $Ah_j = \lambda_j h_j$ and the sequence of eigenvalues $(\lambda_j)_{j=1}^\infty$ consists of positive numbers, is nonstrictly increasing, and approaches ∞. The spectrum of the operator A_0 coincides with the set of all its eigenvalues $\{\lambda_j : j \in \mathbb{N}\}$. The following asymptotic relation is true:

$$\lambda_j \sim c\, j^{\,m/n} \quad \text{as} \quad j \to \infty, \tag{2.27}$$

where c is a positive number depending on A and $n = \dim \Gamma$. An arbitrary distribution $f \in \mathcal{D}'(\Gamma)$ can be expanded in the Fourier series

$$f = \sum_{j=1}^{\infty} c_j(f)\, h_j \quad \text{convergent in } \mathcal{D}'(\Gamma), \tag{2.28}$$

where $c_j(f) := (f, h_j)_\Gamma$ are the Fourier coefficients of f with respect to h_j.

Theorem 2.7. *Let $s \in \mathbb{R}$, and let $\varphi \in \mathcal{M}$. Then*

$$H^{s,\varphi}(\Gamma) = \Big\{ f \in \mathcal{D}'(\Gamma) : \sum_{j=1}^{\infty} j^{\,2s/n}\, \varphi^2(j^{\,1/n})\, |c_j(f)|^2 < \infty \Big\}, \tag{2.29}$$

$$\|f\|_{H^{s,\varphi}(\Gamma)} \asymp \Big(\sum_{j=1}^{\infty} j^{\,2s/n}\, \varphi^2(j^{\,1/n})\, |c_j(f)|^2 \Big)^{1/2}. \tag{2.30}$$

The proof of this theorem is preceded by three lemmas. As above, in these lemmas, $s \in \mathbb{R}$ and $\varphi \in \mathcal{M}$.

Lemma 2.2. *For any function $f \in \mathrm{Dom}\,(\varphi_{s,m}(A_0))$, the following equality is true:*

$$\|\varphi_{s,m}(A_0) f\|^2_{L_2(\Gamma)} = \sum_{j=1}^{\infty} \varphi^2_{s,m}(\lambda_j)\, |c_j(f)|^2. \tag{2.31}$$

Proof. By P_j we denote the orthoprojector of the space $L_2(\Gamma)$ onto its one-dimensional subspace $\{l h_j : l \in \mathbb{C}\}$. Then $P_j f = c_j(f)\, h_j$ for any $f \in L_2(\Gamma)$. Since

$$\mathrm{Spec}\, A_0 = \{\lambda_j : j \in \mathbb{N}\},$$

by virtue of the spectral theorem, we obtain

$$\varphi_{s,m}(A_0)\, f = \sum_{j=1}^{\infty} \varphi_{s,m}(\lambda_j) P_j f = \sum_{j=1}^{\infty} \varphi_{s,m}(\lambda_j) c_j(f)\, h_j$$

for every $f \in \mathrm{Dom}\,(\varphi_{s,m}(A_0))$. These series are convergent in $L_2(\Gamma)$. Thus, in view of the Parseval equality, we get (2.31).

Lemma 2.2 is proved.

Lemma 2.3. *For any $f \in \mathcal{D}'(\Gamma)$, the inclusion $f \in H^{s,\varphi}(\Gamma)$ is equivalent to the inequality*

$$\sum_{j=1}^{\infty} \varphi_{s,m}^2(\lambda_j)\,|c_j(f)|^2 < \infty. \tag{2.32}$$

Proof. Let $f \in H^{s,\varphi}(\Gamma)$. We now prove inequality (2.32). We choose a sequence $(f_k) \subset C^\infty(\Gamma)$ such that $f_k \to f$ in $H^{s,\varphi}(\Gamma)$ as $k \to \infty$. According to Lemma 2.2 and Theorem 2.5, we get

$$\sum_{j=1}^{\infty} \varphi_{s,m}^2(\lambda_j)\,|c_j(f_k)|^2 = \|\varphi_{s,m}(A_0) f_k\|^2_{L_2(\Gamma)}$$

$$\asymp \|f_k\|^2_{H^{s,\varphi}(\Gamma)} \le \big(1 + \|f\|^2_{H^{s,\varphi}(\Gamma)}\big)$$

for any $k \in \mathbb{N}$. In addition, it follows from the continuity of the embedding $H^{s,\varphi}(\Gamma) \hookrightarrow \mathcal{D}'(\Gamma)$ that

$$c_j(f_k) = (f_k, h_j)_\Gamma \to (f, h_j)_\Gamma = c_j(f) \quad \text{as} \quad k \to \infty$$

for any $j \in \mathbb{N}$. Therefore, by the Fatou lemma for positive series (see, e.g., [23, Vol. 1, Chap. 3, Theorem 6.2]), we obtain

$$\sum_{j=1}^{\infty} \varphi_{s,m}^2(\lambda_j)\,|c_j(f)|^2 \le \liminf_{k\to\infty} \sum_{j=1}^{\infty} \varphi_{s,m}^2(\lambda_j)\,|c_j(f_k)|^2 < \infty.$$

Thus, the inclusion $f \in H^{s,\varphi}(\Gamma)$ yields inequality (2.32).

We now suppose that inequality (2.32) holds and prove the inclusion $f \in H^{s,\varphi}(\Gamma)$. By virtue of (2.32), the orthogonal series

$$\sum_{j=1}^{\infty} \varphi_{s,m}(\lambda_j)\, c_j(f)\, h_j =: h \in L_2(\Gamma) \tag{2.33}$$

is convergent in the space $L_2(\Gamma)$. Consider its partial sum

$$w_k := \sum_{j=1}^{k} \varphi_{s,m}(\lambda_j)\, c_j(f)\, h_j \in C^\infty(\Gamma).$$

In view of (2.33), we conclude that

$$w_k \to h \quad \text{in} \quad L_2(\Gamma) \quad \text{as} \quad k \to \infty. \tag{2.34}$$

This means that the sequence $(\varphi_{s,m}^{-1}(A_0)\, w_k)_{k=1}^{\infty}$ is fundamental in the space $H^{s,\varphi}(\Gamma)$. Indeed, since $A_0 h_j = \lambda_j h_j$, we get

$$\varphi_{s,m}^{-1}(A_0)\, h_j = \varphi_{s,m}^{-1}(\lambda_j)\, h_j$$

and

$$\begin{aligned}\varphi_{s,m}^{-1}(A_0)\, w_k &= \sum_{j=1}^{k} \varphi_{s,m}(\lambda_j)\, c_j(f)\, \varphi_{s,m}^{-1}(A_0) h_j \\ &= \sum_{j=1}^{k} c_j(f)\, h_j \in C^\infty(\Gamma).\end{aligned}$$

Hence, by virtue of Theorem 2.5 and relation (2.34), we find

$$\|\varphi_{s,m}^{-1}(A_0)\, w_k - \varphi_{s,m}^{-1}(A_0)\, w_p\|_{H^{s,\varphi}(\Gamma)} \asymp \|w_k - w_p\|_{L_2(\Gamma)} \to 0 \quad \text{as} \quad k, p \to \infty,$$

i.e., the sequence $(\varphi_{s,m}^{-1}(A_0)\, w_k)_{k=1}^{\infty}$ is fundamental in the space $H^{s,\varphi}(\Gamma)$. Its limit is denoted by g. This enables us to write

$$g = \lim_{k\to\infty} \varphi_{s,m}^{-1}(A_0)\, w_k = \sum_{j=1}^{\infty} c_j(f)\, h_j \quad \text{in} \quad H^{s,\varphi}(\Gamma), \tag{2.35}$$

whence, in view of inequality (2.28), we conclude that $f = g \in H^{s,\varphi}(\Gamma)$. Thus, inequality (2.32) implies the inclusion $f \in H^{s,\varphi}(\Gamma)$.

Lemma 2.3 is proved.

Remark 2.4. It follows from relation (2.35) that, for each $f \in H^{s,\varphi}(\Gamma)$, the series (2.28) converges to f in the space $H^{s,\varphi}(\Gamma)$.

Lemma 2.4. *The following equivalence of norms is true:*

$$\|f\|_{H^{s,\varphi}(\Gamma)} \asymp \Big(\sum_{j=1}^{\infty} \varphi_{s,m}^2(\lambda_j)\, |c_j(f)|^2 \Big)^{1/2}, \quad f \in H^{s,\varphi}(\Gamma). \tag{2.36}$$

Proof. We use the reasoning and notation from the proof of Lemma 2.3. Let $f \in H^{s,\varphi}(\Gamma)$ and let g be defined by relation (2.35). Then $f = g$ and

$$\|f\|_{H^{s,\varphi}(\Gamma)} = \lim_{k\to\infty} \|\varphi_{s,m}^{-1}(A_0)\, w_k\|_{H^{s,\varphi}(\Gamma)}. \tag{2.37}$$

Here, we recall that $\varphi_{s,m}^{-1}(A_0)\, w_k \in C^\infty(\Gamma)$. Therefore, by Theorem 2.5, there exists a number $c \geq 1$ independent of f and such that

$$c^{-1}\, \|w_k\|_{L_2(\Gamma)} \leq \|\varphi_{s,m}^{-1}(A_0)\, w_k\|_{H^{s,\varphi}(\Gamma)} \leq c\, \|w_k\|_{L_2(\Gamma)} \tag{2.38}$$

for all $k \in \mathbb{N}$. By virtue of (2.33) and (2.34), we get

$$\lim_{k\to\infty} \|w_k\|_{L_2(\Gamma)} = \|h\|_{L_2(\Gamma)} = \Big(\sum_{j=1}^{\infty} \varphi_{s,m}^2(\lambda_j)\, |c_j(f)|^2 \Big)^{1/2}. \tag{2.39}$$

It remains to pass to the limit as $k \to \infty$ in inequality (2.38) and apply equalities (2.37) and (2.39). We immediately obtain the equivalence of norms (2.36).

Lemma 2.4 is proved.

Proof of Theorem 2.7. It follows from the asymptotic relation (2.27) and the inclusion $\varphi \in \mathcal{M}$ that

$$\varphi_{s,m}(\lambda_j) \asymp j^{\,s/n}\varphi(j^{\,1/n}) \quad \text{as functions of} \quad j \geq 1. \tag{2.40}$$

Indeed, according to relation (2.20), we get

$$\varphi_{s,m}(\lambda_j) = \lambda_j^{s/m}\,\varphi(\lambda_j^{1/m}) \quad \text{for} \quad \lambda_j \geq 1.$$

Since $\varphi \in \mathcal{M} \subset \mathbf{QSV}$, there exists a positive function $\varphi_1 \in \mathbf{SV}$ such that $\varphi(t) \asymp \varphi_1(t)$ whenever $t \gg 1$. Hence,

$$\varphi_{s,m}(\lambda_j) \asymp \lambda_j^{s/m}\,\varphi_1(\lambda_j^{1/m}) \quad \text{provided that} \quad j \gg 1. \tag{2.41}$$

Applying Proposition 1.2 (Uniform Convergence Theorem) to the function $\varphi_1 \in$ **SV**, in view of relation (2.27), we conclude that

$$\lim_{j\to\infty} \frac{\varphi_1(\lambda_j^{1/m})}{\varphi_1(j^{\,1/n})} = \lim_{j\to\infty} \frac{\varphi_1((\lambda_j^{1/m}\, j^{\,-1/n})\, j^{\,1/n})}{\varphi_1(j^{\,1/n})} = 1,$$

whence, by virtue of relations (2.41) and (2.27), we obtain

$$\varphi_{s,m}(\lambda_j) \asymp j^{\,s/n}\varphi_1(j^{\,1/n}) \asymp j^{\,s/n}\varphi(j^{\,1/n}) \quad \text{whenever} \quad j \gg 1. \tag{2.42}$$

Since the functions φ and $1/\varphi$ are bounded on each segment $[1, b]$, $1 < b < \infty$, relation (2.42) implies (2.40).

It remains to note that Theorem 2.7 immediately follows from Lemmas 2.3 and 2.4 and relation (2.40).

Theorem 2.7 is proved.

Example 2.1. Let Γ be a circle of radius 1 and let $A := 1 - d^2/dt^2$, where t specifies a natural parametrization on Γ. The eigenfunctions

$$h_j(t) := (2\pi)^{-1}\mathrm{e}^{ijt}, \quad j \in \mathbb{Z},$$

of the operator A form an orthonormal basis in $L_2(\Gamma)$. They correspond to the eigenvalues $\lambda_j = 1 + j^{\,2}$. Assume that $s \in \mathbb{R}$ and $\varphi \in \mathcal{M}$. By Theorem 2.5, we obtain

$$\|f\|^2_{H^{s,\varphi}(\Gamma)} \asymp \|\varphi_{s,2}(A_0)f\|^2_{L_2(\Gamma)} = \sum_{j=-\infty}^{\infty} (1 + j^{\,2})^s\, \varphi^2((1 + j^{\,2})^{1/2})\, |c_j(f)|^2.$$

Note that we can now use the basis formed by the real-valued eigenfunctions

$$h_0(t) := (2\pi)^{-1}, \quad h_j(t) := \pi^{-1}\cos jt, \quad \text{and} \quad h_{-j}(t) := \pi^{-1}\sin jt$$

with $j \in \mathbb{N}$. Then

$$\|f\|^2_{H^{s,\varphi}(\Gamma)} \asymp |a_0(f)|^2 + \sum_{j=1}^{\infty} j^{2s}\,\varphi^2(j)\,\big(|a_j(f)|^2 + |b_j(f)|^2\big),$$

where $a_0(f)$, $a_j(f)$, and $b_j(f)$ are the Fourier coefficients of f with respect to these eigenfunctions. In this case, the space $H^{s,\varphi}(\Gamma)$ is closely connected with the spaces of periodic real functions considered by A. I. Stepanets [248, Chapt. 1, § 7], [249, Part 1, Chapt. 3, Subsection 7.1].

2.1.4 Embedding theorem

We now prove an important theorem on the embedding of the space $H^{s,\varphi}(\Gamma)$ in the space $C^k(\Gamma)$.

Theorem 2.8. *Assume that a function $\varphi \in \mathcal{M}$ and an integer number $k \geq 0$ are given. Then condition* (1.37) *is equivalent to the embedding*

$$H^{k+n/2,\varphi}(\Gamma) \hookrightarrow C^k(\Gamma). \tag{2.43}$$

This embedding is compact.

Proof. Assume that relation (1.37) holds. Thus, by Theorem 1.15(iii), we get the continuous embedding

$$H^{k+n/2,\varphi}(\mathbb{R}^n) \hookrightarrow C^k_{\mathrm{b}}(\mathbb{R}^n).$$

This immediately yields the continuous embedding (2.43) if we use the local definition 2.1 of the space $H^{k+n/2,\varphi}(\Gamma)$. Indeed, for any $f \in H^{k+n/2,\varphi}(\Gamma)$, we get

$$(\chi_j f)\circ\alpha_j \in H^{k+n/2,\varphi}(\mathbb{R}^n) \hookrightarrow C^k_{\mathrm{b}}(\mathbb{R}^n) \quad \text{for each} \quad j \in \{1,\dots,r\}.$$

Hence,

$$f = \sum_{j=1}^{r} \chi_j f \in C^k(\Gamma).$$

Moreover,

$$\begin{aligned} \|f\|_{C^k(\Gamma)} &\leq \sum_{j=1}^{r} \|\chi_j f\|_{C^k(\Gamma)} \leq c_1 \sum_{j=1}^{r} \|(\chi_j f)\circ\alpha_j\|_{C^k_{\mathrm{b}}(\mathbb{R}^n)} \\ &\leq c_2 \sum_{j=1}^{r} \|(\chi_j f)\circ\alpha_j\|_{H^{k+n/2,\varphi}(\mathbb{R}^n)} \leq c_3\,\|f\|_{H^{k+n/2,\varphi}(\Gamma)}. \end{aligned}$$

Here, the positive numbers c_1, c_2, and c_3 are independent of f. Embedding (2.43) and its continuity are proved.

We now show that this embedding is compact. In view of Theorem 1.12(i), without loss of generality, we can assume that the function $\varphi \in \mathcal{M}$ is continuous. By virtue of (1.37), the function $\psi_1 := \varphi^2$ satisfies the condition of Lemma 1.4. Let ψ_0 be the function appearing in this lemma. Then the function $\varphi_0 = \sqrt{\psi_0} \in \mathcal{M}$ satisfies both the condition $\varphi_0(t)/\varphi(t) \to 0$ as $t \to \infty$ and inequality (1.37) with φ_0 instead of φ. Hence, by virtue of the result proved above and Theorem 2.3(iv), we find

$$H^{k+n/2,\varphi}(\Gamma) \hookrightarrow H^{k+n/2,\varphi_0}(\Gamma) \hookrightarrow C^k(\Gamma).$$

Here, the first embedding is compact and the second embedding is continuous. Hence, embedding (2.43) is compact.

It remains to show that the inclusion $H^{k+n/2,\varphi}(\Gamma) \subseteq C^k(\Gamma)$ implies condition (1.37). We again apply the local definition 2.1 and use the reasoning from the proof of assertion (iv) in Theorem 2.3. For any distribution $u \in H^{k+n/2,\varphi}(\mathbb{R}^n)$ with $\operatorname{supp} u \subset \alpha_1^{-1}(U)$, we obtain

$$\Theta(u \circ \alpha_1^{-1}) \in H^{k+n/2,\varphi}(\Gamma) \subset C^k(\Gamma).$$

Thus, $u \in C^k(\mathbb{R}^n)$. Hence, by virtue of Proposition 1.5, we arrive at the inclusion

$$\langle\xi\rangle^k \, \mu^{-1}(\xi) \in L_2(\mathbb{R}^n_\xi),$$

where

$$\mu(\xi) := \langle\xi\rangle^{k+n/2} \, \varphi(\langle\xi\rangle).$$

In view of (1.41), this inclusion is equivalent to (1.37).

Theorem 2.8 is proved.

2.2 Elliptic operators on closed manifolds

In this section, we study elliptic PsDOs given on infinitely smooth and closed (i.e., compact and without boundary) manifolds. We prove that these PsDOs are bounded and Fredholm in appropriate pairs of spaces from the refined Sobolev scale. We also study the global and local smoothness of solutions to the elliptic equations. Moreover, we consider a class of parameter-elliptic PsDOs specifying isomorphisms on the refined scale provided that the complex-valued parameter is sufficiently large in the absolute value.

2.2.1 Pseudodifferential operators on closed manifolds

For the sake of convenience, we now recall the definition of PsDO on a closed manifold Γ and the related notions required for our presentation. As above, we use the terminology and notation taken from the survey [10, § 2].

Definition 2.5. Let $m \in \mathbb{R}$. A linear operator $A : C^\infty(\Gamma) \to C^\infty(\Gamma)$ is called a PsDO (on Γ) from the class $\Psi^m(\Gamma)$ if the following conditions are satisfied:

(i) For all functions $\varphi, \psi \in C^\infty(\Gamma)$ whose supports are disjoint, the mapping $u \mapsto \varphi A(\psi u)$, $u \in C^\infty(\Gamma)$, is extended by continuity to an operator of order $-\infty$ on the Sobolev scale, i.e., to a bounded operator from $H^s(\Gamma)$ into $H^{s+r}(\Gamma)$ for all $s \in \mathbb{R}$ and $r > 0$;

(ii) We now arbitrarily choose a local C^∞-chart $\alpha : \mathbb{R}^n \to \Gamma_\alpha$ on Γ and an open set Ω whose closure is contained in the coordinate neighborhood $\Gamma_\alpha \subset \Gamma$. Then there exists a PsDO $A_\Omega \in \Psi^m(\mathbb{R}^n)$ such that

$$(\varphi A(\psi u)) \circ \alpha = (\varphi \circ \alpha) A_\Omega((\psi u) \circ \alpha) \quad \text{for any} \quad u \in C^\infty(\Gamma)$$

and all functions $\varphi, \psi \in C^\infty(\Gamma)$ whose supports lie in Ω.

In connection with item (i) of this definition, we note that the condition "to be an operator of order $-\infty$ on the Sobolev scale" used in (i) is equivalent to the condition "to be an integral operator

$$u(y) \mapsto \int_\Gamma K(x, y)\, u(y)\, dy, \quad u \in C^\infty(\Gamma), \tag{2.44}$$

on Γ with infinitely smooth kernel $K(x, y)$." Hence, the presented notion of PsDO on the manifold Γ is not, in fact, related to any scale of Hilbert spaces.

We set

$$\Psi^{-\infty}(\Gamma) := \bigcap_{m \in \mathbb{R}} \Psi^m(\Gamma), \quad \Psi^{\infty}(\Gamma) := \bigcup_{m \in \mathbb{R}} \Psi^m(\Gamma).$$

The class $\Psi^{-\infty}(\Gamma)$ coincides with the class of all integral operators (2.44) with infinitely smooth kernels. Each PsDO $A \in \Psi^\infty(\Gamma)$ is continuous on $C^\infty(\Gamma)$ and can be uniquely extended to a linear continuous operator on $\mathcal{D}'(\Gamma)$. This operator is also denoted by A.

Definition 2.6. Let $m \in \mathbb{R}$. A PsDO $A \in \Psi^m(\Gamma)$ is called polyhomogeneous (or classical) of order m on Γ if each A_Ω in Definition 2.5 belongs to $\Psi^m_{\mathrm{ph}}(\mathbb{R}^n)$.

By $\Psi^m_{\mathrm{ph}}(\Gamma)$ we denote the class of all polyhomogeneous PsDOs of order m on Γ.

Definition 2.7. Let a PsDO $A \in \Psi^m_{\mathrm{ph}}(\Gamma)$ be given. Its principal symbol $a_0(x,\xi)$ is defined as a function of the arguments $x \in \Gamma$ and $\xi \in T^*_x\Gamma$, where $\xi \neq 0$, which coincides (locally in x) with the principal symbol of the corresponding PsDO $A_\Omega \in \Psi^m_{\mathrm{ph}}(\mathbb{R}^n)$.

Here, as usual, $T^*_x\Gamma$ stands for the cotangent space to the manifold Γ at a point $x \in \Gamma$. It is important that the principal symbol of a polyhomogeneous PsDO on Γ does not depend on the choice of local charts on Γ and is an infinitely smooth function of x and ξ positively homogeneous in ξ of order m.

Definition 2.8. A linear operator $A^+ : C^\infty(\Gamma) \to C^\infty(\Gamma)$ is said to be *formally adjoint* to a PsDO $A \in \Psi^\infty(\Gamma)$ if

$$(Au, v)_\Gamma = (u, A^+ v)_\Gamma \quad \text{for all} \quad u, v \in C^\infty(\Gamma).$$

If $A = A^+$, then the PsDO A is called *formally self-adjoint.*

We note that the notion of formally adjoint PsDO is introduced with respect to the density dx given on Γ because $(\cdot,\cdot)_\Gamma$ is the inner product in $L_2(\Gamma, dx)$. If $A \in \Psi^m(\Gamma)$ for some $m \in \mathbb{R}$, then $A^+ \in \Psi^m(\Gamma)$. If, in addition, A is a polyhomogeneous PsDO with the principal symbol a_0, then A^+ is also a polyhomogeneous PsDO with the complex conjugate principal symbol $\overline{a_0}$.

Definition 2.9. A PsDO $A \in \Psi^m_{\mathrm{ph}}(\Gamma)$ and its principal symbol $a_0(x,\xi)$ are called *elliptic* on Γ if $a_0(x,\xi) \neq 0$ for any point $x \in \Gamma$ and each covector $\xi \in T^*_x\Gamma \setminus \{0\}$.

If a PsDO $A \in \Psi^m_{\mathrm{ph}}(\Gamma)$ is elliptic on Γ, then A has a parametrix $B \in \Psi^{-m}_{\mathrm{ph}}(\Gamma)$, which is also elliptic on Γ; i.e., the analog of Proposition 1.6 is true . However, we do not use this analog.

As an important example of elliptic PsDO on Γ, we can mention the Beltrami–Laplace operator Δ_Γ. We now recall its definition. Let a Riemannian metric be introduced on the manifold Γ; i.e., let an infinitely smooth covariant real tensor field $g(x)$, $x \in \Gamma$, be given, where $g(x) = (g_{j,k}(x))^n_{j,k=1}$ is a symmetric positive definite matrix. The Riemannian metric defines a density $dx := (\det g(x))^{1/2} dx_1 \ldots dx_1$ on the local coordinates $x = (x_1, \ldots, x_n)$ in Γ. By definition, the action of the Beltrami–Laplace operator upon the function $u \in C^2(\Gamma)$ is given by the formula

$$(\Delta_\Gamma u)(x) := (\det g(x))^{-1/2} \sum_{j,k=1}^{n} \partial_{x_j} \big((\det g(x))^{1/2}\, g^{j,k}(x)\, \partial_{x_k} u(x) \big),$$

where $g^{-1}(x) = (g^{j,k}(x))^n_{j,k=1}$ is the matrix inverse to $g(x)$. The function $\Delta_\Gamma u$ is independent of the choice of local coordinates on Γ. The principal symbol of the differential operator Δ_Γ is equal to

$$\sum_{j,k=1}^{n} g^{j,k}(x)\,\xi_j\,\xi_k,$$

and, therefore, the ellipticity of Δ_Γ follows from the positive definiteness of the matrix $g^{-1}(x)$. The operator Δ_Γ is formally self-adjoint with respect to the density dx.

2.2.2 Elliptic operators on the refined scale

Consider a given PsDO $A \in \Psi^m_{\mathrm{ph}}(\Gamma)$ of order $m \in \mathbb{R}$. In Subsections 2.2.2 and 2.2.3, we assume that A is elliptic on Γ.

We set

$$\mathcal{N} := \{\, u \in C^\infty(\Gamma) : \, Au = 0 \ \text{ on } \ \Gamma \,\}, \tag{2.45}$$

$$\mathcal{N}^+ := \{\, v \in C^\infty(\Gamma) : \, A^+ v = 0 \ \text{ on } \ \Gamma \,\}. \tag{2.46}$$

Since both PsDOs A and A^+ are elliptic on Γ, the spaces $\mathcal{N}$ and $\mathcal{N}^+$ are finite-dimensional [10, Theorem 2.3.3].

We now investigate the properties of the PsDO A on the refined Sobolev scale over Γ. First, we prove the following assertion about the action of each PsDO on this scale (an analog of Lemma 1.6).

Lemma 2.5. *Let $G \in \Psi^r(\Gamma)$ for some $r \in \mathbb{R}$. Then the restriction of the mapping $u \mapsto Gu$, $u \in \mathcal{D}'(\Gamma)$, to the space $H^{\sigma,\varphi}(\Gamma)$ is a linear bounded operator*

$$G : H^{\sigma,\varphi}(\Gamma) \to H^{\sigma-r,\varphi}(\Gamma) \tag{2.47}$$

for all parameters $\sigma \in \mathbb{R}$ and $\varphi \in \mathcal{M}$.

Proof. In the Sobolev case where $\varphi \equiv 1$, this lemma is well known [10, Theorem 2.1.2]. We arbitrarily choose $\sigma \in \mathbb{R}$ and $\varphi \in \mathcal{M}$. Consider the following bounded linear operators:

$$G : H^{\sigma\mp 1}(\Gamma) \to H^{\sigma\mp 1-r}(\Gamma).$$

Further, we apply the interpolation with the same functional parameter ψ as in Theorem 2.2 for $\varepsilon = \delta = 1$. We get the following bounded operator:

$$G : \bigl[\, H^{\sigma-1}(\Gamma),\, H^{\sigma+1}(\Gamma)\,\bigr]_\psi \to \bigl[\, H^{\sigma-r-1}(\Gamma),\, H^{\sigma-r+1}(\Gamma)\,\bigr]_\psi.$$

This and Theorem 2.2 imply that operator (2.47) is well defined and bounded.

Lemma 2.5 is proved.

In view of Lemma 1.6, we have the bounded operator

$$A : H^{s+m,\varphi}(\Gamma) \to H^{s,\varphi}(\Gamma) \tag{2.48}$$

for any parameters $s \in \mathbb{R}$ and $\varphi \in \mathcal{M}$. We now study the properties of this operator.

Theorem 2.9. *The bounded operator* (2.48) *is a Fredholm operator for all* $s \in \mathbb{R}$ *and* $\varphi \in \mathcal{M}$. *Its kernel coincides with* $\mathcal{N}$. *Its domain has the form*

$$\{f \in H^{s,\varphi}(\Gamma) : (f, w)_\Gamma = 0 \ \textit{for all} \ w \in \mathcal{N}^+\}. \tag{2.49}$$

The index of operator (2.48) *is equal to* $\dim \mathcal{N} - \dim \mathcal{N}^+$ *and does not depend on* s *and* φ.

Proof. For $\varphi \equiv 1$ (the Sobolev scale), this theorem is known (see, e.g., [10, Theorems 2.2.6, 2.3.3, and 2.3.12] or [86, Theorem 19.2.1]). Thus, the general case $\varphi \in \mathcal{M}$ can be obtained with the help of interpolation with a functional parameter. Namely, let $s \in \mathbb{R}$. We have the following bounded Fredholm operators:

$$A : H^{s\mp 1+m}(\Gamma) \to H^{s\mp 1}(\Gamma) \tag{2.50}$$

with common kernel $\mathcal{N}$ and identical indices $\varkappa := \dim \mathcal{N} - \dim \mathcal{N}^+$. In this case,

$$A\big(H^{s\mp 1+m}(\Gamma)\big) = \{f \in H^{s\mp 1}(\Gamma) : (f, w)_\Gamma = 0 \ \text{for all} \ w \in \mathcal{N}^+\}. \tag{2.51}$$

We now apply interpolation with the functional parameter ψ from Theorem 2.2 to (2.50), where $\varepsilon = \delta = 1$. This yields the bounded operator

$$A : \big[H^{s-1+m}(\Gamma), H^{s+1+m}(\Gamma)\big]_\psi \to \big[H^{s-1}(\Gamma), H^{s+1}(\Gamma)\big]_\psi,$$

which coincides, by virtue of Theorem 2.2, with operator (2.48). Hence, by Theorem 1.7, operator (2.48) is a Fredholm operator with kernel $\mathcal{N}$ and index $\varkappa = \dim \mathcal{N} - \dim \mathcal{N}^+$. The domain of this operator is

$$H^{s,\varphi}(\Gamma) \cap A\big(H^{s-1+m}(\Gamma)\big).$$

Hence, in view of (2.51), we conclude that it is equal to (2.49).

Theorem 2.9 is proved.

By virtue of this theorem, $\mathcal{N}^+$ is the defect subspace of operator (2.48). Note that, in view of Theorem 2.3(v), the operator

$$A^+ : H^{-s,1/\varphi}(\Gamma) \to H^{-s-m,1/\varphi}(\Gamma) \tag{2.52}$$

is adjoint to operator (2.48). Since PsDO A^+ is elliptic on Γ, the bounded operator (2.52) is Fredholm. By virtue of Theorem 2.9, it has the kernel $\mathcal{N}^+$ and the defect subspace $\mathcal{N}$. We also note that the indices of operators (2.48) and (2.52) are equal to 0 provided that $\dim \Gamma \geq 2$ (see [18], [10, Sec. 2.3 f]).

If the spaces $\mathcal{N}$ and $\mathcal{N}^+$ are trivial, then Theorem 2.9 and the Banach theorem on inverse operator imply that operator (2.48) is an isomorphism. In the general case, it is convenient to specify the isomorphism with the help of the following projectors.

We represent the spaces of action of operator (2.48) in the form of the direct sums of closed subspaces:

$$H^{s+m,\varphi}(\Gamma) = \mathcal{N} \dotplus \{u \in H^{s+m,\varphi}(\Gamma) : (u,v)_\Gamma = 0 \text{ for all } v \in \mathcal{N}\},$$

$$H^{s,\varphi}(\Gamma) = \mathcal{N}^+ \dotplus \{f \in H^{s,\varphi}(\Gamma) : (f,w)_\Gamma = 0 \text{ for all } w \in \mathcal{N}^+\}.$$

The indicated decompositions in direct sums exist because the intersection of their terms is trivial and the finite dimension of the first term is equal to the codimension of the second term. (Indeed, e.g., in the first sum, the quotient space of the space $H^{s+m,\varphi}(\Gamma)$ by the second term is the space dual to the subspace $\mathcal{N}$ of the space $H^{-s-m,1/\varphi}(\Gamma)$). By $\mathcal{P}$ and $\mathcal{P}^+$, we denote the oblique projectors of the spaces $H^{s+m,\varphi}(\Gamma)$ and $H^{s,\varphi}(\Gamma)$ onto the second terms of the indicated sums parallel to the first terms, respectively. These projectors are independent of s and φ.

Theorem 2.10. *For any $s \in \mathbb{R}$ and $\varphi \in \mathcal{M}$, the restriction of operator* (2.48) *to the subspace $\mathcal{P}(H^{s+m,\varphi}(\Gamma))$ is an isomorphism*

$$A : \mathcal{P}(H^{s+m,\varphi}(\Gamma)) \leftrightarrow \mathcal{P}^+(H^{s,\varphi}(\Gamma)). \tag{2.53}$$

Proof. By Theorem 2.9, $\mathcal{N}$ is the kernel and $\mathcal{P}^+(H^{s,\varphi}(\Gamma))$ is the domain of operator (2.48). Hence, operator (2.53) is a bijection. In addition, this operator is bounded. Therefore, it is an isomorphism by the Banach theorem on inverse operator. Theorem 2.10 is proved.

Theorem 2.10 yields the following *a priori* estimate for the solution of the elliptic equation $Au = f$ on Γ (an analog of Theorem 1.16 for a closed manifold).

Theorem 2.11. *Let $s \in \mathbb{R}$, $\sigma > 0$, and $\varphi \in \mathcal{M}$. There exists a number $c = c(s,\sigma,\varphi) > 0$ such that, for any distribution $u \in H^{s+m,\varphi}(\Gamma)$, the a priori estimate*

$$\|u\|_{H^{s+m,\varphi}(\Gamma)} \leq c\left(\|Au\|_{H^{s,\varphi}(\Gamma)} + \|u\|_{H^{s+m-\sigma,\varphi}(\Gamma)}\right) \tag{2.54}$$

is true.

Proof. Since $\mathcal{N}$ is a finite-dimensional subspace of the spaces $H^{s+m,\varphi}(\Gamma)$ and $H^{s+m-\sigma,\varphi}(\Gamma)$, their norms are equivalent on $\mathcal{N}$. In particular, the distribution $u - \mathcal{P}u \in \mathcal{N}$ satisfies the relation

$$\|u - \mathcal{P}u\|_{H^{s+m,\varphi}(\Gamma)} \leq c_1 \|u - \mathcal{P}u\|_{H^{s+m-\sigma,\varphi}(\Gamma)}$$

with a constant $c_1 > 0$ independent of $u \in H^{s+m,\varphi}(\Gamma)$. This yields

$$\begin{aligned}\|u\|_{H^{s+m,\varphi}(\Gamma)} &\leq \|u - \mathcal{P}u\|_{H^{s+m,\varphi}(\Gamma)} + \|\mathcal{P}u\|_{H^{s+m,\varphi}(\Gamma)} \\ &\leq c_1 \|u - \mathcal{P}u\|_{H^{s+m-\sigma,\varphi}(\Gamma)} + \|\mathcal{P}u\|_{H^{s+m,\varphi}(\Gamma)} \\ &\leq c_1\, c_2 \|u\|_{H^{s+m-\sigma,\varphi}(\Gamma)} + \|\mathcal{P}u\|_{H^{s+m,\varphi}(\Gamma)},\end{aligned}$$

where c_2 is the norm of the projector $1 - \mathcal{P}$ acting in the space $H^{s+m-\sigma,\varphi}(\Gamma)$. Thus,

$$\|u\|_{H^{s+m,\varphi}(\Gamma)} \leq c_1\, c_2 \|u\|_{H^{s+m-\sigma,\varphi}(\Gamma)} + \|\mathcal{P}u\|_{H^{s+m,\varphi}(\Gamma)}. \tag{2.55}$$

We now use the condition $Au = f$. Since $\mathcal{N}$ is the kernel of operator (2.48), and $u - \mathcal{P}u \in \mathcal{N}$, we have $A\mathcal{P}u = f$. Hence, $\mathcal{P}u$ is the preimage of the distribution f under isomorphism (2.53). Therefore,

$$\|\mathcal{P}u\|_{H^{s+m,\varphi}(\Gamma)} \leq c_3 \|f\|_{H^{s,\varphi}(\Gamma)},$$

where c_3 is the norm of the operator inverse to (2.53). This result and inequality (2.55) immediately yield estimate (2.54).

Theorem 2.11 is proved.

Note that if $\mathcal{N} = \{0\}$, i.e., the equation $Au = f$ has at most one solution, then the norm $\|u\|_{H^{s+m-\sigma}(\Gamma)}$ on the right-hand side of estimate (2.54) can be omitted. However, if $\mathcal{N} \neq \{0\}$, then this quantity can be made arbitrarily small for each distribution u by choosing a sufficiently large number σ.

2.2.3 Smoothness of solutions to the elliptic equation

Let Γ_0 be a nonempty open subset of the manifold Γ. We study the local smoothness of solutions to the elliptic equation $Au = f$ on Γ_0 in the refined scale. First, we consider the case where $\Gamma_0 = \Gamma$.

Theorem 2.12. *Assume that the distribution $u \in \mathcal{D}'(\Gamma)$ is a solution of the equation $Au = f$ on Γ, where $f \in H^{s,\varphi}(\Gamma)$ for some parameters $s \in \mathbb{R}$ and $\varphi \in \mathcal{M}$. Then $u \in H^{s+m,\varphi}(\Gamma)$.*

Proof. Since the manifold Γ is compact, the space $\mathcal{D}'(\Gamma)$ is the union of the Sobolev spaces $H^{\sigma}(\Gamma)$, where $\sigma \in \mathbb{R}$. Hence, for the distribution $u \in \mathcal{D}'(\Gamma)$, there exists a number $\sigma < s$ such that $u \in H^{\sigma+m}(\Gamma)$. By virtue of Theorems 2.9 and 2.3(iii), the equality

$$(H^{s,\varphi}(\Gamma)) \cap A(H^{\sigma+m}(\Gamma)) = A(H^{s+m,\varphi}(\Gamma))$$

holds. Hence, the condition $f \in H^{s,\varphi}(\Gamma)$ implies that

$$f = Au \in A(H^{s+m,\varphi}(\Gamma)).$$

Thus, on the manifold Γ, the equalities $Au = f$ and $Av = f$ are true for some distribution $v \in H^{s+m,\varphi}(\Gamma)$. Therefore, $A(u-v) = 0$ on Γ and, by Theorem 2.9, we arrive at the inclusion

$$w := u - v \in \mathcal{N} \subset C^{\infty}(\Gamma) \subset H^{s+m,\varphi}(\Gamma).$$

Hence, $u = v + w \in H^{s+m,\varphi}(\Gamma)$.

Theorem 2.12 is proved.

We now consider the case $\Gamma_0 \neq \Gamma$. Denote

$$H^{\sigma,\varphi}_{\mathrm{loc}}(\Gamma_0) := \{h \in \mathcal{D}'(\Gamma) : \chi h \in H^{\sigma,\varphi}(\Gamma) \text{ for all } \chi \in C^{\infty}(\Gamma),\ \operatorname{supp}\chi \subset \Gamma_0\}.$$

Here, $\sigma \in \mathbb{R}$, $\varphi \in \mathcal{M}$.

Theorem 2.13. *Assume that the distribution $u \in \mathcal{D}'(\Gamma)$ is a solution of the equation $Au = f$ on Γ_0, where $f \in H^{s,\varphi}_{\mathrm{loc}}(\Gamma_0)$ for some parameters $s \in \mathbb{R}$ and $\varphi \in \mathcal{M}$. Then $u \in H^{s+m,\varphi}_{\mathrm{loc}}(\Gamma_0)$.*

Theorem 2.13 is proved by analogy with Theorem 1.18. In this case, it is necessary to apply Theorem 2.12 instead of Theorem 1.17.

As applied to the scale of spaces $H^{s,\varphi}(\Gamma)$, Theorem 2.13 improves the well-known assertions on the increase in the local smoothness of solutions to the elliptic equation on a manifold in the Sobolev scale [52, Chap. 2, Lemma 3.2], [232, Theorem 7.2]. It is easy to see that the local refined smoothness φ of the right-hand side of the elliptic equation is inherited by its solution.

Theorems 2.13 and 2.8 immediately yield the following sufficient condition for the continuity of derivatives of the solution u to the equation $Au = f$.

Corollary 2.3. *Let an integer $r \geq 0$ and a function $\varphi \in \mathcal{M}$ satisfying condition (1.37) be given. Assume that the distribution $u \in \mathcal{D}'(\Gamma)$ is a solution of the equation $Au = f$ on the set Γ_0, where $f \in H^{r-m+n/2,\varphi}_{\mathrm{loc}}(\Gamma_0)$. Then $u \in C^r(\Gamma_0)$.*

Note that condition (1.37) is not only sufficient for the validity of the inclusion $u \in C^r(\Gamma_0)$ but also necessary in the class of all analyzed solutions of the equation $Au = f$; see Remark 1.9.

2.2.4 Parameter-elliptic operators

Various classes of parameter-elliptic operators were studied by S. Agmon and L. Nirenberg [6], M. S. Agranovich and M. I. Vishik [13], A. N. Kozhevnikov [99] and their followers (see the survey [10] and the references therein). It was established that if the modulus of the complex-valued parameter is sufficiently large, then the elliptic operator specifies an isomorphism between suitable Sobolev spaces. Moreover, the norm of the operator admits a two-sided estimate with constants independent of the parameter. We consider a broad class of parameter-elliptic PsDOs on the manifold Γ and extend the indicated result to the refined Sobolev scale.

The definition of parameter-elliptic PsDO is taken from the survey [10, Sec. 4.1]. We now fix arbitrary numbers $q \in \mathbb{N}$ and $m > 0$. Consider a PsDO $A(\lambda) \in \Psi^{mq}(\Gamma)$ depending on a complex-valued parameter λ as follows:

$$A(\lambda) := \sum_{r=0}^{q} \lambda^{q-r} A^{(r)}. \tag{2.56}$$

Here, $A^{(r)}$, $r = 0, \ldots, q$, are polyhomogeneous PsDOs on Γ of orders $\operatorname{ord} A^{(r)} \leq mr$ and A_0 is the operator of multiplication by a function $a_0 \in C^\infty(\Gamma)$.

Let K be a fixed closed angle in the complex plane with vertex at the origin (we do not exclude the case where K is degenerated into a ray).

Definition 2.10. A PsDO $A(\lambda)$ is called *parameter-elliptic* in the angle K if

$$\sum_{r=0}^{q} \lambda^{q-r} a^{r,0}(x,\xi) \neq 0 \tag{2.57}$$

for any $x \in \Gamma$, $\xi \in T^*_x\Gamma$, and $\lambda \in K$ such that $(\xi, \lambda) \neq 0$. Here, $a^{r,0}(x,\xi)$ is the principal symbol of the PsDO $A^{(r)}$ in the case where $\operatorname{ord} A^{(r)} = mr$ or $a^{r,0}(x,\xi) \equiv 0$ if $\operatorname{ord} A^{(r)} < mr$. Moreover, we assume that $a^{0,0}(x,\xi) \equiv a_0(\xi)$ and that the functions $a^{1,0}(x,\xi)$, $a^{2,0}(x,\xi), \ldots$ are equal to 0 for $\xi = 0$ (the last assumption is explained by the fact that the principal symbols are not initially defined for $\xi = 0$).

Example 2.2. Consider a PsDO $A - \lambda I$, where $A \in \Psi^m_{\mathrm{ph}}(\Gamma)$, and I is the identity operator. For this PsDO, the condition of parameter-ellipticity in the angle K simply means that $a_0(x,\xi) \notin K$ for any $\xi \neq 0$. Here, as above, $a_0(x,\xi)$ is the principal symbol of A. This example is of importance for the spectral theory of elliptic operators.

Further, in this subsection, we always assume that the PsDO $A(\lambda)$ satisfies Definition 2.10.

It follows from this definition that the PsDO $A(\lambda)$ is elliptic on Γ for any fixed $\lambda \in \mathbb{C}$. Indeed, the principal symbol of $A(\lambda)$ is $a^{q,0}(x,\xi)$ for any λ. As follows from (2.57), for $\lambda = 0$, it satisfies the inequality $a^{q,0}(x,\xi) \neq 0$ for any $x \in \Gamma$ and $\xi \in T^*_x\Gamma \setminus \{0\}$. This fact means that the PsDO $A(\lambda)$ is elliptic on Γ. By Theorem 2.9, we get the following bounded Fredholm operator:

$$A(\lambda) : H^{s+mq,\varphi}(\Gamma) \to H^{s,\varphi}(\Gamma) \tag{2.58}$$

for any $s \in \mathbb{R}$, $\varphi \in \mathcal{M}$, and $\lambda \in \mathbb{C}$. Moreover, since $A(\lambda)$ is a parameter-elliptic PsDO in the angle K, it has important additional properties.

Theorem 2.14. *The following assertions are true:*

(i) *There exists a number $\lambda_0 > 0$ such that, for every value of the parameter $\lambda \in K$ satisfying the condition $|\lambda| \geq \lambda_0$, the PsDO $A(\lambda)$ establishes, for any $s \in \mathbb{R}$ and $\varphi \in \mathcal{M}$, the isomorphism*

$$A(\lambda) : H^{s+mq,\varphi}(\Gamma) \leftrightarrow H^{s,\varphi}(\Gamma). \tag{2.59}$$

(ii) *For arbitrary parameters $s \in \mathbb{R}$ and $\varphi \in \mathcal{M}$, there exists a number $c = c(s,\varphi) \geq 1$ such that the two-sided estimate*

$$\begin{aligned} c^{-1}\,\|A(\lambda)u\|_{H^{s,\varphi}(\Gamma)} &\leq \big(\,\|u\|_{H^{s+mq,\varphi}(\Gamma)} + |\lambda|^q\,\|u\|_{H^{s,\varphi}(\Gamma)}\,\big) \\ &\leq c\,\big\|A(\lambda)u\big\|_{H^{s,\varphi}(\Gamma)} \end{aligned} \tag{2.60}$$

holds for any $\lambda \in K$ with $|\lambda| \geq \lambda_0$ and any $u \in H^{s+mq,\varphi}(\Gamma)$.

For $\varphi \equiv 1$ (Sobolev spaces), this theorem is known [10, Theorem 4.1.2]; in this case, the left inequality in the two-sided estimate (2.60) holds without the assumption about the parameter-ellipticity of $A(\lambda)$.

We now separately prove assertions (i) and (ii) of Theorem 2.14. The general case of $\varphi \in \mathcal{M}$ is derived from the Sobolev case $\varphi \equiv 1$.

Proof of Theorem 2.14(i). Let $s \in \mathbb{R}$ and let $\varphi \in \mathcal{M}$. Since, for every $\lambda \in \mathbb{C}$, the PsDO $A(\lambda)$ is elliptic on Γ, the bounded Fredholm operator (2.58) has the kernel $\mathcal{N}(\lambda)$ and the defect subspace $\mathcal{N}^+(\lambda)$, which are finite-dimensional and independent of s and φ (by virtue of Theorem 2.9). Then we use the fact that Theorem 2.14 holds for $\varphi \equiv 1$. Thus, there exists a number $\lambda_0 > 0$ such that, for any $\lambda \in K$ satisfying the condition $|\lambda| \geq \lambda_0$, the PsDO $A(\lambda)$ establishes an isomorphism

$$A(\lambda) : H^{s+mq}(\Gamma) \leftrightarrow H^{s+m}(\Gamma).$$

Hence, for the indicated λ, the spaces $\mathcal{N}(\lambda)$ and $\mathcal{N}^{+}(\lambda)$ are trivial, i.e., the linear bounded operator (2.58) is a bijection. By the Banach theorem on inverse operator, this yields isomorphism (2.59). Assertion (i) of Theorem 2.14 is proved.

Assertion (ii) of Theorem 2.14 is proved with the help of interpolation with functional parameter. For this purpose, we need the following space:

Let a function $\varphi \in \mathcal{M}$ and numbers $\sigma \in \mathbb{R}$, $\varrho > 0$, $\theta > 0$ be given. By $H^{\sigma,\varphi}(\Gamma, \varrho, \theta)$ we denote the space $H^{\sigma,\varphi}(\Gamma)$ endowed with the norm depending on the number parameters ϱ and θ in the following way:

$$\|h\|_{H^{\sigma,\varphi}(\Gamma,\varrho,\theta)} := \Big(\|h\|^2_{H^{\sigma,\varphi}(\Gamma)} + \varrho^2 \, \|h\|^2_{H^{\sigma-\theta,\varphi}(\Gamma)} \Big)^{1/2}.$$

This definition is correct in view of the continuous embedding $H^{\sigma,\varphi}(\Gamma) \hookrightarrow H^{\sigma-\theta,\varphi}(\Gamma)$. This means that the norms in the spaces $H^{\sigma,\varphi}(\Gamma, \varrho, \theta)$ and $H^{\sigma,\varphi}(\Gamma)$ are equivalent. The norm in the space $H^{\sigma,\varphi}(\Gamma, \varrho, \theta)$ is generated by the inner product

$$(h_1, h_2)_{H^{\sigma,\varphi}(\Gamma,\varrho,\theta)} := (h_1, h_2)_{H^{\sigma,\varphi}(\Gamma)} + \varrho^2 \, (h_1, h_2)_{H^{\sigma-\theta,\varphi}(\Gamma)}.$$

Hence, this is a Hilbert space. As above, we omit the index φ in the notation used in the Sobolev case $\varphi \equiv 1$. Returning to the formulation of assertion (ii) of Theorem 2.14, we note that

$$\begin{aligned}\|u\|_{H^{s+mq,\varphi}(\Gamma,|\lambda|^q,mq)} &\le \|u\|_{H^{s+mq,\varphi}(\Gamma)} + |\lambda|^q \, \|u\|_{H^{s,\varphi}(\Gamma)} \\ &\le \sqrt{2} \, \|u\|_{H^{s+mq,\varphi}(\Gamma,|\lambda|^q,mq)}.\end{aligned}$$

By virtue of Theorem 2.2, the spaces

$$\big[H^{\sigma-\varepsilon}(\Gamma, \varrho, \theta), \, H^{\sigma+\delta}(\Gamma, \varrho, \theta) \big]_{\psi} \quad \text{and} \quad H^{\sigma,\varphi}(\Gamma, \varrho, \theta)$$

coincide up to equivalence of norms. Here, the numbers ε and δ are positive, and the functional parameter ψ is the same as in Theorem 2.2. It turns out that the constants in the two-sided estimates for the norms in these spaces can be made independent of the parameter ϱ.

Lemma 2.6. *Let $\sigma \in \mathbb{R}$, let $\varphi \in \mathcal{M}$, and let θ, ε, and δ be given positive numbers. Then there exists a number $c_0 \ge 1$ such that the following two-sided estimate of norms holds for any $\varrho > 0$ and $h \in H^{\sigma,\varphi}(\Gamma)$:*

$$c_0^{-1} \, \|h\|_{H^{\sigma,\varphi}(\Gamma,\varrho,\theta)} \le \|h\|_{[H^{\sigma-\varepsilon}(\Gamma,\varrho,\theta), \, H^{\sigma+\delta}(\Gamma,\varrho,\theta)]_{\psi}} \le c_0 \, \|h\|_{H^{\sigma,\varphi}(\Gamma,\varrho,\theta)}. \qquad (2.61)$$

Here, ψ is the interpolation parameter from Theorem 2.2 and the number c_0 is independent of ϱ and h.

Proof. Assume that $\varrho > 0$. First, we prove an analog of estimate (2.61) for the spaces of distributions in $\mathbb{R}^n$. Then, by using the operators of "flattening" and "sewing," we pass to the spaces of distributions on the manifold Γ (cf. the proof of Theorem 2.2). By $H^{\sigma,\varphi}(\mathbb{R}^n, \varrho, \theta)$ we denote the space $H^{\sigma,\varphi}(\mathbb{R}^n)$ endowed with the Hilbert norm

$$\|w\|_{H^{\sigma,\varphi}(\mathbb{R}^n,\varrho,\theta)} := \Big(\|w\|^2_{H^{\sigma,\varphi}(\mathbb{R}^n)} + \varrho^2 \, \|w\|^2_{H^{\sigma-\theta,\varphi}(\mathbb{R}^n)} \Big)^{1/2}$$

$$= \Big(\int\limits_{\mathbb{R}^n} \langle\xi\rangle^{2\sigma} \left(1 + \varrho^2 \langle\xi\rangle^{-2\theta}\right) \varphi^2(\langle\xi\rangle) \, |\widehat{w}(\xi)|^2 \, d\xi \Big)^{1/2}. \tag{2.62}$$

For any fixed $\varrho > 0$, this norm is equivalent to the norm in the space $H^{\sigma,\varphi}(\mathbb{R}^n)$. Hence, $H^{\sigma,\varphi}(\mathbb{R}^n, \varrho, \theta)$ is a Hilbert space. Similarly, we can define the spaces $H^{\sigma-\varepsilon}(\mathbb{R}^n, \varrho, \theta)$ and $H^{\sigma+\delta}(\mathbb{R}^n, \varrho, \theta)$. By virtue of Theorem 1.14, the spaces

$$\left[H^{\sigma-\varepsilon}(\mathbb{R}^n, \varrho, \theta), \, H^{\sigma+\delta}(\mathbb{R}^n, \varrho, \theta) \right]_{\psi}, \tag{2.63}$$

and $H^{\sigma,\varphi}(\mathbb{R}^n, \varrho, \theta)$ are equal up to equivalence of norms for any fixed $\varrho > 0$. We now show that the equality of norms is indeed realized.

We calculate the norm in space (2.63). By J we denote a pseudodifferential operator in the space $\mathbb{R}^n$ with the symbol $\langle\xi\rangle^{\varepsilon+\delta}$, where the argument $\xi \in \mathbb{R}^n$. It can be directly verified that J is a generating operator for the pair of spaces in relation (2.63). With the help of the isometric isomorphism

$$\mathcal{F} : H^{\sigma-\varepsilon}(\mathbb{R}^n, \varrho, \theta) \leftrightarrow L_2\big(\mathbb{R}^n, \langle\xi\rangle^{2(\sigma-\varepsilon)}(1 + \varrho^2 \langle\xi\rangle^{-2\theta}) \, d\xi\big),$$

where $\mathcal{F}$ is the Fourier transformation, the operators J and $\psi(J)$ are reduced to the operators of multiplication by the functions $\langle\xi\rangle^{\varepsilon+\delta}$ and

$$\psi(\langle\xi\rangle^{\varepsilon+\delta}) = \langle\xi\rangle^{\varepsilon} \varphi(\langle\xi\rangle),$$

respectively. Hence,

$$\|w\|^2_{[H^{\sigma-\varepsilon}(\mathbb{R}^n,\varrho,\theta), \, H^{\sigma+\delta}(\mathbb{R}^n,\varrho,\theta)]_{\psi}} = \|\psi(J)\, w\|^2_{H^{\sigma-\varepsilon}(\mathbb{R}^n,\varrho,\theta)}$$

$$= \int\limits_{\mathbb{R}^n} \langle\xi\rangle^{2(\sigma-\varepsilon)} \left(1 + \varrho^2 \langle\xi\rangle^{-2\theta}\right) \left|\langle\xi\rangle^{\varepsilon} \varphi(\langle\xi\rangle) \, \widehat{w}(\xi)\right|^2 d\xi$$

$$= \|w\|^2_{H^{\sigma,\varphi}(\mathbb{R}^n,\varrho,\theta)}.$$

Thus, we get

$$\|w\|_{[H^{\sigma-\varepsilon}(\mathbb{R}^n,\varrho,\theta), \, H^{\sigma+\delta}(\mathbb{R}^n,\varrho,\theta)]_{\psi}} = \|w\|_{H^{\sigma,\varphi}(\mathbb{R}^n,\varrho,\theta)} \tag{2.64}$$

for any $w \in H^{\sigma,\varphi}(\mathbb{R}^n)$ and $\varrho > 0$. We now deduce the two-sided inequality (2.61) from (2.64). To do this, we use the local Definition 2.1 of the spaces $H^{s,\varphi}(\Gamma)$ with $s \in \mathbb{R}$ and $\varphi \in \mathcal{M}$ for a fixed finite atlas and the resolution of identity on Γ. We consider the linear mapping of "flattening" of the manifold Γ :

$$T : h \mapsto \big((\chi_1 h) \circ \alpha_1, \ldots, (\chi_r h) \circ \alpha_r\big), \quad h \in \mathcal{D}'(\Gamma).$$

It can be directly verified that this mapping specifies the isometric operators

$$T : H^{\sigma,\varphi}(\Gamma, \varrho, \theta) \to (H^{\sigma,\varphi}(\mathbb{R}^n, \varrho, \theta))^r, \tag{2.65}$$

$$T : H^{s}(\Gamma, \varrho, \theta) \to (H^{s}(\mathbb{R}^n, \varrho, \theta))^r, \quad s \in \{\sigma - \varepsilon, \sigma + \delta\}. \tag{2.66}$$

Applying the interpolation with the parameter ψ to operators (2.66), we obtain the bounded operator

$$T : \big[\, H^{\sigma-\varepsilon}(\Gamma, \varrho, \theta),\, H^{\sigma+\delta}(\Gamma, \varrho, \theta)\,\big]_{\psi}$$

$$\to \big[\, (H^{\sigma-\varepsilon}(\mathbb{R}^n, \varrho, \theta))^r,\, (H^{\sigma+\delta}(\mathbb{R}^n, \varrho, \theta))^r\,\big]_{\psi}. \tag{2.67}$$

Since the presented pairs of spaces are normal, the norm of operator (2.67) does not exceed some number $c_1 := c(\psi, 1)$ independent of the parameter ϱ by virtue of Theorem 1.8. This result, Theorem 1.5, and equality (2.64) yield the bounded operator

$$T : \big[\, H^{\sigma-\varepsilon}(\Gamma, \varrho, \theta),\, H^{\sigma+\delta}(\Gamma, \varrho, \theta)\,\big]_{\psi} \to (H^{\sigma,\varphi}(\mathbb{R}^n, \varrho, \theta))^r \tag{2.68}$$

whose norm $\leq c_1$.

Parallel with the mapping T, we consider the linear mapping of "sewing"

$$K : (w_1, \ldots, w_r) \mapsto \sum_{j=1}^{r} \Theta_j\big((\eta_j w_j) \circ \alpha_j^{-1}\big),$$

where $w_1, \ldots, w_r$ are distributions in $\mathbb{R}^n$. Here, the function $\eta_j \in C^\infty(\mathbb{R}^n)$ is finite and equal to 1 on the set $\alpha_j^{-1}(\operatorname{supp} \chi_j)$ and Θ_j is the operator of extension by zero onto Γ. By virtue of (2.8), we get the bounded operators

$$K : (H^{s}(\mathbb{R}^n))^r \to H^{s}(\Gamma), \quad s \in \mathbb{R}, \tag{2.69}$$

$$K : (H^{s,\varphi}(\mathbb{R}^n))^r \to H^{s,\varphi}(\Gamma), \quad s \in \mathbb{R}. \tag{2.70}$$

Let c_2 be the maximum of the norms of operators (2.69), where

$$s \in \{\sigma - \varepsilon,\, \sigma - \varepsilon - \theta,\, \sigma + \delta,\, \sigma + \delta - \theta\},$$

and operators (2.70), where $s \in \{\sigma, \sigma-\theta\}$. The number c_2 is independent of the parameter ϱ. It can be directly verified that the norms of the operators

$$K:\ (H^{\sigma,\varphi}(\mathbb{R}^n,\varrho,\theta))^r \to H^{\sigma,\varphi}(\Gamma,\varrho,\theta), \tag{2.71}$$

$$K:\ (H^{s}(\mathbb{R}^n,\varrho,\theta))^r \to H^{s}(\Gamma,\varrho,\theta), \quad s\in\{\sigma-\varepsilon,\sigma+\delta\}, \tag{2.72}$$

do not exceed the number c_2. Applying the interpolation with the parameter ψ to (2.72), we obtain the bounded operator

$$\begin{aligned} K:\ &\big[\,(H^{\sigma-\varepsilon}(\mathbb{R}^n,\varrho,\theta))^r,\ (H^{\sigma+\delta}(\mathbb{R}^n,\varrho,\theta))^r\,\big]_\psi \\ &\to \big[\,H^{\sigma-\varepsilon}(\Gamma,\varrho,\theta),\ H^{\sigma+\delta}(\Gamma,\varrho,\theta)\,\big]_\psi \end{aligned}$$

whose norm does not exceed the number c_1c_2 by virtue of Theorem 1.8. Hence, in view of Theorem 1.5 and equality (2.64), we arrive at the bounded operator

$$K:\ (H^{\sigma,\varphi}(\mathbb{R}^n,\varrho,\theta))^r \to \big[\,H^{\sigma-\varepsilon}(\Gamma,\varrho,\theta),\ H^{\sigma+\delta}(\Gamma,\varrho,\theta)\,\big]_\psi \tag{2.73}$$

whose norm $\le c_1c_2$.

By virtue of (2.7), the product $KT=I$ is the identity operator. Hence, in view of of relations (2.65) (isometric operator) and (2.73), we get the bounded operator

$$I=KT:\ H^{\sigma,\varphi}(\Gamma,\varrho,\theta) \to \big[\,H^{\sigma-\varepsilon}(\Gamma,\varrho,\theta),\ H^{\sigma+\delta}(\Gamma,\varrho,\theta)\,\big]_\psi$$

whose norm $\le c_1c_2$.

In addition, in view of (2.68) and (2.71) (the norm of the second operator does not exceed the number c_2), we arrive at one more bounded operator

$$I=KT:\ \big[\,H^{\sigma-\varepsilon}(\Gamma,\varrho,\theta),\ H^{\sigma+\delta}(\Gamma,\varrho,\theta)\,\big]_\psi \to H^{\sigma,\varphi}(\Gamma,\varrho,\theta)$$

whose norm $\le c_1c_2$. This immediately yields the two-sided estimate (2.61), where the number $c_0:=\max\{1,c_1c_2\}$ is independent of the parameter ϱ.

Lemma 2.6 is proved.

Proof of Theorem 2.14(ii). Let $s\in\mathbb{R}$ and let $\varphi\in\mathcal{M}$. We recall that Theorem 2.14 holds in the Sobolev case where $\varphi\equiv 1$. Therefore, there exists a number $\lambda_0>0$ such that, for every value of the parameter $\lambda\in K$ satisfying the condition $|\lambda|\ge\lambda_0$, the isomorphisms

$$A(\lambda):\ H^{s\mp 1+mq}(\Gamma,|\lambda|^q,mq) \leftrightarrow H^{s\mp 1}(\Gamma) \tag{2.74}$$

hold and the norm of operator (2.74), together with the norm of the inverse operator, are bounded uniformly in the parameter λ. Let ψ be the interpolation parameter from Theorem 2.2, where we set $\varepsilon=\delta=1$. Applying the interpolation with this parameter to (2.74), we obtain an isomorphism

$$A(\lambda) : \bigl[H^{s-1+mq}(\Gamma, |\lambda|^q, mq),\, H^{s+1+mq}(\Gamma, |\lambda|^q, mq) \bigr]_\psi$$

$$\leftrightarrow \bigl[H^{s-1}(\Gamma),\, H^{s+1}(\Gamma) \bigr]_\psi. \tag{2.75}$$

In this case, by virtue of Theorem 1.8, the norm of operator (2.75) and the norm of the inverse operator are bounded uniformly in the parameter λ. (The admissible pairs of spaces in relation (2.75) are normal.) It remains to use Lemma 2.6, where we set

$$\sigma := s + mq, \quad \varrho := |\lambda|^q, \quad \theta := mq, \quad \varepsilon = \delta = 1,$$

and Theorem 2.2. According to these assertions, relation (2.75) leads to the isomorphism

$$A(\lambda) : H^{s+mq,\varphi}(\Gamma, |\lambda|^q, mq) \leftrightarrow H^{s,\varphi}(\Gamma) \tag{2.76}$$

such that the norm of operator (2.76), together with the norm of the inverse operator, are bounded uniformly in the parameter λ. This yields the two-sided estimate (2.60), where the number c is independent of the parameter λ and the distribution $u \in H^{s+mq,\varphi}(\Gamma)$. Assertion (ii) of Theorem 2.14 is proved.

Theorem 2.14(i) yields the following assertion about the indices of parameter-elliptic PsDOs (it should be compared with [13, § 6, Subsec. 4]).

Corollary 2.4. *Assume that a PsDO $A(\lambda)$ is parameter elliptic on a closed ray $K := \{\lambda \in \mathbb{C} : \arg\lambda = \mathrm{const}\}$. Then the index of operator* (2.58) *is equal to zero for any $\lambda \in \mathbb{C}$.*

Proof. For any fixed $\lambda \in \mathbb{C}$, the PsDO $A(\lambda)$ is elliptic on Γ. Therefore, by virtue of Theorem 2.9, the index of operator (2.58) is finite and independent of $s \in \mathbb{R}$ and $\varphi \in \mathcal{M}$. Moreover, this index is also independent of the parameter λ. Indeed, λ affects solely the lowest terms in sum (2.56): $A(\lambda) - A(0) \in \Psi^{m(q-1)}(\Gamma)$. Hence, in view of Lemma 2.5, we have the bounded operator

$$A(\lambda) - A(0) : H^{s+mq,\varphi}(\Gamma) \to H^{s+m,\varphi}(\Gamma).$$

At the same time, by virtue of Theorem 2.3(iii) and the condition $m > 0$, we get the compact embedding $H^{s+m,\varphi}(\Gamma) \hookrightarrow H^{s,\varphi}(\Gamma)$. This means that the operator

$$A(\lambda) - A(0) : H^{s+mq,\varphi}(\Gamma) \to H^{s,\varphi}(\Gamma)$$

is compact. Thus (see, e.g., [86, Corollary 19.1.8]), the Fredholm operators $A(\lambda)$ and $A(0)$ have the same index, i.e., this index is independent of the parameter λ. Then, by Theorem 2.14(i), isomorphism (2.59) is true for the values of the parameter $\lambda \in K$ sufficiently large in the absolute value.

Hence, the index of the operator $A(\lambda)$ is equal to zero for $\lambda \in K$, $|\lambda| \gg 1$ and, therefore, for any $\lambda \in \mathbb{C}$.

Corollary 2.4 is proved.

2.3 Convergence of spectral expansions

In this subsection, we discuss the applications of the refined Sobolev scale to the investigation of convergence of spectral expansions almost everywhere or in the spaces C^k with integer $k \geq 0$.

2.3.1 Convergence almost everywhere for general orthogonal series

First, we present the classical results on convergence almost everywhere for arbitrary orthogonal series required in what follows. Let Γ be a measurable space with finite measure μ and let $(h_j)_{j=1}^{\infty}$ be an orthonormal system of functions in $L_2(\Gamma, \mu)$. (The functions h_j are, generally speaking, complex-valued.) We use the symbol log to denote the logarithmic function for any fixed base $a > 1$.

Proposition 2.1 (Men'shov–Rademacher theorem). *Let a sequence $(a_j)_{j=1}^{\infty}$ of complex numbers be such that*

$$L := \sum_{j=1}^{\infty} |a_j|^2 \log^2 (j+1) < \infty. \tag{2.77}$$

Then the orthogonal series

$$\sum_{j=1}^{\infty} a_j \, h_j(x) \tag{2.78}$$

converges μ-almost everywhere on Γ. In addition, if

$$S^*(x) := \sup_{1 \leq k < \infty} \left| \sum_{j=1}^{k} a_j \, h_j(x) \right|,$$

is the majorant of the partial sums of series (2.78), *then*

$$\|S^*\|_{L_2(\Gamma)} \leq C \sqrt{L}, \tag{2.79}$$

where the number $C > 0$ depends only on Γ and μ.

This result was proved independently by D. E. Men'shov [134] and H. Rademacher [197] in the case where Γ is a finite interval on the axis $\mathbb{R}$, μ is the Lebesgue measure, and the functions h_j are real-valued. The proof of the Men'shov–Rademacher theorem presented in the monograph [94, Chap. 8, § 1] remains valid in the analyzed general case.

Note that the Men'shov–Rademacher theorem is exact. Men'shov [134] constructed an example of the orthonormal system $(h_j)_{j=1}^{\infty}$ in $L_2((0,1))$ such that, for any number sequence $(\omega_j)_{j=1}^{\infty}$ satisfying the conditions

$$1 = \omega_1 \leq \omega_2 \leq \omega_3 \leq \ldots \quad \text{and} \quad \lim_{j\to\infty} \frac{\omega_j}{\log^2 j} = 0,$$

there exists a series of the form (2.78) divergent almost everywhere whose coefficients satisfy the inequality

$$\sum_{j=1}^{\infty} |a_j|^2 \, \omega_j < \infty.$$

(The presentation of this result can be found, e.g., in [14, Sec. 2.4.1] or [94, Chap. 8, § 1]).

Note that (see [94, Chap. 8, § 2]) the convergence of series (2.78) μ-almost everywhere does not, generally speaking, imply that this series converges unconditionally μ-almost everywhere. We recall that series (2.78) is called unconditionally convergent μ-almost everywhere on Γ if the series

$$\sum_{j=1}^{\infty} a_{\sigma(j)} \, h_{\sigma(j)} \tag{2.80}$$

converges μ-almost everywhere for any permutation of the series of positive integers $\sigma = (\sigma(j))_{j=1}^{\infty}$. (In this case, the set of measure zero of all points at which series (2.80) diverges may depend on the permutation σ.)

Proposition 2.2 (Orlicz–Ul'yanov theorem). *Let a sequence $(a_j)_{j=1}^{\infty}$ of complex numbers and a (nonstrictly) increasing sequence $(\omega_j)_{j=1}^{\infty}$ of positive numbers satisfy the conditions*

$$\sum_{j=2}^{\infty} |a_j|^2 \, (\log^2 j) \, \omega_j < \infty, \tag{2.81}$$

$$\sum_{j=2}^{\infty} \frac{1}{j \, (\log j) \, \omega_j} < \infty. \tag{2.82}$$

Then series (2.78) converges unconditionally μ-almost everywhere on Γ.

This is an equivalent formulation of the Orlicz theorem [178] proposed by P. L. Ul'yanov [261, § 4] (see also [262, § 9, Subsec. 1]). As shown by K. Tandori [251], the Orlicz theorem is the best possible assertion in a sense that condition (2.81) for the sequence $(\omega_j)_{j=1}^{\infty}$ cannot be weakened.

W. Orlicz and P. L. Ul'yanov restricted themselves to the case where Γ is a finite interval on the axis $\mathbb{R}$, μ is the Lebesgue measure, and the functions h_j are real valued. In the general case considered in our monograph, Proposition 2.2 remains true. This follows from the more general Tandori theorem [251] whose proof can be found in the monograph [94, Chap. 8, § 2, Theorem 5] and remains valid in the analyzed general situation.

2.3.2 Convergence almost everywhere for spectral expansions

We now study the convergence almost everywhere for the spectral expansions in eigenfunctions of elliptic PsDOs. In this subsection, Γ is an infinitely smooth closed (compact) manifold of dimension $n \geq 1$ with a fixed density dx. Let A be a classic elliptic PsDO of positive order on Γ. Assume that A is an (unbounded) normal operator in the Hilbert space $L_2(\Gamma, dx)$. Let $(h_j)_{j=1}^{\infty}$ be the complete orthonormal system of the eigenfunctions of this operator. They are enumerated so that the absolute values of the corresponding eigenvalues form a monotonically nondecreasing sequence. For any function $f \in L_2(\Gamma, dx)$, we have

$$f = \sum_{j=1}^{\infty} c_j(f)\, h_j \quad \text{in } L_2(\Gamma, dx), \tag{2.83}$$

where $c_j(f) := (f, h_j)_\Gamma$ are the Fourier coefficients of f in the basis $(h_j)_{j=1}^{\infty}$.

Series (2.83) is called convergent in the indicated sense on some functional class $X(\Gamma)$ if, for any function $f \in X(\Gamma)$, this series converges to f in a proper way.

We now study the problem of convergence almost everywhere for series (2.83) in the Hörmander spaces. Note that if, for some function $f \in L_2(\Gamma, dx)$, series (2.83) converges almost everywhere on Γ, then f is the sum of this series almost everywhere on Γ. This follows from the fact that the convergence in $L_2(\Gamma, dx)$ and the convergence almost everywhere on Γ yield the convergence in the measure generated by the density dx (see, e.g., [23, Chap. I, Sec. 4; Chap. 6, Sec. 8.3]).

Consider the majorant of the partial sums of series (2.83) for the function $f \in L_2(\Gamma, dx)$:

$$S^*(f,x) := \sup_{1 \leq k < \infty} \Bigl| \sum_{j=1}^{k} c_j(f)\, h_j(x) \Bigr|.$$

For $A = \Delta_\Gamma$, Meaney [133] proved that series (2.83) converges almost everywhere on Γ for the functional class

$$H^{0+}(\Gamma) := \bigcup_{\varepsilon>0} H^{\varepsilon}(\Gamma).$$

Moreover, for any $\varepsilon > 0$ and $f \in H^{\varepsilon}(\Gamma)$,

$$\|S^*(f,\cdot)\|_{L_2(\Gamma,dx)} \le C_\varepsilon \|f\|_{H^\varepsilon(\Gamma)},$$

where the number C_ε is independent of f.

We generalize and improve this result by using Hörmander spaces. Denote $\log^* t := \max\{1, \log t\}$.

Theorem 2.15. *In the functional class $H^{0,\log^*}(\Gamma)$, series (2.83) converges almost everywhere on Γ. Moreover, the following estimate is true for any function $f \in H^{0,\log^*}(\Gamma)$:*

$$\|S^*(f,\cdot)\|_{L_2(\Gamma,dx)} \le C\,\|f\|_{H^{0,\log^*}(\Gamma)}, \tag{2.84}$$

where the number $C > 0$ is independent of f.

Proof. Let $f \in H^{0,\log^*}(\Gamma)$. If the operator A is self-adjoint and positive definite in $L_2(\Gamma, dx)$, then, by virtue of Theorem 2.7,

$$\|f\|^2_{H^{0,\log^*}(\Gamma)} \asymp \sum_{j=1}^{\infty} (\log^*(j^{\,1/n}))^2\,|c_j(f)|^2,$$

where $\log^*(j^{\,1/n}) \asymp \log(j+1)$ with $j \ge 1$.

Hence,

$$\sum_{j=1}^{\infty} |c_j(f)|^2 \log^2(j+1) \asymp \|f\|^2_{H^{0,\log^*}(\Gamma)} < \infty.$$

Therefore, by the Men'shov–Rademacher theorem (Proposition 2.1), series (2.83) converges almost everywhere on Γ (to f) and estimate (2.84) is true.

The general case, where the operator A is normal, is reduced to the analyzed case by the transition to the self-adjoint positive definite operator $B := 1 + A^*A$. In this case, one must take into account the fact that $(h_j)_{j=1}^{\infty}$ is the system of eigenfunctions both of the normal operator A and the self-adjoint operator B. This system is complete in $L_2(\Gamma, dx)$ and the numbering of eigenfunctions is consistent.

Theorem 2.15 is proved.

This result is refined by the following theorem on unconditional convergence:

Theorem 2.16. *Assume that an increasing function $\varphi \in \mathcal{M}$ satisfies the condition*

$$\int_2^\infty \frac{dt}{t\,(\log t)\,\varphi^2(t)} < \infty. \tag{2.85}$$

Then, in the functional class $H^{0,\varphi \log^*}(\Gamma)$, series (2.83) is unconditionally convergent almost everywhere on Γ.

Proof. Let $f \in H^{0,\varphi \log^*}(\Gamma)$. If the operator A is self-adjoint and positive definite in $L_2(\Gamma, dx)$, then, by Theorem 2.7, we can write

$$\|f\|^2_{H^{0,\varphi\log^*}(\Gamma)} \asymp \sum_{j=1}^\infty \varphi^2(j^{\,1/n})(\log^*(j^{\,1/n}))^2\,|c_j(f)|^2. \tag{2.86}$$

Consider an increasing sequence of numbers $\omega_j := \varphi^2(j^{\,1/n})$, $j \in \mathbb{N}$. By virtue of (2.86), we have

$$\sum_{j=2}^\infty |c_j(f)|^2\,(\log^2 j)\,\omega_j < \infty. \tag{2.87}$$

In addition, according to condition (2.85),

$$\sum_{j=3}^\infty \frac{1}{j\,(\log j)\,\omega_j} \le \int_2^\infty \frac{d\tau}{\tau\,(\log\tau)\,\varphi^2(\tau^{1/n})} = \int_{2^{1/n}}^\infty \frac{n\,t^{n-1}\,dt}{t^n\,n\,(\log t)\,\varphi^2(t)} < \infty. \tag{2.88}$$

By virtue of the Orlicz–Ul'yanov theorem (Proposition 2.2), inequalities (2.87) and (2.88) yield the unconditional convergence almost everywhere of series (2.83) in Γ.

As in the proof of Theorem 2.15, the general case, where the operator A is normal, is reduced to the already analyzed case by the transition to the self-adjoint operator $B := 1 + A^*A$.

Theorem 2.16 is proved.

Theorems 2.15 and 2.16 belong to the theory of general orthogonal series. However, the conditions imposed on the function f are formulated in constructive terms of the smoothness of functions.

2.3.3 Convergence of spectral expansions in the metric of the space C^k

At the end of Section 2.3, we prove a criterion of convergence of series (2.83) in the spaces $C^k(\Gamma)$ with integer $k \ge 0$ on the classes $H^{s,\varphi}(\Gamma)$.

Theorem 2.17. *Let an integer $k \geq 0$ and a function $\varphi \in \mathcal{M}$ be given. Series* (2.83) *converges in the space $C^k(\Gamma)$ on the class $H^{k+n/2,\varphi}(\Gamma)$ if and only if the function φ satisfies condition* (1.37).

Proof. *Sufficiency.* Assume that φ satisfies condition (1.37). Let $f \in H^{k+n/2,\varphi}(\Gamma)$. As indicated in Remark 2.4, series (2.83) converges to f in the space $H^{k+n/2,\varphi}(\Gamma)$. By Theorem 2.8, inequality (1.37) yields the continuity of the embedding $H^{k+n/2,\varphi}(\Gamma) \hookrightarrow C^k(\Gamma)$. Hence, series (2.83) converges to f in the space $C^k(\Gamma)$.

Sufficiency is proved.

Necessity. Assume that, for any function $f \in H^{k+n/2,\varphi}(\Gamma)$, series (2.83) converges (to f) in the space $C^k(\Gamma)$. Then $H^{k+n/2,\varphi}(\Gamma) \subseteq C^k(\Gamma)$, which yields condition (1.37) by virtue of Theorem 2.8.

Necessity is proved.

Theorem 2.17 is proved.

2.4 RO-varying functions and Hörmander spaces

In the present section, we describe *all* (up to equivalence of norms) interpolation Hilbert spaces for the pairs of inner product Sobolev spaces $H^{s_0}(\mathbb{R}^n)$ and $H^{s_1}(\mathbb{R}^n)$, where $s_0, s_1 \in \mathbb{R}$ and $s_0 < s_1$. It is shown that the class of these interpolation spaces is formed solely by isotropic Hörmander spaces with weight functions, RO-varying at infinity in a sense of Avakumović. We also consider the indicated spaces over smooth closed manifolds and discuss their possible applications.

2.4.1 RO-varying functions in the sense of Avakumović

We now present a definition, which is of fundamental importance for our presentation.

Definition 2.11. Let RO be the set of all Borel measurable functions

$$\varphi : [1, \infty) \to (0, \infty)$$

for which one can find numbers $a > 1$ and $c \geq 1$ such that

$$c^{-1} \leq \frac{\varphi(\lambda t)}{\varphi(t)} \leq c \quad \text{for any} \quad t \geq 1 \quad \text{and} \quad \lambda \in [1, a] \tag{2.89}$$

(the constants a and c may depend on φ). These functions are called *RO-varying* at infinity.

The class of RO-varying (or OR-varying) functions was introduced by V. G. Avakumović [19] in 1936 and studied fairly comprehensively (see, e.g., [26, Sec. 2.0–2.2] or [235, Appendix]).

We now recall some well-known properties of functions from the class **RO**. It is clear that the number a for the function $\varphi \in \mathbf{RO}$ in relation (2.89) can be made arbitrarily large. This yields the following property:

Proposition 2.3. *If $\varphi \in \mathbf{RO}$, then the functions φ and $1/\varphi$ are bounded on every segment $[1, b]$, where $1 < b < \infty$.*

Proposition 2.4. *The following description of the class* **RO** *is true:*

$$\varphi \in \mathbf{RO} \Leftrightarrow \varphi(t) = \exp\left(\beta(t) + \int_1^t \frac{\varepsilon(\tau)}{\tau}\, d\tau\right), \quad t \geq 1,$$

where β and ε are real-valued Borel measurable functions bounded on the semiaxis $[1, \infty)$.

Proposition 2.5. *For any function $\varphi : [1, \infty) \to (0, \infty)$, condition (2.89) is equivalent to the following fact: there exist numbers $s_0, s_1 \in \mathbb{R}$, $s_0 \leq s_1$, and $c \geq 1$ such that*

$$t^{-s_0}\varphi(t) \leq c\,\tau^{-s_0}\varphi(\tau) \quad \textit{and} \quad \tau^{-s_1}\varphi(\tau) \leq c\,t^{-s_1}\varphi(t) \tag{2.90}$$

for all $t \geq 1$ and $\tau \geq t$.

Condition (2.90) indicates that the function $t^{-s_0}\varphi(t)$ is equivalent to an increasing function and the function $t^{-s_1}\varphi(t)$ is equivalent to a decreasing function on the semiaxis $[1, \infty)$. In this case, the property of equivalence is understood in a weak sense, whereas the properties of increasing and decreasing are understood in the nonstrict sense.

Setting $\lambda := \tau/t$ in condition (2.90), we rewrite it in the equivalent form

$$c^{-1}\lambda^{s_0} \leq \frac{\varphi(\lambda t)}{\varphi(t)} \leq c\lambda^{s_1} \quad \text{for all} \quad t \geq 1 \quad \text{and} \quad \lambda \geq 1. \tag{2.91}$$

For any function $\varphi \in \mathbf{RO}$, we denote

$$\sigma_0(\varphi) := \sup\{s_0 \in \mathbb{R} : \ (2.91) \text{ is true}\},$$

$$\sigma_1(\varphi) := \inf\{s_1 \in \mathbb{R} : \ (2.91) \text{ is true}\}.$$

It is clear that

$$-\infty < \sigma_0(\varphi) \leq \sigma_1(\varphi) < \infty.$$

The numbers $\sigma_0(\varphi)$ and $\sigma_1(\varphi)$ are the lower and upper Matuszewska indices of φ (see [130] and [26, Sec. 2.1.2]).

We now additionally define the function $\varphi(t) := \varphi^2(1)/\varphi(t^{-1})$ for $0 < t < 1$. This gives a function φ positive on the semiaxis $(0, \infty)$ and such that (2.91) yields the condition

$$c^{-2}\lambda^{s_0} \le \frac{\varphi(\lambda t)}{\varphi(t)} \le c^2\lambda^{s_1} \quad \text{for all} \quad t > 0 \quad \text{and} \quad \lambda \ge 1.$$

This means that the numbers $\sigma_0(\varphi)$ and $\sigma_1(\varphi)$ are equal to the lower and upper stretching indices of the function φ, respectively [109, Chap. II, § 1, Subsec. 2]. They are given by the formulas

$$\sigma_0(\varphi) = \lim_{\lambda\to 0+} \frac{\log m_\varphi(\lambda)}{\log \lambda}, \quad \sigma_1(\varphi) = \lim_{\lambda\to\infty} \frac{\log m_\varphi(\lambda)}{\log \lambda}, \tag{2.92}$$

where

$$m_\varphi(\lambda) := \sup_{t>0} \frac{\varphi(\lambda t)}{\varphi(t)}, \quad \lambda > 0,$$

is the stretching function for φ. Note that the right-hand sides of relations (2.92) are equal, by definition, to the lower and upper Boyd indices of the function m_φ (see [27]).

In an important case where

$$\sigma_0(\varphi) = \sigma_1(\varphi) =: \sigma,$$

the number σ is called the *order of variation* of the function φ. Note that every Borel measurable function $\varphi : [1, \infty) \to (0, \infty)$ belongs to the class **RO** and has the order of variation σ provided that it is a quasiregularly varying function of order σ at ∞ and that both φ and $1/\varphi$ are bounded on each segment $[1, b]$, where $1 < b < \infty$. This follows from [235, Sec. 1.5, Subsec. 4] and Proposition 2.5. In particular, $\mathcal{M} \subset$ **RO**.

2.4.2 Interpolation spaces for a pair of Sobolev spaces

In this subsection, we describe all interpolation Hilbert spaces for the pairs of inner product Sobolev spaces. For this purpose, we first select the following class of inner product Hörmander spaces:

Definition 2.12. Let $\varphi \in$ **RO**. By definition, the linear space $H^\varphi(\mathbb{R}^n)$ is formed by all distributions $u \in \mathcal{S}'(\mathbb{R}^n)$ such that their Fourier transform $\widehat{u}$ is locally Lebesgue summable in $\mathbb{R}^n$ and satisfies the inequality

$$\int\limits_{\mathbb{R}^n} \varphi^2(\langle\xi\rangle)\, |\widehat{u}(\xi)|^2\, d\xi < \infty.$$

In the space $H^{\varphi}(\mathbb{R}^n)$, the inner product of distributions u_1 and u_2 is defined by the formula

$$(u_1, u_2)_{H^{\varphi}(\mathbb{R}^n)} := \int\limits_{\mathbb{R}^n} \varphi^2(\langle\xi\rangle)\, \widehat{u_1}(\xi)\, \overline{\widehat{u_2}(\xi)}\, d\xi.$$

The inner product specifies the norm in a standard way.

For every functional parameter $\varphi \in \mathbf{RO}$, the space $H^{\varphi}(\mathbb{R}^n)$ is a special (isotropic) Hilbert case of Hörmander spaces: $H^{\varphi}(\mathbb{R}^n) = B_{2,\mu}(\mathbb{R}^n)$, where $\mu(\xi) := \varphi(\langle\xi\rangle)$, $\xi \in \mathbb{R}^n$. In this case, we can show that μ is a weight function.

Lemma 2.7. *Let $\varphi \in \mathbf{RO}$. Then the function $\mu(\xi) := \varphi(\langle\xi\rangle)$ of the argument $\xi \in \mathbb{R}^n$ is a weight function in a sense of Definition* 1.8.

Proof. Let $\xi, \eta \in \mathbb{R}^n$. The inequality $|\langle\xi\rangle - \langle\eta\rangle| \le \big|\,|\xi| - |\eta|\,\big|$ is checked by raising it to the square. Thus, for $\langle\xi\rangle \ge \langle\eta\rangle$, we find

$$\frac{\langle\xi\rangle}{\langle\eta\rangle} = 1 + \frac{\langle\xi\rangle - \langle\eta\rangle}{\langle\eta\rangle} \le 1 + |\xi| - |\eta| \le 1 + |\xi - \eta|.$$

Hence, by virtue of Proposition 2.5, we get

$$\frac{\varphi(\langle\xi\rangle)}{\varphi(\langle\eta\rangle)} \le c \left(\frac{\langle\xi\rangle}{\langle\eta\rangle}\right)^{s_1} \le c\,(1 + |\xi - \eta|)^{\max\{0, s_1\}}.$$

At the same time, if $\langle\eta\rangle \ge \langle\xi\rangle$, then

$$\frac{\varphi(\langle\xi\rangle)}{\varphi(\langle\eta\rangle)} \le c \left(\frac{\langle\xi\rangle}{\langle\eta\rangle}\right)^{s_0} = c \left(\frac{\langle\eta\rangle}{\langle\xi\rangle}\right)^{-s_0} \le c\,(1 + |\xi - \eta|)^{\max\{0, -s_0\}}.$$

Thus, for any $\xi, \eta \in \mathbb{R}^n$, we obtain

$$\frac{\mu(\xi)}{\mu(\eta)} = \frac{\varphi(\langle\xi\rangle)}{\varphi(\langle\eta\rangle)} \le c\,(1 + |\xi - \eta|)^{l},$$

where the numbers $c \ge 1$ and $l := \max\{0, -s_0, s_1\}$ are independent of ξ and η. This means that μ is a weight function in a sense of Definition 1.8.

Lemma 2.7 is proved.

Note that if $\varphi_0 \in \mathcal{M}$ and $s \in \mathbb{R}$, then the function $\varphi_s(t) := t^s \varphi_0(t)$ belongs to the class $\mathbf{RO}$. Hence, the class of spaces $\{H^{\varphi}(\mathbb{R}^n) : \varphi \in \mathbf{RO}\}$ contains the refined Sobolev scale.

We now indicate the properties of the spaces $H^{\varphi}(\mathbb{R}^n)$, which follow from the corresponding properties of the Hörmander spaces [81, Sec. 2.2].

Proposition 2.6. *Let $\varphi, \varphi_1 \in \mathrm{RO}$. Then*

(i) *$H^{\varphi}(\mathbb{R}^n)$ is a separable Hilbert space.*

(ii) *The set $C_0^{\infty}(\mathbb{R}^n)$ is dense in $H^{\varphi}(\mathbb{R}^n)$.*

(iii) *The function $\varphi(t)/\varphi_1(t)$ is bounded in the vicinity of ∞ if and only if $H^{\varphi_1}(\mathbb{R}^n) \hookrightarrow H^{\varphi}(\mathbb{R}^n)$. This embedding is continuous and dense.*

(iv) *For any real numbers $s_0 < \sigma_0(\varphi)$ and $s_1 > \sigma_1(\varphi)$, the following continuous and dense embeddings are true: $H^{s_1}(\mathbb{R}^n) \hookrightarrow H^{\varphi}(\mathbb{R}^n) \hookrightarrow H^{s_0}(\mathbb{R}^n)$.*

(v) *The spaces $H^{\varphi}(\mathbb{R}^n)$ and $H^{1/\varphi}(\mathbb{R}^n)$ are mutually dual with respect to the extension by continuity of the inner product in the space $L_2(\mathbb{R}^n)$.*

(vi) *For every fixed integer $k \geq 0$, the condition*

$$\int_1^{\infty} t^{2k+n-1}\, \varphi^{-2}(t)\, dt < \infty \tag{2.93}$$

is equivalent to the embedding $H^{\varphi}(\mathbb{R}^n) \hookrightarrow C_{\mathrm{b}}^k(\mathbb{R}^n)$. This embedding is continuous.

We now make some comments to Proposition 2.6. Assertion (iv) follows from assertion(iii) and inequality (2.91) in which we set $t := 1$. Since

$$\varphi \in \mathrm{RO} \Leftrightarrow 1/\varphi \in \mathrm{RO},$$

the space $H^{1/\varphi}(\mathbb{R}^n)$ from assertion (v) is well defined. Assertion (vi) follows from the Hörmander embedding theorem (Proposition 1.5). In this case, it should be taken into account that (1.33) $\Leftrightarrow$ (2.93) for the function $\mu(\xi) := \varphi(\langle\xi\rangle)$ of the argument $\xi \in \mathbb{R}^n$ if we pass to the spherical coordinates.

We now study the interpolation properties of the Hörmander spaces $H^{\varphi}(\mathbb{R}^n)$, where the function parameter $\varphi \in \mathrm{RO}$.

Theorem 2.18. *For given functions $\varphi_0, \varphi_1 \in \mathrm{RO}$ and $\psi \in \mathcal{B}$, the function φ_0/φ_1 is assumed to be bounded in the vicinity of ∞. Let ψ be an interpolation parameter and let*

$$\varphi(t) := \varphi_0(t)\, \psi(\varphi_1(t)/\varphi_0(t)) \quad \text{for} \quad t \geq 1.$$

Then $\varphi \in \mathrm{RO}$ and

$$[H^{\varphi_0}(\mathbb{R}^n), H^{\varphi_1}(\mathbb{R}^n)]_{\psi} = H^{\varphi}(\mathbb{R}^n) \quad \text{with equality of the norms.} \tag{2.94}$$

Proof. First, we show that $\varphi \in \mathbf{RO}$. By definition, the function φ is Borel measurable on the semiaxis $[1,\infty)$. We now show that it satisfies condition (2.89). Since $\varphi_0, \varphi_1 \in \mathbf{RO}$, there exist numbers $a > 1$ and $c > 1$ such that

$$c^{-1} \leq \frac{\varphi_j(\lambda t)}{\varphi_j(t)} \leq c \text{ for all } t \geq 1, \ \lambda \in [1,a], \text{ and } j \in \{0, 1\}. \tag{2.95}$$

The boundedness of the function φ_0/φ_1 in the vicinity of ∞ and Proposition 2.3 imply that there exists a number $\varepsilon > 0$ such that

$$\frac{\varphi_1(t)}{\varphi_0(t)} > \varepsilon \quad \text{for any} \quad t \geq 1. \tag{2.96}$$

Further, since ψ is an interpolation parameter, the function ψ is pseudoconcave in the vicinity of ∞ by Theorem 1.9. Thus, by virtue of Lemmas 1.1 and 1.2, the function ψ is weakly equivalent to a concave function on the semiaxis (ε,∞). This is equivalent to the following condition: there exists a number $c_0 > 1$ such that

$$\frac{\psi(\tau)}{\psi(t)} \leq c_0 \max\left\{1, \frac{\tau}{t}\right\} \quad \text{for all} \quad \tau > \varepsilon \quad \text{and} \quad t > \varepsilon. \tag{2.97}$$

This yields the inequality

$$\frac{\psi(t)}{\psi(\tau)} \geq c_0^{-1} \min\left\{1, \frac{t}{\tau}\right\} \quad \text{for all} \quad \tau > \varepsilon \quad \text{and} \quad t > \varepsilon. \tag{2.98}$$

Relations (2.95), (2.96), and (2.97) now imply that, for any $t \geq 1$ and $\lambda \in [1,a]$, we get

$$\frac{\varphi(\lambda t)}{\varphi(t)} = \frac{\varphi_0(\lambda t)}{\varphi_0(t)} \cdot \frac{\psi(\varphi_1(\lambda t)/\varphi_0(\lambda t))}{\psi(\varphi_1(t)/\varphi_0(t))} \leq c \cdot c_0 \max\left\{1, \frac{\varphi_1(\lambda t)/\varphi_0(\lambda t)}{\varphi_1(t)/\varphi_0(t)}\right\} \leq c^3 c_0.$$

Similarly, relations (2.95), (2.96), and (2.98) yield

$$\frac{\varphi(\lambda t)}{\varphi(t)} \geq c^{-1} c_0^{-1} \min\left\{1, \frac{\varphi_1(\lambda t)/\varphi_0(\lambda t)}{\varphi_1(t)/\varphi_0(t)}\right\} \geq c^{-3} c_0^{-1}.$$

Thus, the function φ satisfies condition (2.89) and, therefore, $\varphi \in \mathbf{RO}$.

We now prove equality (2.94). By virtue of Proposition 2.6(iii), the pair $[H^{\varphi_0}(\mathbb{R}^n), H^{\varphi_1}(\mathbb{R}^n)]$ is admissible. The pseudodifferential operator with the symbol $\varphi_1(\langle\xi\rangle)/\varphi_0(\langle\xi\rangle)$, where $\xi \in \mathbb{R}^n$, is the generating operator J for this pair. By using the Fourier transformation

$$\mathcal{F} : H^{\varphi_0}(\mathbb{R}^n) \leftrightarrow L_2(\mathbb{R}^n, \varphi_0^2(\langle\xi\rangle)\, d\xi),$$

the generating operator J is reduced to the form of multiplication by the function $\varphi_1(\langle\xi\rangle)/\varphi_0(\langle\xi\rangle)$. Hence, the operator $\psi(J)$ is reduced to the form of multiplication by the function

$$\psi\left(\frac{\varphi_1(\langle\xi\rangle)}{\varphi_0(\langle\xi\rangle)}\right) = \frac{\varphi(\langle\xi\rangle)}{\varphi_0(\langle\xi\rangle)}.$$

Thus, for any function $u \in C_0^\infty(\mathbb{R}^n)$, we can write

$$\|u\|^2_{[H^{\varphi_0}(\mathbb{R}^n),H^{\varphi_1}(\mathbb{R}^n)]_\psi} = \|\psi(J)u\|^2_{H^{\varphi_0}(\mathbb{R}^n)} = \int\limits_{\mathbb{R}^n} \varphi_0^2(\langle\xi\rangle)\, |(\widehat{\psi(J)u})(\xi)|^2\, d\xi$$

$$= \int\limits_{\mathbb{R}^n} \varphi^2(\langle\xi\rangle)\, |\widehat{u}(\xi)|^2\, d\xi = \|u\|^2_{H^{\varphi}(\mathbb{R}^n)}.$$

This yields the equality of spaces (2.94) because the set $C_0^\infty(\mathbb{R}^n)$ is dense in each of these spaces. This fact follows from Proposition 2.6(ii) and Theorem 1.1 according to which the space $H^{\varphi_1}(\mathbb{R}^n)$ is continuously and densely embedded in $[H^{\varphi_0}(\mathbb{R}^n), H^{\varphi_1}(\mathbb{R}^n)]_\psi$.

Theorem 2.18 is proved.

Theorem 2.19. *Assume that a function $\varphi \in \mathrm{RO}$ and real numbers s_0 and s_1 such that $s_0 < \sigma_0(\varphi)$ and $s_1 > \sigma_1(\varphi)$ are given. Also let*

$$\psi(t) := \begin{cases} t^{-s_0/(s_1-s_0)}\, \varphi(t^{1/(s_1-s_0)}) & \text{for } t \geq 1, \\ \varphi(1) & \text{for } 0 < t < 1. \end{cases}$$

Then the function $\psi \in \mathcal{B}$ is an interpolation parameter and

$$[H^{s_0}(\mathbb{R}^n), H^{s_1}(\mathbb{R}^n)]_\psi = H^{\varphi}(\mathbb{R}^n) \quad \text{with equality of the norms.} \tag{2.99}$$

Proof. Since

$$\varphi(t) = t^{s_0}\, \psi(t^{s_1}/t^{s_0})$$

for $t \geq 1$, Theorem 2.19 follows from Theorems 1.9 and 2.18 if we prove that the function ψ belongs to the set $\mathcal{B}$ and is pseudoconcave in the vicinity of ∞. By virtue of Proposition 2.5, the function ψ satisfies condition (2.97) for $\varepsilon = 1$.

Indeed, if $t \geq 1$ and $\tau \geq 1$, then

$$\frac{\psi(\tau)}{\psi(t)} = \left(\frac{\tau}{t}\right)^{-s_0/(s_1-s_0)} \frac{\varphi(\tau^{1/(s_1-s_0)})}{\varphi(t^{1/(s_1-s_0)})}$$

$$\leq \left(\frac{\tau}{t}\right)^{-s_0/(s_1-s_0)} c \max\left\{\left(\frac{\tau}{t}\right)^{s_1/(s_1-s_0)}, \left(\frac{\tau}{t}\right)^{s_0/(s_1-s_0)}\right\}$$

$$= c \max\left\{\frac{\tau}{t}, 1\right\}.$$

In particular, this implies that the function $1/\psi$ is bounded on the semiaxis $[1,\infty)$ and, hence, $\psi \in \mathcal{B}$ in view of Proposition 2.3. It remains to use Lemma 1.2 according to which condition (2.97) is equivalent to the requirement that the function ψ is pseudoconcave on the semiaxis (ε,∞).

Theorem 2.19 is proved.

We can now prove the following fundamental property of the class of Hörmander spaces:

$$\{H^{\varphi}(\mathbb{R}^n) : \varphi \in \mathrm{RO}\}. \tag{2.100}$$

Theorem 2.20. *The class of spaces* (2.100) *coincides (up to equivalence of norms) with the set of all Hilbert interpolation spaces for the pairs of inner product Sobolev spaces*

$$[H^{s_0}(\mathbb{R}^n), H^{s_1}(\mathbb{R}^n)], \tag{2.101}$$

where $s_0, s_1 \in \mathbb{R}$ *and* $s_0 < s_1$.

Proof. By virtue of Theorem 2.19, each space $H^{\varphi}(\mathbb{R}^n)$, where $\varphi \in \mathrm{RO}$, is an interpolation space for the pair of spaces (2.101) provided that $s_0 < \sigma_0(\varphi)$ and $s_1 > \sigma_1(\varphi)$. Conversely, if H is an interpolation Hilbert space for the pair of spaces (2.101), where $s_0, s_1 \in \mathbb{R}$ and $s_0 < s_1$, then, by Proposition 1.1, we get

$$H = [H^{s_0}(\mathbb{R}^n), H^{s_1}(\mathbb{R}^n)]_{\psi} \quad \text{with equivalence of the norms}$$

for a function $\psi \in \mathcal{B}$ pseudoconcave in the vicinity of ∞. By Theorem 1.9, this function is an interpolation parameter. According to Theorem 2.18, this yields $H = H^{\varphi}(\mathbb{R}^n)$, where the function $\varphi \in \mathrm{RO}$ is given by the formula

$$\varphi(t) := t^{s_0}\psi(t^{s_1-s_0}) \quad \text{for} \quad t \geq 1.$$

Theorem 2.20 is proved.

In view of this theorem, the class of spaces (2.100) is called the extended Sobolev scale (by means of the Hilbert interpolation spaces).

Remark 2.5. By virtue of Theorems 2.18 and 1.9 and Proposition 1.1, the class of spaces (2.100) is closed relative to the interpolation as a result of which we obtain a Hilbert space.

We now consider the extended Sobolev scale over an infinitely smooth closed manifold Γ of dimension $n \geq 1$. The following theorem gives equivalent definitions of the space $H^{\varphi}(\Gamma)$, $\varphi \in \mathrm{RO}$ (cf. Subsection 2.1.1). We use the notation from Subsection 2.1.1.

Theorem 2.21. *Let* $\varphi \in \mathrm{RO}$. *The following definitions give the same Hilbert space* $H^{\varphi}(\Gamma)$ *up to equivalence of norms:*

(i) *The linear space* $H^{\varphi}(\Gamma)$ *consists, by the definition, of all distributions* $f \in \mathcal{D}'(\Gamma)$ *such that* $(\chi_j f) \circ \alpha_j \in H^{\varphi}(\mathbb{R}^n)$ *for any* $j \in \{1, \ldots, r\}$*. The space* $H^{\varphi}(\Gamma)$ *is endowed with the inner product given by the formula*

$$(f_1, f_2)_{H^{\varphi}(\Gamma)} := \sum_{j=1}^{r} ((\chi_j f_1) \circ \alpha_j, (\chi_j f_2) \circ \alpha_j)_{H^{\varphi}(\mathbb{R}^n)}$$

and the corresponding Hilbert norm.

(ii) *Let* k_0 *and* k_1 *be integers such that* $k_0 < \sigma_0(\varphi)$ *and* $k_1 > \sigma_1(\varphi)$*. By the definition,*

$$H^{\varphi}(\Gamma) := [H^{k_0}(\Gamma), H^{k_1}(\Gamma)]_{\psi},$$

where the interpolation parameter ψ *is given by the formula*

$$\psi(t) = \begin{cases} t^{-k_0/(k_1-k_0)}\,\varphi(t^{1/(k_1-k_0)}) & \text{for } t \geq 1, \\ \varphi(1) & \text{for } 0 < t < 1. \end{cases}$$

(iii) *By the definition, the space* $H^{\varphi}(\Gamma)$ *is the completion of the linear manifold* $C^{\infty}(\Gamma)$ *with respect to the Hilbert norm*

$$f \mapsto \|\varphi((1 - \Delta_{\Gamma})^{1/2})\, f\|_{L_2(\Gamma)}, \quad f \in C^{\infty}(\Gamma).$$

This theorem is a special case of the following two theorems. As the basic definition of the space $H^{\varphi}(\Gamma)$, we take assertion (i) of Theorem 2.21.

Theorem 2.22. *The interpolation theorems* 2.18 *and* 2.19 *remain true if, in their formulations,* $\mathbb{R}^n$ *is replaced by* Γ *and the equality of norms is replaced by their equivalence.*

Theorem 2.22 is proved by analogy with Theorem 2.2.

Let A be an elliptic PsDO of order $m > 0$. Assume that A is an (unbounded) self-adjoint positive-definite operator in the Hilbert space $L_2(\Gamma)$. We additionally define the function $\varphi \in \mathbf{RO}$ by the equality $\varphi(t) := \varphi(1)$ for $0 < t < 1$.

Theorem 2.23. *For any* $\varphi \in$ RO, *the norm in the space* $H^{\varphi}(\Gamma)$ *is equivalent to the norm*

$$f \mapsto \|\varphi(A^{1/m})\, f\|_{L_2(\Gamma)} \tag{2.102}$$

in the dense set $C^{\infty}(\Gamma)$*. Thus, the space* $H^{\varphi}(\Gamma)$ *coincides with the completion of the set* $C^{\infty}(\Gamma)$ *with respect to norm* (2.102).

Theorem 2.23 is proved by analogy with Theorem 2.5.

At the end of this subsection, we present an analog of Theorem 2.20 for the class of spaces

$$\{H^{\varphi}(\Gamma) : \varphi \in \mathbf{RO}\}. \tag{2.103}$$

Theorem 2.24. *The class of spaces* (2.103) *coincides (up to equivalence of norms) with the set of all Hilbert interpolation spaces for the pairs of inner product Sobolev spaces* $[H^{s_0}(\Gamma), H^{s_1}(\Gamma)]$, *where* $s_0, s_1 \in \mathbb{R}$ *and* $s_0 < s_1$.

This theorem is proved by analogy with Theorem 2.20.

The class of spaces (2.103) is called the extended Sobolev scale over the manifold Γ.

2.4.3 Applications to elliptic operators

We now discuss some applications of the extended Sobolev scale to elliptic PsDOs.

First, we consider the PsDOs in $\mathbb{R}^n$. It is useful to compare the results presented in what follows with the results from Section 1.4. We preliminarily consider the action of PsDOs on the extended Sobolev scale (2.100). We set $\varrho(t) := t$ for $t \geq 1$.

Lemma 2.8. *Let* A *be a PsDO from the class* $\Psi^m(\mathbb{R}^n)$, *where* $m \in \mathbb{R}$. *Then the restriction of the mapping* $u \mapsto Au$, $u \in \mathcal{S}'(\mathbb{R}^n)$, *to the space* $H^{\varphi\varrho^m}(\mathbb{R}^n)$ *is a linear bounded operator*

$$A : H^{\varphi\varrho^m}(\mathbb{R}^n) \to H^{\varphi}(\mathbb{R}^n) \quad \textit{for any} \quad \varphi \in \mathrm{RO}. \tag{2.104}$$

This lemma is proved with the help of the interpolation theorem 2.19 by analogy with Lemma 1.6.

Assume that the PsDO $A \in \Psi^m_{\mathrm{ph}}(\mathbb{R}^n)$ is uniformly elliptic in $\mathbb{R}^n$. Then mapping (2.104) has the following properties:

Theorem 2.25. *Assume that a function* $\varphi \in \mathrm{RO}$ *and a number* $\sigma > 0$ *are given. There exists a number* $c = c(\varphi, \sigma) > 0$ *such that, for any distribution* $u \in H^{\varphi\varrho^m}(\mathbb{R}^n)$, *the following a priori estimate is true:*

$$\|u\|_{H^{\varphi\varrho^m}(\mathbb{R}^n)} \leq c\,\big(\,\|Au\|_{H^{\varphi}(\mathbb{R}^n)} + \|u\|_{H^{\varphi\varrho^{m-\sigma}}(\mathbb{R}^n)}\big).$$

This theorem is proved by analogy with Theorem 1.16.

Let V be an arbitrary nonempty open subset of the space $\mathbb{R}^n$. For $\varphi \in \mathrm{RO}$, we set

$$H^{\varphi}_{\mathrm{int}}(V) := \big\{w \in H^{-\infty}(\mathbb{R}^n) : \chi w \in H^{\varphi}(\mathbb{R}^n)$$
$$\text{for all } \chi \in C^{\infty}_{\mathrm{b}}(\mathbb{R}^n),\ \operatorname{supp}\chi \subset V,\ \operatorname{dist}(\operatorname{supp}\chi, \partial V) > 0\big\}.$$

Theorem 2.26. *Assume that* $u \in H^{-\infty}(\mathbb{R}^n)$ *is a solution of the equation* $Au = f$ *on the set* V, *where* $f \in H^{\varphi}_{\mathrm{int}}(V)$ *for a certain parameter* $\varphi \in \mathrm{RO}$. *Then* $u \in H^{\varphi\varrho^m}_{\mathrm{int}}(V)$.

This theorem is proved by analogy with Theorem 1.18.

Theorem 2.26 and Proposition 2.6(vi) yield the following sufficient condition (sharper than Theorem 1.19) for the existence of continuous derivatives of the solutions to the equation $Au = f$.

Theorem 2.27. *Let an integer $r \geq 0$ and a function $\varphi \in \mathcal{M}$ satisfying the condition*

$$\int_1^\infty t^{2(r-m)+n-1} \varphi^{-2}(t)\, dt < \infty,$$

be given. Assume that a distribution $u \in H^{-\infty}(\mathbb{R}^n)$ is a solution of the equation $Au = f$ on the open set $V \subseteq \mathbb{R}^n$ and that $f \in H^{\varphi}_{\mathrm{int}}(V)$. Then the assertion of Theorem 1.19 *is true.*

We now briefly consider the application of the spaces $H^{\varphi}(\Gamma)$, $\varphi \in \mathrm{RO}$, to elliptic PsDOs on the infinitely smooth closed manifold Γ.

Let a PsDO $A \in \Psi^m_{\mathrm{ph}}(\Gamma)$, where $m \in \mathbb{R}$, be elliptic on Γ. For the operator A, the finite-dimensional spaces $\mathcal{N}$ and $\mathcal{N}^+$ are given by relations (2.45) and (2.46).

Theorem 2.28. *For any parameter $\varphi \in \mathrm{RO}$, the restriction of the mapping $u \mapsto Au$, $u \in \mathcal{D}'(\Gamma)$, to the space $H^{\varphi\varrho^m}(\Gamma)$ is a bounded operator*

$$A : H^{\varphi\varrho^m}(\Gamma) \to H^{\varphi}(\Gamma). \tag{2.105}$$

This operator is Fredholm. Its kernel coincides with $\mathcal{N}$ and the domain is given by the formula

$$\{f \in H^{\varphi}(\Gamma) : (f, w)_\Gamma = 0 \ \text{ for all } \ w \in \mathcal{N}^+\}.$$

The index of operator (2.105) *is equal to* $\dim \mathcal{N} - \dim \mathcal{N}^+$. *It is independent of φ.*

This theorem is proved by analogy with Theorem 2.9. Theorem 2.28 yields analogs of Theorems 2.10–2.13 for the extended Sobolev scale over Γ. This can be established by using the same reasoning as in Subsections 2.2.2 and 2.2.3. We omit the formulations of these analogs.

2.5 Remarks and comments

Section 2.1. The proposed equivalent definitions of the Hörmander spaces on closed (compact) smooth manifolds are similar to the definitions used for the Sobolev spaces; see, e.g., the monographs by J.-L. Lions and E. Magenes [121, Sec. 7.3] and M. Taylor [253, Chap. I, § 5].

G. Shlenzak [231] used a certain class of locally introduced Hörmander spaces over the boundary of an Euclidean domain. However, the independence of these spaces and their topologies of the choice of local charts is not proved in [231].

The real Hörmander spaces over a circle, where the functions are given by trigonometric Fourier series, were used by J. Pöschel [194], P. Djakov and B. S. Mityagin [44, 45], and V. A. Mikhailets and V. Molyboga [140, 141, 142]. These spaces are closely related to the spaces of periodic real functions introduced by A. I. Stepanets [248, 249]. They are extensively used in the approximation theory.

All results presented in Section 2.1 (except Theorem 2.7) were obtained by the authors in [153, Sec. 3.3 and 3.4]. The proof of Theorem 2.7 is presented here for this first time. These results were partially announced in [159, Sec. 7.2]. Numerous results, including Theorems 2.2, 2.3, and 2.8, remain valid for the compact smooth manifolds with boundary [145, Sec. 3].

Section 2.2. The analysis of classical PsDOs on smooth manifolds without boundary was developed by L. Hörmander [82]. The systematic presentation of the theory of elliptic PsDOs on these manifolds can be found, e.g., in his monograph [86, Chap. 19] and in the survey by M. S. Agranovich [10, § 2]. The classical results presented in these works concerning the Fredholm property of elliptic PsDOs and the regularity of solutions on the Sobolev scale have found various applications to the theory of elliptic boundary-value problems for differential equations, spectral theory, the theory of function spaces, etc. (see also the monographs by M. Taylor [253], F. Treves [254, 255], and M. A. Shubin [232].

The elliptic PsDOs realizing isomorphisms on the Sobolev scale are of substantial independent interest. A broad class of operators of this kind, namely, the elliptic operators with parameter, was selected and studied by S. Agmon and L. Nirenberg [2, 6] and M. S. Agranovich and M. I. Vishik [13]; see also the survey by M. S. Agranovich [10, § 4]. These operators found important applications to the spectral theory and to the theory of parabolic equations.

All theorems in Section 2.2 were proved in [165].

Section 2.3. For any orthogonal series, the classical theorem on convergence almost everywhere was proved independently by D. E. Men'shov [134] and H. Rademacher [197]. The theorems on unconditional convergence were established by W. Orlicz [178] and K. Tandori [251]. We use the equivalent formulation of the Orlicz theorem proposed by P. L. Ul'yanov [261, § 4] (see also [262, § 9, Subsec. 1]). All these theorems were proved by the mentioned authors for real-valued series given on a finite interval of the axis $\mathbb{R}$ and for the Lebesgue measure. The presentation of these results can be found, e.g., in the monographs by G. Alexits [14] and B. S. Kashin and A. A. Saakyan [94].

Apparently, the most general versions of the Men'shov–Rademacher, Orlicz–Ul'yanov, and Tandori theorems are considered in [157] (see also [158]). The complete characterization of the coefficients of arbitrary orthogonal series convergent almost everywhere was given by A. Paszkiewicz [185].

Theorems 2.15 and 2.16 were announced in [156, Sec. 5]. They generalize and significantly improve the result obtained by C. Meaney [133] who used the Sobolev scale and considered the expansions in eigenfunctions of the Beltrami–Laplace operator. Theorem 2.17 was presented in [159]

Section 2.4. The class of RO-varying functions was introduced by V. G. Avakumović [19] in 1936 and studied fairly comprehensively; see, e.g., the monographs by N. H. Bingham, C. M. Goldie, and J. L. Teugels [26, Sec. 2.0–2.2] and E. Seneta [235, Appendix].

The Hörmander spaces parametrized by RO-varying functions, the extended Sobolev scale formed by these spaces, and their applications to scalar elliptic operators were studied in [152, 156, 160]. All theorems presented in Section 2.4 were proved in the cited works. The parameter-elliptic operators were investigated on this scale by A. A. Murach and T. N. Zinchenko in [174].

Chapter 3

Semihomogeneous elliptic boundary-value problems

3.1 Regular elliptic boundary-value problems

In the present section, we give a definition of the regular elliptic boundary-value problem in a bounded Euclidean domain and formulate some notions related to this class of problems.

3.1.1 Definition of the problem

In Chapters 3 and 4, we assume that Ω is an arbitrary bounded domain in the Euclidean space $\mathbb{R}^n$ $(n \geq 2)$ with a boundary Γ, which is supposed to be an infinitely smooth manifold of dimension $n-1$ that has no boundary. The domain Ω is locally located on one side of Γ. We set $\overline{\Omega} := \Omega \cup \Gamma$ and $\widehat{\Omega} := \mathbb{R}^n \setminus \Omega$. For a point $x \in \Gamma$, let $\nu(x)$ denote the unit vector of the inner normal to the boundary Γ at x. Also let ν denote the infinitely smooth vector field of these unit vectors.

In the domain Ω, we consider the boundary-value problem

$$Lu \equiv \sum_{|\mu| \leq 2q} l_\mu(x)\, D^\mu u = f \;\text{ in }\; \Omega, \tag{3.1}$$

$$B_j\, u \equiv \sum_{|\mu| \leq m_j} b_{j,\mu}(x)\, D^\mu u = g_j \;\text{ on }\; \Gamma, \;\; j = 1, \ldots, q, \tag{3.2}$$

where $L = L(x, D)$ is a linear differential expression of even order $2q \geq 2$ on $\overline{\Omega}$ and

$$B_j = B_j(x, D), \quad j = 1, \ldots, q,$$

are boundary linear differential expressions of orders $\operatorname{ord} B_j = m_j \leq 2q - 1$ on Γ. All coefficients of differential expressions L and B_j are supposed to be complex-valued functions that are infinitely smooth:

$$l_\mu \in C^\infty(\overline{\Omega}) \quad \text{and} \quad b_{j,\mu} \in C^\infty(\Gamma).$$

We set $B := (B_1, \ldots, B_q)$.

In relations (3.1) and (3.2) (and in what follows), we use the standard notation:

$$\mu := (\mu_1, \ldots, \mu_n) \;\text{ is a multiindex,}$$

$$|\mu| := \mu_1 + \ldots + \mu_n,$$

$$D^{\mu} := D_1^{\mu_1} \dots D_n^{\mu_n},$$

$$D_k := i\partial/\partial x_k \quad \text{for} \quad k = 1, \dots, n,$$

i is the imaginary unit, and $x = (x_1, \dots, x_n)$ is a point in the space $\mathbb{R}^n$.

In addition, we assume that (3.1), (3.2) is a regular elliptic boundary-value poblem in the domain Ω.

Let us recall the corresponding definition (see, e.g., [121, Chap. 2, Sec. 1] or [258, Sec. 5.2.1]).

The above-mentioned differential expressions $L(x, D)$ for fixed $x \in \overline{\Omega}$ and $B_j(x, D)$ for fixed $x \in \Gamma$ are associated with the following homogeneous characteristic polynomials, respectively:

$$L^{(0)}(x,\xi) := \sum_{|\mu|=2q} l_{\mu}(x)\,\xi^{\mu} \quad \text{and} \quad B_j^{(0)}(x,\xi) := \sum_{|\mu|=m_j} b_{j,\mu}(x)\,\xi^{\mu}.$$

These polynomials are also called principal symbols of expressions L and B_j. Here, the variable $\xi = (\xi_1, \dots, \xi_n) \in \mathbb{C}^n$ and $\xi^{\mu} := \xi_1^{\mu_1} \dots \xi_n^{\mu_n}$.

Definition 3.1. The boundary-value problem (3.1), (3.2) is called *regular elliptic* in the domain Ω if the following conditions are satisfied:

(i) The differential expression L is *properly elliptic* on $\overline{\Omega}$; i.e., for any point $x \in \overline{\Omega}$ and any linearly independent vectors $\xi', \xi'' \in \mathbb{R}^n$, the polynomial $L^{(0)}(x, \xi' + \tau\xi'')$ in the variable τ has exactly q roots $\tau_j^{+}(x; \xi', \xi'')$, $j = 1, \dots, q$, with positive imaginary parts and exactly q roots with negative imaginary parts (with regard for multiplicities of the roots).

(ii) The system of boundary expressions $\{B_1, \dots, B_q\}$ satisfies the *complementing condition* with respect to L on Γ, i.e., for any point $x \in \Gamma$ and any vector $\xi \neq 0$ tangent to Γ at x, the polynomials $B_j^{(0)}(x, \xi + \tau\nu(x))$, $j = 1, \dots, q$, in the variable τ are linearly independent modulo the polynomial

$$\prod_{j=1}^{q} (\tau - \tau_j^{+}(x; \xi, \nu(x))).$$

(iii) The system of boundary expressions $\{B_1, \dots, B_q\}$ is *normal*, i.e., their orders m_j, $j = 1, \dots, q$, are mutually distinct and $B_j^{(0)}(x, \nu(x)) \neq 0$ for any $x \in \Gamma$.

In what follows, we always assume (except for Subsection 4.1.3) that the boundary-value problem (3.1), (3.2) is regular elliptic in its domain Ω.

Remark 3.1. Condition (i) of Definition 3.1 yields

$$L^{(0)}(x,\xi) \neq 0 \quad \text{for all} \quad x \in \overline{\Omega} \quad \text{and} \quad \xi \in \mathbb{R}^n \setminus \{0\}, \tag{3.3}$$

i.e., the differential expression L is *elliptic* on $\overline{\Omega}$. If $n \geq 3$, then conditions (i) and (3.3) are equivalent (see, e.g., [121, Chap. 2, Sec. 1.1] or [258, Sec. 5.2.1]). If all coefficients of the expression L are real, then this equivalence holds also for $n = 2$.

Remark 3.2. The complementing condition (Definition 3.1(ii)) was first formulated by Ya. B. Lopatinskii [123], [124] and, in particular cases, by Z. Ya. Shapiro [230]. Other equivalent forms of this condition are also known [11, Sec. 1.3]. The condition of normality for the system of boundary expressions (Definition 3.1(iii)) was independently introduced by N. Aronszajn and A. N. Milgram [16] and M. Schechter [222].

Example 3.1. Let $k \in \mathbb{Z}$ and $0 \leq k \leq q$. The system of boundary expressions

$$B_j u := \partial^{k+j-1} u / \partial \nu^{k+j-1}, \quad j = 1, \dots, q,$$

is normal and satisfies the complementing condition on Γ with respect to any differential expression L, which is properly elliptic on $\overline{\Omega}$ (see, e.g., [258, Sec. 5.2.1, Remark 4]). If $k = 0$, then we have the Dirichlet boundary-value problem for the equation $Lu = f$.

3.1.2 Formally adjoint problem

Along with problem (3.1), (3.2), we consider the boundary-value problem

$$L^+ v \equiv \sum_{|\mu| \leq 2q} D^\mu (\overline{l_\mu(x)}\, v) = \omega \ \text{ in } \Omega, \tag{3.4}$$

$$B_j^+ v = h_j \ \text{ on } \Gamma, \ j = 1, \dots, q. \tag{3.5}$$

It is formally adjoint to the problem (3.1), (3.2) with respect to Green's formula

$$(Lu, v)_\Omega + \sum_{j=1}^{q} (B_j u,\, C_j^+ v)_\Gamma = (u, L^+ v)_\Omega + \sum_{j=1}^{q} (C_j u,\, B_j^+ v)_\Gamma, \tag{3.6}$$

which is valid for all functions $u, v \in C^\infty(\overline{\Omega})$ (see, e.g., [121, Chap. 2, Sec. 2.2] or [258, Sec. 5.4.2]). Here, $\{B_j^+\}$, $\{C_j\}$, and $\{C_j^+\}$ are normal systems of boundary linear differential expressions with coefficients from $C^\infty(\Gamma)$. Their orders satisfy the condition

$$\operatorname{ord} B_j + \operatorname{ord} C_j^+ = \operatorname{ord} C_j + \operatorname{ord} B_j^+ = 2q - 1. \tag{3.7}$$

Additionally, we should note that here and hereinafter $(\cdot,\cdot)_\Omega$ and $(\cdot,\cdot)_\Gamma$ are used to indicate as inner products in the spaces $L_2(\Omega)$ and $L_2(\Gamma)$, respectively, as well as natural extensions of these inner products by continuity.

The differential expression L^+ is called formally adjoint to the expression L, and the system of boundary expressions $\{B_1^+, \dots, B_q^+\}$ is called adjoint to the system $\{B_1, \dots, B_q\}$ with respect to L. The adjoint system is not uniquely defined but all adjoint systems are equivalent in the following sense [227, Sec. 10-4]:

If we have a distinct system $\{\widetilde{B}_1^+, \dots, \widetilde{B}_q^+\}$, which is adjoint to $\{B_1, \dots, B_q\}$ with respect to L, then

$$\{v \in C^\infty(\overline{\Omega}) : \; B_j^+ v = 0 \;\text{ on }\; \Gamma \;\text{ for every }\; j = 1, \dots, q\}$$

$$= \{v \in C^\infty(\overline{\Omega}) : \; \widetilde{B}_j^+ v = 0 \;\text{ on }\; \Gamma \;\text{ for every }\; j = 1, \dots, q\}. \qquad (3.8)$$

It is known that a boundary-value problem is regular elliptic if and only if the problem, which is formally adjoint to it, is regular elliptic (see, e.g., [121, Chap. 2, Sec. 2.5] or [227, Sec. 10-3]).

We set

$$N := \{u \in C^\infty(\overline{\Omega}) : \; Lu = 0 \text{ in } \Omega, \;\; B_j u = 0 \text{ on } \Gamma, \;\; j = 1, \dots, q\},$$

$$N^+ := \{v \in C^\infty(\overline{\Omega}) : \; L^+ v = 0 \text{ in } \Omega, \;\; B_j^+ v = 0 \text{ on } \Gamma, \;\; j = 1, \dots, q\}.$$

Due to equality (3.8), the set N^+ does not depend on the choice of the adjoint system of boundary expressions $\{B_1^+, \dots, B_q^+\}$. Since problems (3.1), (3.2) and (3.4), (3.5) are regular elliptic, spaces N and N^+ are finite-dimensional (see, e.g., [121, Chap. 2, Sec. 5.4])

Example 3.2. Consider the Dirichlet problem for the differential equation $Lu = f$, where the expression L is properly elliptic on $\overline{\Omega}$. For this boundary-value problem, its adjoint system is the Dirichlet problem for the equation $L^+ v = \omega$. In this case, $\dim N = \dim N^+$ [121, Chap. 2, Sec. 2.5, 8.5].

In the present chapter, we study *semihomogeneous* regular elliptic boundary-value problems; i.e., for (3.1) and (3.2) we suppose that either $f = 0$ in the domain Ω or all $g_j = 0$ on the boundary Γ. These two important subclasses of problems are studied separately in Sections 3.3 and 3.4.

3.2 Hörmander spaces for Euclidean domains

In this section, we study the classes of Hörmander spaces separately for open and closed Euclidean domains. In the first case, the space consists of distributions *defined in* an open domain, whereas in the second case, it is formed by distributions in $\mathbb{R}^n$, which are *supported on* a closed domain [269, § 3, Subcec. 1 and 2]. For open and closed domains, we consider Ω and $\overline{\Omega}$, respectively. We use the indicated spaces in order to investigate the boundary-value problem (3.1), (3.2). We discover a relation between these spaces and the spaces over the boundary Γ. As usual, $\mathcal{D}'(\Omega)$ is a topological linear space of all distributions defined in the domain Ω.

3.2.1 Spaces for open domains

Let $s \in \mathbb{R}$ and $\varphi \in \mathcal{M}$.

Definition 3.2. The linear space $H^{s,\varphi}(\Omega)$ consists of all distributions $w \in H^{s,\varphi}(\mathbb{R}^n)$ restricted to the domain Ω. In the space $H^{s,\varphi}(\Omega)$, the norm of any distribution $u \in H^{s,\varphi}(\Omega)$ is given by the formula

$$\|u\|_{H^{s,\varphi}(\Omega)} := \inf\big\{\, \|w\|_{H^{s,\varphi}(\mathbb{R}^n)} : w \in H^{s,\varphi}(\mathbb{R}^n),\ \ w = u \ \text{ on } \ \Omega \,\big\}. \tag{3.9}$$

Theorem 3.1. *The space $H^{s,\varphi}(\Omega)$ is a separable Hilbert space with respect to norm* (3.9).

Proof. In accordance with the definition, $H^{s,\varphi}(\Omega)$ is the factor space of the Hilbert space $H^{s,\varphi}(\mathbb{R}^n)$ by its subspace

$$\big\{w \in H^{s,\varphi}(\mathbb{R}^n) : \operatorname{supp} w \subseteq \widehat{\Omega}\,\big\}. \tag{3.10}$$

(Linear manifold (3.10) is closed in the topology of the space $H^{s,\varphi}(\mathbb{R}^n)$ due to the continuous embedding $H^{s,\varphi}(\mathbb{R}^n) \hookrightarrow \mathcal{D}'(\mathbb{R}^n)$ (see Theorem 3.6 below)). Hence, $H^{s,\varphi}(\Omega)$ is a Hilbert space with respect to the inner product

$$(u_1, u_2)_{H^{s,\varphi}(\Omega)} := (w_1 - \Pi w_1, w_2 - \Pi w_2)_{H^{s,\varphi}(\mathbb{R}^n)}. \tag{3.11}$$

Here, $u_j \in H^{s,\varphi}(\Omega)$, $w_j \in H^{s,\varphi}(\mathbb{R}^n)$, $u_j = w_j$ in Ω for $j \in \{1,2\}$, and Π is the orthoprojector of $H^{s,\varphi}(\mathbb{R}^n)$ onto subspace (3.10). (The right-hand side of equality (3.11) is independent of the choice of w_1 and w_2.) The norm defined by formula (3.9) is generated by this inner product. Due to Definition 3.2, the separability property of the space $H^{s,\varphi}(\mathbb{R}^n)$ implies the separability property for $H^{s,\varphi}(\Omega)$.

Theorem 3.1 is proved.

In the particular case where $\varphi \equiv 1$, the space $H^{s,\varphi}(\Omega)$ is also denoted by $H^{s}(\Omega)$. This is the Sobolev space of order $s \in \mathbb{R}$ over the domain Ω [258, Sec. 4.2.].

The class of spaces

$$\{H^{s,\varphi}(\Omega) : s \in \mathbb{R}, \varphi \in \mathcal{M}\} \tag{3.12}$$

is called the refined Sobolev scale over the domain Ω or, shortly, the refined scale. Let us study the properties of this scale.

Theorem 3.2. *Let a function $\varphi \in \mathcal{M}$ and positive numbers ε and δ be given. Then, for any $s \in \mathbb{R}$,*

$$[H^{s-\varepsilon}(\Omega), H^{s+\delta}(\Omega)]_{\psi} = H^{s,\varphi}(\Omega) \tag{3.13}$$

with equivalence of norms. Here, ψ is the interpolation parameter from Theorem 1.14.

Proof. The pair of Sobolev spaces on the left-hand side of (3.13) is admissible. Consider the operator R_{Ω} of the restriction of distributions $u \in \mathcal{D}'(\mathbb{R}^n)$ to the domain Ω. We have the following bounded and surjective linear operators:

$$R_{\Omega} : H^{s-\varepsilon}(\mathbb{R}^n) \to H^{s-\varepsilon}(\Omega), \quad R_{\Omega} : H^{s+\delta}(\mathbb{R}^n) \to H^{s+\delta}(\Omega), \tag{3.14}$$

$$R_{\Omega} : H^{s,\varphi}(\mathbb{R}^n) \to H^{s,\varphi}(\Omega). \tag{3.15}$$

By considering the interpolation with parameter ψ applied to (3.14), we obtain the bounded operator

$$R_{\Omega} : [H^{s-\varepsilon}(\mathbb{R}^n), H^{s+\delta}(\mathbb{R}^n)]_{\psi} \to [H^{s-\varepsilon}(\Omega), H^{s+\delta}(\Omega)]_{\psi}.$$

In view of Theorem 1.14, this implies the boundedness of the operator

$$R_{\Omega} : H^{s,\varphi}(\mathbb{R}^n) \to [H^{s-\varepsilon}(\Omega), H^{s+\delta}(\Omega)]_{\psi}.$$

Since operator (3.15) is surjective, this yields the inclusion

$$H^{s,\varphi}(\Omega) \subseteq [H^{s-\varepsilon}(\Omega), H^{s+\delta}(\Omega)]_{\psi}. \tag{3.16}$$

Let us prove that the inverse inclusion is also true and continuous. It is shown in monograph [258, Theorem 4.2.2] that for each $k \in \mathbb{N}$ one can construct a linear mapping T_k, which extends each distribution $u \in H^{-k}(\Omega)$ onto the space $\mathbb{R}^n$, such that

$$T_k : H^{\sigma}(\Omega) \to H^{\sigma}(\mathbb{R}^n) \quad \text{for} \quad |\sigma| < k \tag{3.17}$$

is a bounded operator. Let us pick $k \in \mathbb{N}$ such that $|s - \varepsilon| < k$ and $|s + \delta| < k$ and consider bounded operators (3.17) for $\sigma = s - \varepsilon$ and $\sigma = s + \delta$. Since ψ is an interpolation parameter, we have the bounded operator

$$T_k : [H^{s-\varepsilon}(\Omega), H^{s+\delta}(\Omega)]_{\psi} \to [H^{s-\varepsilon}(\mathbb{R}^n), H^{s+\delta}(\mathbb{R}^n)]_{\psi}.$$

Hence, in view of Theorem 1.14, we obtain the bounded operator

$$T_k : [H^{s-\varepsilon}(\Omega), H^{s+\delta}(\Omega)]_{\psi} \to H^{s,\varphi}(\mathbb{R}^n). \tag{3.18}$$

The product of bounded operators (3.15) and (3.18) generates the bounded identity operator

$$I = R_{\Omega} T_k : [H^{s-\varepsilon}(\Omega), H^{s+\delta}(\Omega)]_{\psi} \to H^{s,\varphi}(\Omega).$$

Thus, along with inclusion (3.16), its inverse continuous embedding holds too. Hence, the equality of spaces (3.13) takes place. Moreover, by the Banach inverse operator theorem, norms in these spaces are equivalent.

Theorem 3.2 is proved.

Let us recall that

$$C_0^\infty(\Omega) := \{u \in C^\infty(\Omega) : \operatorname{supp} u \subset \Omega\}.$$

We identify functions $u \in C_0^\infty(\Omega)$ with their extensions by zero onto $\mathbb{R}^n$. In what follows, it should be clear from context in which one of the domains (Ω or $\mathbb{R}^n$) the function $u \in C_0^\infty(\Omega)$ is defined.

Theorem 3.3. *Let $s \in \mathbb{R}$ and $\varphi, \varphi_1 \in \mathcal{M}$. Then the following assertions hold:*

(i) *The set $C^\infty(\overline{\Omega})$ is dense in the space $H^{s,\varphi}(\Omega)$.*

(ii) *If $s < 1/2$, then the set $C_0^\infty(\Omega)$ is dense in the space $H^{s,\varphi}(\Omega)$.*

(iii) *For any number $\varepsilon > 0$, the dense compact embedding $H^{s+\varepsilon,\varphi_1}(\Omega) \hookrightarrow H^{s,\varphi}(\Omega)$ takes place.*

(iv) *The function φ/φ_1 is bounded in a neighborhood of ∞ if and only if $H^{s,\varphi_1}(\Omega) \hookrightarrow H^{s,\varphi}(\Omega)$. This embedding is dense and continuous. It is compact if and only if $\varphi(t)/\varphi_1(t) \to 0$ as $t \to \infty$.*

Proof. Assertion (i) follows immediately from the fact that the set $C_0^\infty(\mathbb{R}^n)$ is dense in the space $H^{s,\varphi}(\mathbb{R}^n)$ [269, Lemma 3.1].

Assertion (ii) holds in the Sobolev case where $\varphi \equiv 1$ (see, e.g., [258, Theorem 4.7.1 (d)]). From this result, the assertion can be proved for any $\varphi \in \mathcal{M}$ by using Theorems 3.2 and 1.1.

Let $s < 1/2$. Due to the above-mentioned theorems, the dense continuous embedding $H^{s+\delta}(\Omega) \hookrightarrow H^{s,\varphi}(\Omega)$ takes place for any $\delta > 0$. Moreover, if $s+\delta < 1/2$, then the set $C_0^\infty(\Omega)$ is dense in the Sobolev space $H^{s+\delta}(\Omega)$. Hence, this set is also dense in $H^{s,\varphi}(\Omega)$. Assertion (ii) is proved.

Assertions (iii) and (iv) are contained in Theorems 7.4 and 8.1 from [269]. The density of these embeddings is implied by assertion (i).

Theorem 3.3 is proved.

Theorem 3.4. *Let a function $\varphi \in \mathcal{M}$ and an integer $k \geq 0$ be given. Then condition (1.37) is equivalent to the embedding*

$$H^{k+n/2,\varphi}(\Omega) \hookrightarrow C^k(\overline{\Omega}). \tag{3.19}$$

This embedding is compact.

Proof. Assume that the function φ satisfies condition (1.37). Then, by Theorem 1.15(iii), the continuous embedding

$$H^{k+n/2,\varphi}(\mathbb{R}^n) \hookrightarrow C^k_{\mathrm{b}}(\mathbb{R}^n) \tag{3.20}$$

takes place. This yields continuous embedding (3.19). Indeed, for arbitrary functions $u \in H^{k+n/2,\varphi}(\Omega)$ and $w \in H^{k+n/2,\varphi}(\mathbb{R}^n)$ such that $u = w$ in Ω, we have $w \in C^k_{\mathrm{b}}(\mathbb{R}^n)$, $u \in C^k(\overline{\Omega})$, and

$$\|u\|_{C^k(\overline{\Omega})} \leq \|w\|_{C^k_{\mathrm{b}}(\mathbb{R}^n)} \leq c\,\|w\|_{H^{k+n/2,\varphi}(\mathbb{R}^n)},$$

where c is the norm of the embedding operator (3.20). Passing to the infimum over the functions w in this inequality, we obtain the estimate

$$\|u\|_{C^k(\overline{\Omega})} \leq c\,\|u\|_{H^{k+n/2,\varphi}(\Omega)}.$$

The continuity of embedding (3.19) is proved. Its compactness can be proved similarly to the proof of compactness for embedding (2.43) in Theorem 2.8. It remains to note that inclusion (2.43) leads to property (1.37) due to Proposition 1.5 (for $p = q = 2$, $V := \Omega$) and equivalence (1.41).

Theorem 3.4 is proved.

Remark 3.3. For $k = 0$, the statement of the theorem can be found in Theorem 7.5 in the paper [269] by L. R. Volevich and B. P. Paneah.

Let us consider the problem of existence of traces on the boundary Γ for arbitrary distributions $u \in H^{s,\varphi}(\Omega)$ and properties of these traces. Since Γ is a closed compact infinitely smooth manifold of dimension $n-1$, the refined Sobolev scale over Γ is defined.

Theorem 3.5. *For arbitrary parameters $s > 1/2$ and $\varphi \in \mathcal{M}$, the linear mapping*

$$u \mapsto u \restriction \Gamma, \quad u \in C^\infty(\overline{\Omega}), \tag{3.21}$$

($u \restriction \Gamma$ is the trace of a function u on Γ) can be uniquely extended (by continuity) to the bounded surjective operator

$$R_\Gamma : H^{s,\varphi}(\Omega) \to H^{s-1/2,\varphi}(\Gamma). \tag{3.22}$$

This operator has a right inverse operator

$$S_\Gamma : H^{s-1/2,\varphi}(\Gamma) \to H^{s,\varphi}(\Omega), \tag{3.23}$$

which is linear and bounded and such that the mapping S_Γ does not depend on s and φ.

Proof. In the case where $\varphi \equiv 1$ (Sobolev spaces), this theorem is known (see, e.g., the monograph by H. Triebel [258, Lemma 5.4.4]). By using an interpolation, we extend this case onto the general case of any $\varphi \in \mathcal{M}$.

Let us choose a number $\varepsilon > 0$ such that $s - \varepsilon > 1/2$. We have the following bounded linear operators:

$$R_\Gamma : H^{s\mp\varepsilon}(\Omega) \to H^{s\mp\varepsilon-1/2,\varphi}(\Gamma),$$

$$S_\Gamma : H^{s\mp\varepsilon-1/2,\varphi}(\Gamma) \to H^{s\mp\varepsilon,\varphi}(\Omega).$$

Applying the interpolation with the parameter ψ from Theorem 1.14, where we set $\delta := \varepsilon$, in view of Theorems 2.2 and 3.2 we obtain bounded operators (3.22) and (3.23). Since $R_\Gamma S_\Gamma h = h$ for any $h \in H^{s-\varepsilon-1/2}(\Gamma)$, operator (3.23) is a right inverse operator to operator (3.22), and the latter one is surjective.

Theorem 3.5 is proved.

Remark 3.4. Theorems on traces for hyperplanes and flat pieces of a boundary were proved for general Hörmander spaces in [81, Theorem 2.2.8] and [269, Theorems 6.1, 6.2, and 7.6].

Corollary 3.1. *Let* $\sigma > 0$ *and* $\varphi \in \mathcal{M}$. *Then*

$$H^{\sigma,\varphi}(\Gamma) = \{R_\Gamma f : f \in H^{\sigma+1/2,\varphi}(\Omega)\}, \tag{3.24}$$

$$\|h\|_{H^{\sigma,\varphi}(\Gamma)} \asymp \inf\big\{ \|f\|_{H^{\sigma+1/2,\varphi}(\Omega)} : f \in H^{\sigma+1/2,\varphi}(\Omega),\ R_\Gamma f = h \big\}. \tag{3.25}$$

Proof. We set $s := \sigma + 1/2$ in Theorem 3.5. Equality (3.24) holds because operator (3.22) is surjective. Let us prove the equivalence of norms (3.25). For arbitrary functions $h \in H^{\sigma,\varphi}(\Gamma)$ and $f \in H^{\sigma+1/2,\varphi}(\Omega)$ that satisfy the equality $R_\Gamma f = h$, we have

$$\|h\|_{H^{\sigma,\varphi}(\Gamma)} \leq c_1 \, \|f\|_{H^{\sigma+1/2,\varphi}(\Omega)},$$

where c_1 is the norm of operator (3.22). Passing to the infimum over the functions f in this inequality, we obtain the following estimate:

$$\|h\|_{H^{\sigma,\varphi}(\Gamma)} \leq c_1 \inf\big\{\|f\|_{H^{\sigma+1/2,\varphi}(\Omega)} : f \in H^{\sigma+1/2,\varphi}(\Omega),\ R_\Gamma f = h\big\}.$$

The inverse estimate is implied by the fact that

$$\|f\|_{H^{\sigma+1/2,\varphi}(\Omega)} \leq c_2 \, \|h\|_{H^{\sigma,\varphi}(\Gamma)}$$

for the function $f := S_\Gamma h$, where c_2 is the norm of operator (3.23).

Corollary 3.1 is proved.

Remark 3.5. If $s < 1/2$, then mapping (3.21) cannot be extended to a continuous operator $R_\Gamma : H^{s,\varphi}(\Omega) \to \mathcal{D}'(\Gamma)$. Indeed, if it were possible, then by Theorem 3.3 (ii) we would have $R_\Gamma f = 0$ for any distribution $f \in H^{s,\varphi}(\Omega)$. However, it is not true, e.g., for the function $f := 1$ on $\overline{\Omega}$. Since the set $C_0^\infty(\Omega)$ is dense in $H^{1/2}(\Omega)$, this remark remains true for Sobolev spaces in the limit case $s = 1/2$ [258, Theorem 4.7.1 (d)].

3.2.2 Spaces for closed domains

Let $s \in \mathbb{R}$ and $\varphi \in \mathcal{M}$. Also let Q be an arbitrary nonempty closed subset of the space $\mathbb{R}^n$.

Definition 3.3. The linear space $H_Q^{s,\varphi}(\mathbb{R}^n)$ consists of all distributions $w \in H^{s,\varphi}(\mathbb{R}^n)$ such that their supports $\operatorname{supp} w \subseteq Q$. The space $H_Q^{s,\varphi}(\mathbb{R}^n)$ is endowed with the inner product and norm from the space $H^{s,\varphi}(\mathbb{R}^n)$.

Theorem 3.6. *The space $H_Q^{s,\varphi}(\mathbb{R}^n)$ is a separable Hilbert space.*

Proof. Let a sequence (w_j) be fundamental in $H_Q^{s,\varphi}(\mathbb{R}^n)$. Since the space $H^{s,\varphi}(\mathbb{R}^n)$ is complete, the sequence has a limit w in this space. In view of continuity of the embedding $H^{s,\varphi}(\mathbb{R}^n) \hookrightarrow \mathcal{D}'(\mathbb{R}^n)$, we can conclude that $w_j \to w$ in $\mathcal{D}'(\mathbb{R}^n)$ as $j \to \infty$. This result along with inclusions $\operatorname{supp} w_j \subseteq Q$ yield $\operatorname{supp} w \subseteq Q$. Thus, the sequence (w_j) has a limit w in the space $H_Q^{s,\varphi}(\mathbb{R}^n)$. This proves the completeness of the space. The space is separable because it is a subspace of the separable space $H^{s,\varphi}(\mathbb{R}^n)$.

Theorem 3.6 is proved.

In the Sobolev case ($\varphi \equiv 1$), we omit the index φ in the notation of the space $H_Q^{s,\varphi}(\mathbb{R}^n)$ and other spaces considered in this chapter.

It is important for us to consider the case where $Q := \overline{\Omega}$. We are going to study the properties of the space $H_{\overline{\Omega}}^{s,\varphi}(\mathbb{R}^n)$ and its relation to the refined Sobolev scale over Ω.

Theorem 3.7. *Let a function $\varphi \in \mathcal{M}$ and positive numbers ε and δ be given. Then, for any $s \in \mathbb{R}$,*

$$[H_{\overline{\Omega}}^{s-\varepsilon}(\mathbb{R}^n), H_{\overline{\Omega}}^{s+\delta}(\mathbb{R}^n)]_\psi = H_{\overline{\Omega}}^{s,\varphi}(\mathbb{R}^n) \tag{3.26}$$

with equivalence of norms. Here, ψ is the interpolation parameter from Theorem 1.14.

Proof. Since Ω is a bounded domain with infinitely smooth boundary, the set $C_0^\infty(\Omega)$ is dense in the space $H_{\overline{\Omega}}^{\sigma}(\mathbb{R}^n)$ for any $\sigma \in \mathbb{R}$ (see, e.g., [258, Theorem 4.3.2/1 (b)]). Hence, the continuous embedding $H_{\overline{\Omega}}^{s+\delta}(\mathbb{R}^n) \hookrightarrow H_{\overline{\Omega}}^{s-\varepsilon}(\mathbb{R}^n)$

is dense, the pair of spaces $H^{s-\varepsilon}_{\overline{\Omega}}(\mathbb{R}^n)$ and $H^{s+\delta}_{\overline{\Omega}}(\mathbb{R}^n)$ is admissible, and the left-hand side of formula (3.26) is defined.

We will deduce this formula from Theorem 1.14 by using Theorem 1.6 (interpolation of subspaces). To this end, we have to construct a mapping which is a projector for any space $H^{\sigma}(\mathbb{R}^n)$ with $s-\varepsilon \le \sigma \le s+\delta$ onto the subspace $H^{\sigma}_{\overline{\Omega}}(\mathbb{R}^n)$. We construct this mapping as follows. Let us pick a number $r > 0$ such that $|x| < r$ for all $x \in \Omega$ and set

$$G = \{x \in \mathbb{R}^n : |x| < 4r\} \setminus \overline{\Omega}.$$

Let R denote the mapping that corresponds each distribution in $\mathbb{R}^n$ to its restriction in the domain G. In such a way, we obtain the linear bounded operator

$$R\colon H^{\sigma}(\mathbb{R}^n) \to H^{\sigma}(G) \quad \text{for every} \quad \sigma \in \mathbb{R}. \tag{3.27}$$

Note that G is a bounded open domain with infinitely smooth boundary. Then (see, e.g., [258, Theorem 4.2.2]), for any compact set $K \subset \mathbb{R}$, there exists a linear mapping T, which is a bounded operator

$$T: H^{\sigma}(G) \to H^{\sigma}(\mathbb{R}^n) \text{ for every } \sigma \in K \tag{3.28}$$

that extends the distribution from the domain G onto $\mathbb{R}^n$. This means that operator (3.28) is right inverse to operator (3.27). We take $K = [s-\varepsilon, s+\delta]$ and consider the mapping

$$P_0 : u \mapsto u - TRu, \quad u \in H^{s-\varepsilon}(\mathbb{R}^n).$$

Due to the boundedness of operators (3.27) and (3.28), we have the bounded linear operator

$$P_0 : H^{\sigma}(\mathbb{R}^n) \to H^{\sigma}(\mathbb{R}^n) \quad \text{for any} \ \sigma \in [s-\varepsilon, s+\delta].$$

It projects the space $H^{\sigma}(\mathbb{R}^n)$ onto the subspace $H^{\sigma}_{\overline{G}}(\mathbb{R}^n)$ for any $\sigma \in [s-\varepsilon, s+\delta]$. Indeed,

$$u \in H^{\sigma}(\mathbb{R}^n) \Rightarrow RP_0 u = R(u - TRu) = Ru - RTRu = Ru - Ru = 0$$
$$\Rightarrow P_0 u \in H^{\sigma}_{\overline{G}}(\mathbb{R}^n).$$

In addition,

$$u \in H^{\sigma}_{\overline{G}}(\mathbb{R}^n) \Rightarrow Ru = 0 \Rightarrow P_0 u = u - TRu = u.$$

Now we choose a function $\chi \in C^{\infty}(\mathbb{R}^n)$ such that $\chi(x) = 1$ for $|x| \le 2r$ and $\chi(x) = 0$ for $|x| \ge 3r$ and consider the mapping

$$P : u \mapsto \chi \cdot P_0 u, \quad u \in H^{s-\varepsilon}(\mathbb{R}^n).$$

Since the operator of multiplication by the function χ is bounded in any Sobolev space over $\mathbb{R}^n$, we conclude that P is a projector of the space $H^\sigma(\mathbb{R}^n)$ onto the subspace $H^\sigma_{\overline{\Omega}}(\mathbb{R}^n)$ for any $\sigma \in [s-\varepsilon,\, s+\delta]$. Hence, in view of Theorems 1.6 and 1.14, we have the following equalities of the spaces up to equivalence of their norms:

$$\begin{aligned}[H^{s-\varepsilon}_{\overline{\Omega}}(\mathbb{R}^n), H^{s+\delta}_{\overline{\Omega}}(\mathbb{R}^n)]_\psi &= [H^{s-\varepsilon}(\mathbb{R}^n), H^{s+\delta}(\mathbb{R}^n)]_\psi \cap H^{s-\varepsilon}_{\overline{\Omega}}(\mathbb{R}^n)\\ &= H^{s,\varphi}(\mathbb{R}^n) \cap H^{s-\varepsilon}_{\overline{\Omega}}(\mathbb{R}^n)\\ &= H^{s,\varphi}_{\overline{\Omega}}(\mathbb{R}^n).\end{aligned}$$

Theorem 3.7 is proved.

Theorem 3.8. *Let $s \in \mathbb{R}$ and $\varphi \in \mathcal{M}$. Then the following assertions hold:*

(i) *The set $C^\infty_0(\Omega)$ is dense in $H^{s,\varphi}_{\overline{\Omega}}(\mathbb{R}^n)$.*

(ii) *If $|s| < 1/2$, then spaces $H^{s,\varphi}(\Omega)$ and $H^{s,\varphi}_{\overline{\Omega}}(\mathbb{R}^n)$ are equal to each other as completions of $C^\infty_0(\Omega)$ with respect to equivalent norms.*

(iii) *Spaces $H^{s,\varphi}_{\overline{\Omega}}(\mathbb{R}^n)$ and $H^{-s,1/\varphi}(\Omega)$ are mutually dual with respect to the extension of the inner product in $L_2(\Omega)$ by continuity.*

(iv) *Assertions* (iii) *and* (iv) *of Theorem* 3.3 *remain true if one replace spaces $H^{\cdot,\cdot}(\Omega)$ by spaces $H^{\cdot,\cdot}_{\overline{\Omega}}(\mathbb{R}^n)$ with the same upper indices.*

Proof. Assertions (i), (iii), and (iv) are contained, respectively, in Lemma 3.3, Section 3.4, and Theorems 7.1 and 8.1 in [269].

Let us prove assertion (ii). Assume that $|s| < 1/2$. Assertion (ii) for the Sobolev case ($\varphi \equiv 1$) was proved, e.g., in [258, Theorem 4.3.2/1 (a) and (c)]. From this result, we can prove the assertion for any $\varphi \in \mathcal{M}$ with the help of interpolation. Let us pick a number $\varepsilon > 0$ such that $|s \mp \varepsilon| < 1/2$. The identity mapping on the set $C^\infty_0(\Omega)$ can be extended by continuity to isomorphisms

$$I : H^{s\mp\varepsilon}_{\overline{\Omega}}(\mathbb{R}^n) \leftrightarrow H^{s\mp\varepsilon}(\Omega).$$

By applying the interpolation with the parameter ψ from Theorem 1.14 under the choice of $\delta := \varepsilon$, we obtain, in view of Theorems 3.2 and 3.7, an extra isomorphism

$$I : H^{s,\varphi}_{\overline{\Omega}}(\mathbb{R}^n) \leftrightarrow H^{s,\varphi}(\Omega).$$

This proves assertion (ii).

Theorem 3.8 is proved.

3.2.3 Rigging of $L_2(\Omega)$ with Hörmander spaces

In the study of the elliptic problem with homogeneous boundary conditions, the following scale of Hörmander spaces proves to be useful.

Definition 3.4. Let $s \in \mathbb{R}$ and $\varphi \in \mathcal{M}$. If $s \geq 0$, then $H^{s,\varphi,(0)}(\Omega)$ denotes the Hilbert space $H^{s,\varphi}(\Omega)$. If $s < 0$, then $H^{s,\varphi,(0)}(\Omega)$ denotes the Hilbert space $H^{s,\varphi}_{\overline{\Omega}}(\mathbb{R}^n)$, which is dual to the space $H^{-s,1/\varphi}(\Omega)$ with respect to the extension of the inner product in $L_2(\Omega)$ by continuity (due to Theorem 3.8(iii)).

The scale of spaces

$$\{H^{s,\varphi,(0)}(\Omega) : s \in \mathbb{R},\ \varphi \in \mathcal{M}\} \tag{3.29}$$

is two-sided in the parameter s and is refined in the parameter φ. For $s \geq 0$, the positive part of the scale consists of spaces $H^{s,\varphi,(0)}(\Omega) = H^{s,\varphi}(\Omega)$ of distributions defined in the domain Ω. For $s < 0$, the negative part of the scale consists of spaces $H^{s,\varphi,(0)}(\Omega) = H^{s,\varphi}_{\overline{\Omega}}(\mathbb{R}^n)$ of distributions supported on the closure of the domain Ω. Thus, scale (3.29) is formed by the spaces of distributions of various nature.

We identify (and this is natural) the functions from the space $L_2(\Omega) = H^0(\Omega)$ with their extensions by zero over $\mathbb{R}^n$. In this sense, $L_2(\Omega) = H^0_{\overline{\Omega}}(\mathbb{R}^n)$. In view of Theorems 3.3(i), (iii), and 3.8(i), we have dense continuous embeddings

$$H^{s,\varphi,(0)}(\Omega) \hookleftarrow L_2(\Omega) \hookleftarrow H^{-s,1/\varphi,(0)}(\Omega) \quad \text{for all} \quad s < 0,\ \ \varphi \in \mathcal{M}. \tag{3.30}$$

Here, the flanked spaces are mutually dual relative to the extension of the inner product in $L_2(\Omega)$ by continuity. Thus, we obtain a *rigging* of the Hilbert space $L_2(\Omega)$ with spaces of scale (3.29). (For the definition of Hilbert rigging and related concepts, see [21, Chap. 1, Sec. 1.1] and [23, Chap. 14, Sec. 1.1].)

In the Sobolev case where $\varphi \equiv 1$, rigging (3.30) was introduced and studied by Yu. M. Berezansky (see [21, Chap. 1, Sec. 3] or [23, Chap. 14, Sec. 3 and 4]). In this case, we denote the space $H^{s,\varphi,(0)}(\Omega)$ also by $H^{s,(0)}(\Omega)$.

The fact that continuous embeddings (3.30) are dense implies that functions of the class $C^\infty(\overline{\Omega})$ (extended by zero over $\mathbb{R}^n$) form a dense subset in any negative space $H^{s,\varphi,(0)}(\Omega)$, $s < 0$. They are also dense in any positive space $H^{s,\varphi,(0)}(\Omega)$, $s > 0$. This allows us to consider a continuous dense embedding of the form $H^{r,\chi,(0)}(\Omega) \hookrightarrow H^{s,\varphi,(0)}(\Omega)$, where $s, r \in \mathbb{R}$ and $\varphi, \chi \in \mathcal{M}$. Naturally, this means that

$$\|u\|_{H^{s,\varphi,(0)}(\Omega)} \leq \text{const}\, \|u\|_{H^{r,\chi,(0)}(\Omega)} \quad \text{for any} \quad u \in C^\infty(\overline{\Omega}),$$

and moreover, the identity mapping on the set $C^\infty(\overline{\Omega})$ can be extended by continuity to a bounded *injective* operator $I : H^{r,\chi,(0)}(\Omega) \to H^{s,\varphi,(0)}(\Omega)$.

It is called the operator of embedding of the space $H^{r,\chi,(0)}(\Omega)$ into the space $H^{s,\varphi,(0)}(\Omega)$ (on this subject, see [23, Chap. 15, Sec. 7]).

Let us study the properties of scale (3.29).

Theorem 3.9. *Let $s \in \mathbb{R}$ and $\varphi, \varphi_1 \in \mathcal{M}$. Then the following assertions hold:*

(i) *Up to equivalence of norms,*

$$H^{s,\varphi,(0)}(\Omega) = H^{s,\varphi}_{\overline{\Omega}}(\mathbb{R}^n) \quad \text{for } s < 1/2 \tag{3.31}$$

$$H^{s,\varphi,(0)}(\Omega) = H^{s,\varphi}(\Omega) \quad \text{for } s > -1/2. \tag{3.32}$$

(ii) *If $s < 1/2$, then the set $C_0^\infty(\Omega)$ is dense in $H^{s,\varphi,(0)}(\Omega)$.*

(iii) *Spaces $H^{s,\varphi,(0)}(\Omega)$ and $H^{-s,1/\varphi,(0)}(\Omega)$ are mutually dual (with equal norms for $s \neq 0$ and equivalent norms for $s = 0$) relative to the extension of the inner product in $L_2(\Omega)$ by continuity.*

(iv) *For any $\varepsilon > 0$, the compact and dense embedding*

$$H^{s+\varepsilon,\varphi_1,(0)}(\Omega) \hookrightarrow H^{s,\varphi,(0)}(\Omega)$$

takes place.

Proof. (i) Equalities (3.31) and (3.32) are implied immediately by Theorem 3.8(ii) and the definition of the space $H^{s,\varphi,(0)}(\Omega)$.

(ii) This assertion is a consequence of equality (3.31) and Theorem 3.8(i).

(iii) For $s \neq 0$, assertion (iii) is justified in Definition 3.4. For $s = 0$, it follows from Theorem 3.8(iii) and formula (3.31). Namely,

$$(H^{0,\varphi,(0)}(\Omega))' = (H^{0,\varphi}(\Omega))' = H^{0,1/\varphi}_{\overline{\Omega}}(\mathbb{R}^n) = H^{0,1/\varphi,(0)}(\Omega)$$

(the latter equality holds up to equivalence of norms).

(iv) For $s \geq 0$, assertion (iv) coincides with Theorem 3.3(iii). In the case where $s+\varepsilon \leq 0$, it is contained in Theorem 3.8(iv); and if $s+\varepsilon = 0$ in this case, then we use (3.31). In the remaining case where $s < 0 < s + \varepsilon$, the required result is implied by the following dense compact embeddings

$$H^{s+\varepsilon,\varphi_1,(0)}(\Omega) \hookrightarrow H^{0,\varphi_1,(0)}(\Omega) \hookrightarrow H^{s,\varphi,(0)}(\Omega).$$

Theorem 3.9 is proved.

Theorem 3.10. *Let a function $\varphi \in \mathcal{M}$ and positive numbers ε, δ be given. Then, for any $s \in \mathbb{R}$,*

$$[H^{s-\varepsilon,(0)}(\Omega), H^{s+\delta,(0)}(\Omega)]_\psi = H^{s,\varphi,(0)}(\Omega) \tag{3.33}$$

with equivalence of norms. Here, ψ is the interpolation parameter from Theorem 1.14.

Proof. In the case where $s-\varepsilon \geq 0$ or $s+\delta \leq 0$, this theorem follows immediately from interpolation theorems 3.2 and 3.7. Consider the remaining case where $s-\varepsilon < 0 < s+\delta$. We set $\lambda := \min\{\varepsilon/2, \delta/2, 1/4\}$. Since we have either $s \mp \lambda < 1/2$ or $s \mp \lambda > -1/2$, we conclude that

$$[H^{s-\lambda,(0)}(\Omega), H^{s+\lambda,(0)}(\Omega)]_\eta = H^{s,\varphi,(0)}(\Omega) \tag{3.34}$$

in view of Theorem 3.9(i) and the mentioned interpolation theorems. Here, the interpolation parameter η is given by the formula

$$\eta(t) := \begin{cases} t^{1/2}\,\varphi(t^{1/(2\lambda)}) & \text{for} \quad t \geq 1, \\ \varphi(1) & \text{for} \quad 0 < t < 1. \end{cases}$$

Additionally, since $s-\varepsilon < s \mp \lambda < s+\delta$, we can use the result [121, Chap. 1, Theorem 12.5] by J.-L. Lions and E. Magenes in order to prove that

$$\begin{aligned} \big[H^{s-\varepsilon,(0)}(\Omega), H^{s+\delta,(0)}(\Omega)\big]_{\psi_\mp} &= \big[(H^{-s+\varepsilon}(\Omega))', H^{s+\delta}(\Omega)\big]_{\psi_\mp} \\ &= H^{s\mp\lambda,(0)}(\Omega). \end{aligned} \tag{3.35}$$

Here, $\psi_\mp(t) := t^{\theta_\mp}$ and the number $\theta_\mp \in (0,\,1)$ is determined by the condition $s \mp \lambda = (1-\theta_\mp)(s-\varepsilon) + \theta_\mp(s+\delta)$. This implies the equality $\theta_\mp = (\varepsilon \mp \lambda)/(\varepsilon+\delta)$.

In view of the reiteration theorem 1.3, equalities (3.34) and (3.35) (with equivalence of norms) yield the equality

$$\begin{aligned} H^{s,\varphi,(0)}(\Omega) &= [H^{s-\lambda,(0)}(\Omega), H^{s+\lambda,(0)}(\Omega)]_\eta \\ &= \big[[H^{s-\varepsilon,(0)}(\Omega), H^{s+\delta,(0)}(\Omega)]_{\psi_-}, [H^{s-\varepsilon,(0)}(\Omega), H^{s+\delta,(0)}(\Omega)]_{\psi_+}\big]_\eta \\ &= [H^{s-\varepsilon,(0)}(\Omega), H^{s+\delta,(0)}(\Omega)]_\omega. \end{aligned}$$

Here,

$$\begin{aligned} \omega(t) &:= \psi_-(t)\,\eta\Big(\frac{\psi_+(t)}{\psi_-(t)}\Big) \\ &= t^{\theta_-}\,\eta(t^{\theta_+-\theta_-}) = t^{(\varepsilon-\lambda)/(\varepsilon+\delta)}\,\eta(t^{2\lambda/(\varepsilon+\delta)}) \\ &= t^{(\varepsilon-\lambda)/(\varepsilon+\delta)}\,t^{\lambda/(\varepsilon+\delta)}\,\varphi(t^{1/(\varepsilon+\delta)}) \\ &= \psi(t) \quad \text{for} \quad t \geq 1. \end{aligned}$$

Hence (see Remark 1.1), equality (3.34) holds up to equivalence of norms.

Theorem 3.10 is proved.

3.3 Boundary-value problems for homogeneous elliptic equations

Consider the regular elliptic boundary-value problem (3.1), (3.2) in the case where the elliptic equation (3.1) is homogeneous:

$$L\,u = 0 \text{ in } \Omega, \quad B_j\,u = g_j \text{ on } \Gamma \text{ for } j = 1, \dots, q. \tag{3.36}$$

Let us study the operator $B = (B_1, \dots, B_q)$, which corresponds to this problem, on the refined Sobolev scale.

3.3.1 Main result: boundedness and Fredholm property of the operator

We set

$$K_L^\infty(\Omega) := \{\, u \in C^\infty(\overline{\Omega}) : \; L\,u = 0 \quad \text{in} \quad \Omega \,\} \tag{3.37}$$

and associate the linear mapping

$$u \mapsto Bu = (B_1\,u, \dots, B_q\,u), \qquad u \in K_L^\infty(\Omega), \tag{3.38}$$

with problem (3.36).

Let $s \in \mathbb{R}$ and $\varphi \in \mathcal{M}$. We denote

$$K_L^{s,\varphi}(\Omega) := \{\, u \in H^{s,\varphi}(\Omega) : \; L\,u = 0 \text{ in } \Omega \,\}. \tag{3.39}$$

Since the embedding $H^{s,\varphi}(\Omega) \hookrightarrow \mathcal{D}'(\Omega)$ is continuous, $K_L^{s,\varphi}(\Omega)$ is a closed subspace of $H^{s,\varphi}(\Omega)$.

Indeed, let $u \in H^{s,\varphi}(\Omega)$, and let a sequence $(u_j) \subset K_L^{s,\varphi}(\Omega)$ be such that $u_j \to u$ in $H^{s,\varphi}(\Omega)$ as $j \to \infty$. Then $u_j \to u$ in $\mathcal{D}'(\Omega)$, and this yields $0 = Lu_j \to Lu$ in $\mathcal{D}'(\Omega)$ as $j \to \infty$. Hence, $Lu = 0$ in the domain Ω, i.e., $u \in K_L^{s,\varphi}(\Omega)$.

We treat $K_L^{s,\varphi}(\Omega)$ as a Hilbert space with respect to the inner product in $H^{s,\varphi}(\Omega)$.

Let us formulate the main result of Section 3.3.

Theorem 3.11. *For arbitrary parameters $s \in \mathbb{R}$ and $\varphi \in \mathcal{M}$, the set $K_L^\infty(\Omega)$ is dense in $K_L^{s,\varphi}(\Omega)$ and the mapping (3.38) can be uniquely extended (by continuity) to the bounded linear operator*

$$B : K_L^{s,\varphi}(\Omega) \to \bigoplus_{j=1}^{q} H^{s-m_j-1/2,\,\varphi}(\Gamma) =: \mathcal{H}_{s,\varphi}(\Gamma). \tag{3.40}$$

It is a Fredholm operator with the kernel N and the domain

$$\Big\{ (g_1, \dots, g_q) \in \mathcal{H}_{s,\varphi}(\Gamma) : \sum_{j=1}^{q} (g_j, C_j^+\,v)_\Gamma = 0 \;\; \textit{for all} \;\; v \in N^+ \Big\}. \tag{3.41}$$

The index of operator (3.40) does not depend on s and φ.

Note that the term $(g_j, C_j^+ v)_\Gamma$ in formula (3.41) is the value of the antilinear functional g_j at the function $C_j^+ v \in C^\infty(\overline{\Omega})$. This implies that set (3.41) is closed in the space $\mathcal{H}_{s,\varphi}(\Gamma)$. Then, in accordance with Theorem 3.11, the set

$$G := \{ (C_1^+ v, \dots, C_q^+ v) : v \in N^+ \} \tag{3.42}$$

is a defect subspace of operator (3.40): this set is orthogonal to the domain of the operator with respect to the extension of the inner product in $(L_2(\Gamma))^q$ by continuity. The index of operator (3.40) is equal to $\dim N - \dim G$. It is clear that $\dim G \le \dim N^+$, where the strict inequality is also possible. This is implied by [85, Theorem 13.6.15].

In the case where $\varphi \equiv 1$ and $s \notin \{1/2 - k : k \in \mathbb{N}\}$, the statement of Theorem 3.11 is contained in the Lions–Magenes theorem [121, Chap. 2, Sec. 7.3] on solvability of the inhomogeneous problem (3.1), (3.2) in the two-sided scale of Sobolev spaces. The general case, where $\varphi \in \mathcal{M}$ and $s \in \mathbb{R}$, can be deduced from the Lions–Magenes theorem by using the interpolation with a suitable functional parameter and subsequent restriction of the operator of the problem to the space $K_L^{s,\varphi}(\Omega)$. The required steps are represented in Subsection 3.3.4. In Subsections 3.3.2 and 3.3.3, we present necessary results from the theory of interpolations and elliptic boundary-value problems.

Note that Theorem 3.11 provides new results even in the Sobolev case $\varphi \equiv 1$ if the number $s < 0$ is half-integer (see Remark 3.6 in Subsection 3.3.3). This theorem can be regarded as an analog of the Harnack theorem on convergence of sequences of harmonic functions (see, e.g., [161, Cap. 11, Sec. 9]). In this case, we have to use the metric in $H^{s,\varphi}(\Omega)$ instead of the uniform metric.

With respect to Theorem 3.11, we also mention the work [233] by Seeley on Cauchy data for solutions of the homogeneous elliptic equation in the two-sided scale of Sobolev spaces (additionally, see [11, Subsec. 5.4 b]).

3.3.2 A theorem on interpolation of subspaces

In this subsection, we formulate and prove an assertion (which is somewhat awkward) that concerns the interpolation of subspaces related to a linear operator. This assertion plays an important role in the proof of the main result. In the case of (complex) holomorphic interpolations, the assertion was proved by J.-L. Lions and E. Magenes [121, Theorem 14.3]. We show that this result is also true for interpolations of Hilbert spaces with a functional parameter. In this case, unlike the cited monograph, our proof does not use the interpolation functor construction.

First of all, note that we accept the following notation. Let H, Φ, and Ψ be Hilbert spaces and $\Phi \hookrightarrow \Psi$ be continuous. Also let a linear bounded operator $T : H \to \Psi$ be given. We denote

$$(H)_{T,\Phi} = \{u \in H : Tu \in \Phi\}.$$

The space $(H)_{T,\Phi}$ is a Hilbert space with respect to the graphic inner product

$$(u_1,u_2)_{(H)_{T,\Phi}} = (u_1,u_2)_H + (Tu_1,Tu_2)_\Phi$$

and it does not depend on Ψ.

Theorem 3.12. *Let six Hilbert spaces X_0, Y_0, Z_0, X_1, Y_1, and Z_1 and three linear mappings T, R, and S be given and the following conditions be satisfied:*

(i) *Pairs $X=[X_0,X_1]$ and $Y=[Y_0,Y_1]$ are admissible;*

(ii) *Z_0 and Z_1 are subspaces of a linear space E;*

(iii) *For any $j\in\{0,1\}$, the continuous embedding $Y_j\hookrightarrow Z_j$ takes place;*

(iv) *The mapping T is defined on X_0 and, for any $j\in\{0,1\}$, specifies the bounded operator $T: X_j\to Z_j$;*

(v) *The mapping R is defined on E and, for any $j\in\{0,1\}$, specifies the bounded operator $R: Z_j\to X_j$ for each $j\in\{0,1\}$;*

(vi) *The mapping S is defined on E and, for any $j\in\{0,1\}$, specifies the bounded operator $S: Z_j\to Y_j$ for each $j\in\{0,1\}$;*

(vii) *For any $\omega\in E$, the equality $TR\omega=\omega+S\omega$ is valid.*

Then the pair of spaces $[\,(X_0)_{T,Y_0},\,(X_1)_{T,Y_1}\,]$ is admissible, and, for any interpolation parameter $\psi\in\mathcal{B}$, the equality of spaces

$$[\,(X_0)_{T,Y_0},\,(X_1)_{T,Y_1}\,]_\psi = (X_\psi)_{T,Y_\psi}. \tag{3.43}$$

takes place up to equivalence of norms.

Proof. In view of conditions (iii) and (iv), spaces $(X_j)_{T,Y_j}$, $j\in\{0,1\}$, are well defined. We prove that the space on the right-hand side of equality (3.43) is well defined too. Due to condition (i), spaces X_ψ and Y_ψ are defined; for these spaces, continuous embeddings $X_\psi\hookrightarrow X_0$ and $Y_\psi\hookrightarrow Y_0$ take place. The first embedding and condition (iv) for $j=0$ yield the boundedness of the operator $T: X_\psi\to Z_0$. In addition, the second embedding and condition (iii) yield the continuity of the embedding $Y_\psi\hookrightarrow Z_0$. Thus, the space on the right-hand side of equality (3.43) is well defined and it is a Hilbert space likewise spaces $(X_j)_{T,Y_j}$ for $j\in\{0,1\}$ are.

In our proof, we consider the mapping

$$Pu=-RTu+u,\quad u\in X_0. \tag{3.44}$$

For any $j\in\{0,1\}$, the operator $P: X_j\to X_j$ is bounded due to conditions (iv) and (v). Moreover, for any $u\in X_j$, conditions (vi) and (vii) imply that

$$TPu=-TRTu+Tu=-(Tu+STu)+Tu=-STu\in Y_j,$$

i.e., $Pu \in (X_j)_{T,Y_j}$. In addition, the boundedness of the operator $P : X_j \to X_j$ and conditions (iv), (vi) yield the estimate

$$\|Pu\|^2_{(X_j)_{T,Y_j}} = \|Pu\|^2_{X_j} + \|TPu\|^2_{Y_j} = \|Pu\|^2_{X_j} + \| - STu\|^2_{Y_j} \leq c_1 \|u\|^2_{X_j},$$

where the number $c_1 > 0$ does not depend on u. Thus, mapping (3.44) sets bounded operators

$$P : X_j \to (X_j)_{T,Y_j} \quad \text{for each } j \in \{0, 1\}. \tag{3.45}$$

Besides, consider restrictions of the mapping R to Y_j for $j \in \{0, 1\}$. Due to conditions (iii) and (v), we have the bounded operator $R : Y_j \to X_j$. Moreover, for any $\omega \in Y_j$, conditions (vi) and (vii) imply the equality $TR\omega = \omega + S\omega \in Y_j$, i.e., $R\omega \in (X_j)_{T,Y_j}$. In addition, conditions (iii), (vi), and (vii) and the boundedness of the operator $R : Y_j \to X_j$ yield the estimate

$$\begin{aligned}
\|R\omega\|^2_{(X_j)_{T,Y_j}} &= \|R\omega\|^2_{X_j} + \|TR\omega\|^2_{Y_j} = \|R\omega\|^2_{X_j} + \|\omega + S\omega\|^2_{Y_j} \\
&\leq \|R\omega\|^2_{X_j} + \left(\|\omega\|_{Y_j} + \|S\omega\|_{Y_j}\right)^2 \\
&\leq c_2 \|\omega\|^2_{Y_j} + \left(\|\omega\|_{Y_j} + c_3 \|\omega\|_{Z_j}\right)^2 \\
&\leq c_4 \|\omega\|^2_{Y_j},
\end{aligned}$$

where constants c_2, c_3, and c_4 are independent of ω. Thus, operators

$$R : Y_j \to (X_j)_{T,Y_j} \quad \text{for} \quad j \in \{0, 1\} \tag{3.46}$$

are bounded.

We use operators (3.45) and (3.46) in order to prove that the pair of spaces

$$[\,(X_0)_{T,Y_0}, (X_1)_{T,Y_1}\,] \tag{3.47}$$

is admissible. First we prove that the space $(X_j)_{T,Y_j}$ is separable for each $j \in \{0, 1\}$. Due to condition (i), spaces X_j and Y_j are separable. Consider arbitrary countable sets X_j^0 and Y_j^0 that belong to and are dense in X_j and Y_j, respectively. Using these sets, we construct the countable set

$$Q = \{Pu_0 + Rv_0 : \, u_0 \in X_j^0, \, v_0 \in Y_j^0\}$$

and approximate any $u \in (X_j)_{T,Y_j}$ by elements of this set. Since $u \in X_j$ and $Tu \in Y_j$, one can find sequences of elements $u_k \in X_j^0$ and $v_k \in Y_j^0$ such that $u_k \to u$ in X_j and $v_k \to Tu$ in Y_j as $k \to \infty$. Due to bounded operators (3.45) and (3.46) and equality (3.44), we obtain

$$w_k := Pu_k + Rv_k \to Pu + RTu = u \quad \text{in} \quad (X_j)_{T,Y_j} \quad \text{as} \quad k \to \infty, \tag{3.48}$$

with $w_k \in Q$. Hence, the countable set Q is dense in the space $(X_j)_{T,Y_j}$, i.e., the space is separable.

In order to prove that pair (3.47) is admissible, it remains to prove the density of the continuous embedding $(X_1)_{T,Y_1} \hookrightarrow (X_0)_{T,Y_0}$. Fix any $u \in (X_0)_{T,Y_0}$. Then $u \in X_0$ and $Tu \in Y_0$. Due to condition (i), the space X_1 is dense in X_0 and the space Y_1 is dense in Y_0. Hence, there exist sequences of elements $u_k \in X_1$ and $v_k \in Y_1$ such that $u_k \to u$ in X_0 and $v_k \to Tu$ in Y_0 as $k \to \infty$. From here, by using operators (3.45) and (3.46) and equality (3.44), we obtain (3.48) for $j = 0$ and prove that $w_k \in (X_1)_{T,Y_1}$. Thus, $(X_1)_{T,Y_1}$ is dense in $(X_0)_{T,Y_0}$.

Further, let us prove formula (3.43). First we show that the continuous embedding

$$[\,(X_0)_{T,Y_0},\,(X_1)_{T,Y_1}\,]_\psi \hookrightarrow (X_\psi)_{T,Y_\psi} \tag{3.49}$$

takes place. In view of definition of the space $(X_j)_{T,Y_j}$, we have bounded operators

$$I : (X_j)_{T,Y_j} \to X_j \quad \text{and} \quad T : (X_j)_{T,Y_j} \to Y_j \quad \text{for each} \quad j \in \{0,1\},$$

where, as usual, I stands for the identity mapping. For the interpolation parameter ψ, this implies the boundedness of operators

$$I : \,[\,(X_0)_{T,Y_0},\,(X_1)_{T,Y_1}\,]_\psi \to X_\psi,$$

$$T : \,[\,(X_0)_{T,Y_0},\,(X_1)_{T,Y_1}\,]_\psi \to Y_\psi.$$

Hence, if $u \in [\,(X_0)_{T,Y_0},\,(X_1)_{T,Y_1}\,]_\psi$, then $u \in X_\psi$, $Tu \in Y_\psi$ and

$$\|u\|^2_{X_\psi} + \|Tu\|^2_{Y_\psi} \le c\,\|u\|^2_{[(X_0)_{T,Y_0},\,(X_1)_{T,Y_1}]_\psi}$$

for a constant c that does not depend on u. In other words, the continuous embedding (3.49) takes place.

In view of the Banach inverse operator theorem, it remains to prove the inclusion, which is inverse to (3.49). To this end, we apply the interpolation with the parameter ψ to (3.45) and (3.46). We obtain bounded operators

$$P : \,X_\psi \to [\,(X_0)_{T,Y_0},\,(X_1)_{T,Y_1}\,]_\psi,$$

$$R : \,Y_\psi \to [\,(X_0)_{T,Y_0},\,(X_1)_{T,Y_1}\,]_\psi.$$

Hence, if $u \in (X_\psi)_{T,Y_\psi}$ (i.e., $u \in X_\psi$ and $Tu \in Y_\psi$), then, in view of (3.44), we conclude that

$$u = Pu + RTu \in [\,(X_0)_{T,Y_0},\,(X_1)_{T,Y_1}\,]_\psi.$$

Thus, we have the inclusion inverse to (3.49).

Theorem 3.12 is proved.

3.3.3 Elliptic boundary-value problem in Sobolev spaces

To prove the main result, we use the classical theorem on solvability of the inhomogeneous boundary-value problem (3.1), (3.2) in positive Sobolev spaces (see, e.g., [11, Theorems 2.4.1 and 4.3.1] or [121, Chap. 2, Theorem 5.4]).

Proposition 3.1. *The mapping*

$$u \mapsto (Lu, B_1 u, \dots, B_q u), \quad u \in C^\infty(\overline{\Omega}), \tag{3.50}$$

can be uniquely extended (by continuity) to the bounded Fredholm operator

$$(L, B) : H^s(\Omega) \to H^{s-2q}(\Omega) \oplus \mathcal{H}_s(\Gamma) \tag{3.51}$$

for any real $s \geq 2q$. *The kernel of the operator coincides with* N, *and the domain consists of the vectors*

$$(f, g_1, \dots, g_q) \in H^{s-2q}(\Omega) \oplus \mathcal{H}_s(\Gamma)$$

that satisfy the condition

$$(f, v)_\Omega + \sum_{j=1}^{q} (g_j, C_j^+ v)_\Gamma = 0 \quad \textit{for all} \quad v \in N^+. \tag{3.52}$$

The index of operator (3.51) *is equal to* $\dim N - \dim N^+$ *and does not depend on* s.

In the theorem and in what follows, we use the notation

$$\mathcal{H}_s(\Gamma) := \bigoplus_{j=1}^{q} H^{s-m_j-1/2}(\Gamma), \quad s \in \mathbb{R},$$

(cf. formula (3.40))

Proposition 3.1 was extended to the case of any real s by J.-L. Lions and E. Magenes [119, 120, 121] and Ya. A. Roitberg [202, 203, 209] (see also the presentation of Roitberg's results in [21, Chapt. III, § 6]). In that case, the operator (L, B) was studied in spaces that were constructed in different ways on the base of Sobolev spaces of relevant orders. In our proof, we use the construction suggested by J.-L. Lions and E. Magenes in the monograph [121, Chap. 2, Sections 6 and 7] because, unlike the cited papers by Ya. A. Roitberg. the construction by J.-L. Lions and E. Magenes remains applicable within the framework of spaces of distributions in the domain Ω. For the sake of simplicity, we consider here only integer s because it is enough to meet our purposes. The general case is considered in Subsection 4.4.1.

Let a function $\varrho_1 \in C^\infty(\overline{\Omega})$ be positive in Ω and equal to the distance to the boundary Γ in a certain neighborhood of the boundary.

For any integer $\sigma \geq 0$, we set

$$\Xi^\sigma(\Omega) := \{u \in \mathcal{D}'(\Omega) : \varrho_1^{|\mu|} D^\mu u \in L_2(\Omega), \ |\mu| \leq \sigma\}, \tag{3.53}$$

where μ is an n-dimensional multiindex. The space $\Xi^\sigma(\Omega)$ is a Hilbert space with respect to the inner product

$$(u_1, u_2)_{\Xi^\sigma(\Omega)} := \sum_{|\mu| \leq \sigma} \big(\varrho_1^{|\mu|} D^\mu u_1, \, \varrho_1^{|\mu|} D^\mu u_2\big)_{L_2(\Omega)}.$$

We have the following dense continuous embeddings:

$$H_0^\sigma(\Omega) \hookrightarrow \Xi^\sigma(\Omega) \hookrightarrow L_2(\Omega), \tag{3.54}$$

where $H_0^\sigma(\Omega)$ is the closure of the set $C_0^\infty(\Omega)$ in the topology of the space $H^\sigma(\Omega)$.

Let $\Xi^{-\sigma}(\Omega)$ denote the Hilbert space dual to $\Xi^\sigma(\Omega)$ with respect to the inner product in $L_2(\Omega)$. Since spaces $H_0^\sigma(\Omega)$ and $H^{-\sigma}(\Omega)$ are mutually dual with respect to the same inner product [258, Theorem 4.8.2(a)], relation (3.54) yields the continuity of dense embeddings

$$L_2(\Omega) \hookrightarrow \Xi^{-\sigma}(\Omega) \hookrightarrow H^{-\sigma}(\Omega) \tag{3.55}$$

for any integer $\sigma > 0$. The embedding on the right-hand side implies that the space $\Xi^{-\sigma}(\Omega)$ consists of distributions in the domain Ω.

For any integer $s < 2q$, we define the linear space

$$D_L^s(\Omega) := \{u \in H^s(\Omega) : Lu \in \Xi^{s-2q}(\Omega)\}$$

with the graphic inner product

$$(u_1, u_2)_{D_L^s(\Omega)} = (u_1, u_2)_{H^s(\Omega)} + (Lu_1, Lu_2)_{\Xi^{s-2q}(\Omega)}.$$

The space $D_L^s(\Omega)$ is complete with respect to this inner product.

The set $C^\infty(\overline{\Omega})$ is dense in the space $D_L^s(\Omega)$. Due to (3.54) and the boundedness property of the operator $L : H^{2q}(\Omega) \to L_2(\Omega)$, we have dense continuous embeddings

$$H^{2q}(\Omega) \hookrightarrow D_L^s(\Omega) \hookrightarrow H^s(\Omega) \quad \text{for any integer} \quad s < 2q. \tag{3.56}$$

Remark 3.6. In the cited monograph by J.-L. Lions and E. Magenes [121], $H^s(\Omega)$ for $s < 0$ denotes the space dual to $H_0^{-s}(\Omega)$ with respect to the inner product in $L_2(\Omega)$. For all $s < 0$ that are not half-integers, this dual space coincides with our space $H^s(\Omega)$ [258, Theorem 4.8.2(a)]. For half-integer $s < 0$, this dual space does not coincide with $H^s(\Omega)$ (see also Subsection 4.4.1).

The following result was proved by J.-L. Lions and E. Magenes [121, Theorems 6.7 and 7.4].

Proposition 3.2. *Mapping* (3.50) *can be uniquely extended (by continuity) to the bounded Fredholm operator*

$$(L, B) : D^s_L(\Omega) \to \Xi^{s-2q}(\Omega) \oplus \mathcal{H}_s(\Gamma) \tag{3.57}$$

for any integer $s < 2q$. *The kernel of the operator coincides with* N *and its domain consists of the vectors*

$$(f, g_1, \ldots, g_q) \in \Xi^{s-2q}(\Omega) \oplus \mathcal{H}_s(\Gamma)$$

that satisfy condition (3.52). *The index of operator* (3.51) *is equal to* $\dim N - \dim N^+$ *and does not depend on* s.

For our study, we have to establish an assertion on the isomorphism related to the operator that corresponds to a homogeneous boundary-value Dirichlet problem. Fix an integer $r \geq 1$ and consider the r th iteration L^r of the expression L. Let L^{r+} denote the expression, which is formally adjoint to L^r. Look at the linear differential expression $L^r L^{r+} + 1$ of order $4qr$ with coefficients from $C^\infty(\overline{\Omega})$. For any integer $\sigma \geq 2qr$, we set

$$H^\sigma_D(\Omega) := \{u \in H^\sigma(\Omega) : \gamma_j u = 0 \ \text{ on } \ \Gamma, \ \ j = 0, \ldots, 2qr - 1\}.$$

Here and in what follows, $\gamma_j u := (\partial^j u / \partial^j_\nu) \restriction \Gamma$ is the trace operator for the normal derivative of order j on Γ. In this case, the trace is understood in the sense of Theorem 3.5 that states the existence of the bounded operator $\gamma_j : H^\sigma(\Omega) \to H^{\sigma - j - 1/2}(\Gamma)$. Therefore, the linear space $H^\sigma_D(\Omega)$ is complete with respect to the inner product in the space $H^\sigma(\Omega)$.

Lemma 3.1. *Let an integer* $r \geq 1$. *The restriction of the mapping* $u \mapsto L^r L^{r+} u + u$, *where* $u \in \mathcal{D}'(\Omega)$, *defines the isomorphism*

$$L^r L^{r+} + 1 \, : \, H^\sigma_D(\Omega) \leftrightarrow H^{\sigma - 4qr}(\Omega) \tag{3.58}$$

for any integer $\sigma \geq 2qr$.

Proof. The differential expression $L^r L^{r+} + 1$ is properly elliptic in $\overline{\Omega}$ since the expression L is properly elliptic. Consider the inhomogeneous boundary-value Dirichlet problem

$$L^r L^{r+} u + u = f \ \text{ in } \ \Omega,$$

$$\gamma_j \, u = g_j \ \text{ on } \ \Gamma \ \text{ for } \ j = 0, \ldots, 2qr - 1.$$

This is a regular elliptic boundary-value problem, and it was proved in [121, Chap. 2, Theorem 8.3 and Remark 8.5] that the operator of this problem is a bounded Fredholm operator with zero index in the pair of spaces

$$(L^r L^{r+} + 1; \gamma_0, \dots, \gamma_{2qr-1}) : H^\sigma(\Omega) \to H^{\sigma-4qr}(\Omega) \oplus \bigoplus_{j=0}^{2qr-1} H^{\sigma-j-1/2}(\Gamma) \tag{3.59}$$

for any integer $\sigma \geq 2qr$. The kernel N_D of operator (3.59) belongs to $C^\infty(\overline{\Omega})$. Using integration by parts, it is easy to verify that this kernel is trivial:

$$u \in N_D \Rightarrow \quad (u,u)_\Omega = -(L^r L^{r+} u, u)_\Omega = -(L^{r+} u, L^{r+} u)_\Omega \leq 0$$
$$\Rightarrow \quad u = 0.$$

Note that by applying the method of integration by parts and transferring the differential expression L^r of order $2qr$, we obtain expressions of the form $(\,\cdot\,, \gamma_j u)_\Gamma$, where $j = 0, \dots, 2qr-1$. These expressions are equal to zero for $u \in N_D$. Hence, we conclude that operator (3.59) is an isomorphism and its restriction to the subspace $H^\sigma_D(\Omega)$ defines isomorphism (3.58).

Lemma 3.1 is proved.

Remark 3.7. An assertion similar to Lemma 3.1 can be found in Triebel's monograph [258, Sec. 5.7.1, Remark 1].

3.3.4 Proof of the main result

We now prove the main result of Section 3.3, namely, we prove Theorem 3.11.

Proof of Theorem 3.11. Let $s \in \mathbb{R}$ and $\varphi \in \mathcal{M}$. Choose an integer $r \geq 1$ such that

$$2q(1-r) < s < 2qr \tag{3.60}$$

and use Proposition 3.2 for the integer $s = 2q(1-r) \leq 0$ and Proposition 3.1 for $s = 2qr \geq 2q$. We conclude that mapping (3.50) can be extended by continuity to bounded Fredholm operators

$$(L, B) : D_L^{2q(1-r)}(\Omega) \to \Xi^{-2qr}(\Omega) \oplus \mathcal{H}_{2q(1-r)}(\Gamma), \tag{3.61}$$

$$(L, B) : H^{2qr}(\Omega) \to H^{2q(r-1)}(\Omega) \oplus \mathcal{H}_{2qr}(\Gamma) \tag{3.62}$$

with the same index and common kernel N.

Note that pairs of spaces

$$[D_L^{2q(1-r)}(\Omega), H^{2qr}(\Omega)] \quad \text{and} \quad [\Xi^{-2qr}(\Omega), H^{2q(r-1)}(\Omega)] \tag{3.63}$$

are admissible. Indeed, in view of (3.55) and (3.56), the following dense continuous embeddings take place:

$$H^{2qr}(\Omega) \hookrightarrow H^{2q}(\Omega) \hookrightarrow D_L^{2q(1-r)}(\Omega),$$

$$H^{2q(r-1)}(\Omega) \hookrightarrow L_2(\Omega) \hookrightarrow \Xi^{-2qr}(\Omega).$$

Hence, the spaces on the right-hand sides in pairs (3.63) are embedded continuously and densely into their respective spaces on the left-hand sides. This implies that the spaces on the left-hand sides are separable because the (Sobolev) spaces on the right-hand sides are separable. Thus, both pairs in (3.63) are admissible.

In view of (3.60), we set

$$\varepsilon := s - 2q(1-r) > 0 \quad \text{and} \quad \delta := 2qr - s > 0. \tag{3.64}$$

Let ψ be the interpolation parameter from Theorems 3.2 and 2.2 relative to chosen parameters φ, ε, and δ. If we apply the interpolation with the parameter ψ to the action spaces of bounded Fredholm operators (3.61) and (3.62), then by Theorems 1.7 and 1.5 we obtain the bounded Fredholm operator

$$\begin{aligned}(L,B) :\ & [D_L^{2q(1-r)}(\Omega), H^{2qr}(\Omega)]_\psi \\ & \to [\Xi^{-2qr}(\Omega), H^{2q(r-1)}(\Omega)]_\psi \oplus [\mathcal{H}_{2q(1-r)}(\Gamma), \mathcal{H}_{2qr}(\Gamma)]_\psi.\end{aligned} \tag{3.65}$$

By virtue of Theorems 2.2, 1.5 and in view of (3.64), we have

$$\begin{aligned}[\mathcal{H}_{2q(1-r)}(\Gamma), \mathcal{H}_{2qr}(\Gamma)]_\psi &= \left[\bigoplus_{j=1}^{q} H^{2q(1-r)-m_j-1/2}(\Gamma), \bigoplus_{j=1}^{q} H^{2qr-m_j-1/2}(\Gamma)\right]_\psi \\ &= \bigoplus_{j=1}^{q} [H^{2q(1-r)-m_j-1/2}(\Gamma), H^{2qr-m_j-1/2}(\Gamma)]_\psi \\ &= \bigoplus_{j=1}^{q} [H^{s-m_j-1/2-\varepsilon}(\Gamma), H^{s-m_j-1/2+\delta}(\Gamma)]_\psi \\ &= \bigoplus_{j=1}^{q} H^{s-m_j-1/2,\varphi}(\Gamma) \\ &= \mathcal{H}_{s,\varphi}(\Gamma)\end{aligned}$$

up to equivalence of norms. Denoting

$$Z(\Omega) := [\Xi^{-2qr}(\Omega), H^{2q(r-1)}(\Omega)]_\psi, \tag{3.66}$$

we can see that operator (3.65) is a bounded Fredholm operator in the pair of spaces

$$(L,B):\ \bigl[D_L^{2q(1-r)}(\Omega),H^{2qr}(\Omega)\bigr]_{\psi}\to Z(\Omega)\oplus\mathcal{H}_{s,\varphi}(\Gamma). \tag{3.67}$$

This operator has the same kernel N and the same index as operators (3.61) and (3.62) have.

By considering $Z(\Omega)$, let us describe the domain of operator (3.67). We are going to apply Theorem 3.12. To this end, we set

$$X_0=H^{2q(1-r)}(\Omega),\quad Y_0=\Xi^{-2qr}(\Omega),\quad Z_0=E=H^{-2qr}(\Omega),$$

$$X_1=H^{2qr}(\Omega),\quad Y_1=Z_1=H^{2q(r-1)}(\Omega),$$

$$T=L$$

in the conditions of the theorem. Since the second pair in formula (3.63) is admissible, the embedding on the right-hand side of (3.55) implies that conditions (i), (ii), and (iii) of Theorem 3.12 are satisfied. Condition (iv) of this theorem is also valid because the operator $L:H^{\sigma}(\Omega)\to H^{\sigma-2q}(\Omega)$ is bounded for any $\sigma\in\mathbb{R}$.

Additionally, we have to define linear mappings R and S that satisfy conditions (v), (vi), and (vii). We define them as follows. By applying Lemma 3.1 and considering the mapping $(L^rL^{r+}+1)^{-1}$, which is inverse to isomorphism (3.58), we obtain the bounded linear operator

$$(L^rL^{r+}+1)^{-1}:H^{\sigma-4qr}(\Omega)\to H^{\sigma}(\Omega) \tag{3.68}$$

for any integer $\sigma\geq 2qr$. Then we set

$$R=L^{r-1}L^{r+}(L^rL^{r+}+1)^{-1}\quad\text{and}\quad S=-(L^rL^{r+}+1)^{-1}.$$

In view of (3.68) for $\sigma=2qr$ and $\sigma=2q(3r-1)$, we obtain bounded operators

$$R:Z_0=H^{-2qr}(\Omega)\to H^{2qr-2q(2r-1)}(\Omega)=X_0,$$

$$R:Z_1=H^{2q(r-1)}(\Omega)\to H^{2q(3r-1)-2q(2r-1)}(\Omega)=X_1,$$

$$S:Z_0=H^{-2qr}(\Omega)\to H^{2qr}(\Omega)\hookrightarrow H^{0}(\Omega)\hookrightarrow\Xi^{-2qr}(\Omega)=Y_0,$$

$$S:Z_1=H^{2q(r-1)}(\Omega)\to H^{2q(3r-1)}(\Omega)\hookrightarrow H^{2qr}(\Omega)=X_1.$$

In addition, on the set $E=H^{-2qr}(\Omega)$, the following equalities hold:

$$\begin{aligned}TR&=LL^{r-1}L^{r+}(L^rL^{r+}+1)^{-1}\\&=(L^rL^{r+}+1-1)(L^rL^{r+}+1)^{-1}\\&=1-S.\end{aligned}$$

Thus, all conditions of Theorem 3.12 are satisfied. By this theorem, for the interpolation parameter ψ, we have the equality of spaces up to equivalence of norms:

$$\big[(X_0)_{L,Y_0}, (X_1)_{L,Y_1}\big]_\psi = (X_\psi)_{L,Y_\psi}. \tag{3.69}$$

Here,

$$(X_0)_{L,Y_0} = \big\{u \in H^{2q(1-r)}(\Omega) : Lu \in \Xi^{-2qr}(\Omega)\big\} = D_L^{2q(1-r)}(\Omega),$$

and the norms of spaces, which are located at the edges, are equal. Furthermore, due to the boundedness of the operator $L : H^{2qr}(\Omega) \to H^{2q(r-1)}(\Omega)$, we have the equality

$$(X_1)_{L,Y_1} = \big\{u \in H^{2qr}(\Omega) : Lu \in H^{2q(r-1)}(\Omega)\big\} = H^{2qr}(\Omega)$$

with equivalence of norms in spaces, which are located at the edges. In addition, by Theorem 3.2 and in view of (3.64), we have

$$X_\psi = [H^{2q(1-r)}(\Omega), H^{2qr}(\Omega)]_\psi = [H^{s-\varepsilon}(\Omega), H^{s+\delta}(\Omega)]_\psi = H^{s,\varphi}(\Omega).$$

Thus, relation (3.69) takes the form

$$[D_L^{2q(1-r)}(\Omega), H^{2qr}(\Omega)]_\psi = \{u \in H^{s,\varphi}(\Omega) : Lu \in Z(\Omega)\}. \tag{3.70}$$

In the latter space, the graphic inner product is defined and the space is complete with respect to this inner product (we used the equality $Y_\psi = Z(\Omega)$ here in accordance with notation (3.66)).

By substituting equality (3.70) into (3.67), we obtain the bounded operator

$$(L, B) : \{u \in H^{s,\varphi}(\Omega) : Lu \in Z(\Omega)\} \to Z(\Omega) \oplus \mathcal{H}_{s,\varphi}(\Gamma). \tag{3.71}$$

In accordance with the results proved above, it is a Fredholm operator that has the kernel N. Moreover, since operator (3.71) is obtained by using the interpolation procedure applied to Fredholm operators (3.61) and (3.62), by Theorem 1.7 and Proposition 3.2 we conclude that the range of operator (3.71) is equal to

$$Z(\Omega) \oplus \mathcal{H}_{s,\varphi}(\Gamma) \cap (L, B)\big(D_L^{2q(1-r)}(\Omega)\big).$$

In other words, the range consists of the vectors

$$(f, g_1, \dots, g_q) \in Z(\Omega) \oplus \mathcal{H}_{s,\varphi}(\Gamma)$$

that satisfy condition (3.52). The restriction of operator (3.71) to the subspace

$$K_L^{s,\varphi}(\Omega) = \{u \in H^{s,\varphi}(\Omega) : Lu = 0 \text{ in } \Omega\}$$

defines the bounded operator

$$B:\ K_L^{s,\varphi}(\Omega)\to\mathcal{H}_{s,\varphi}(\Gamma). \tag{3.72}$$

Its kernel is equal to $N\cap K_L^{s,\varphi}(\Omega)=N$ (and hence, it is finite-dimensional) and its range coincides with subspace (3.41). Therefore, the range is closed, its codimension is finite and equal to the dimension of the space G defined by (3.42). Thus, operator (3.72) is a Fredholm operator that has the kernel N, range (3.41), and the finite index $\dim N-\dim G$, which does not depend on s,φ.

It remains to show that the set $K_L^{\infty}(\Omega)$ is dense in $K_L^{s,\varphi}(\Omega)$ and operator (3.72) is an extension of mapping (3.38) by continuity. In this connection, let us note the following. Since operator (3.71) is an extension of (3.50), operator (3.72) is an extension of (3.38) in accordance with its definition (3.72). Therefore, to complete the proof, we have to verify that the set $K_L^{\infty}(\Omega)$ is dense in $K_L^{s,\varphi}(\Omega)$. To this end, we consider the isomorphism

$$B:\ K_L^{s,\varphi}(\Omega)/\,N\leftrightarrow\mathcal{R}_{s,\varphi}(\Gamma) \tag{3.73}$$

generated by the Fredholm operator (3.72). Here, $\mathcal{R}_{s,\varphi}(\Gamma)$ denotes the domain (3.41) of operator (3.72). Consider the isomorphism B^{-1}, which is inverse to (3.73). It puts each vector $g=(g_1,\dots,g_q)\in\mathcal{R}_{s,\varphi}(\Gamma)$ in a correspondence with the coset

$$B^{-1}g=[\,u\,]=\{u+w:\ w\in N\}$$

of the element $u\in K_L^{s,\varphi}(\Omega)$ such that $Bu=g$.

As a preliminary result, we prove that mapping (3.73) possesses the following property of increase in smoothness:

$$g\in\mathcal{R}_{s,\varphi}(\Gamma)\cap(C^{\infty}(\Gamma))^{q}$$
$$\Rightarrow\ \bigl(B^{-1}g=[\,u\,]\ \text{ for some }\ u\in K_L^{\infty}(\Omega)\bigr). \tag{3.74}$$

Let

$$g=(g_1,\dots,g_q)\in\mathcal{R}_{s,\varphi}(\Gamma)\cap(C^{\infty}(\Gamma))^{q}.$$

Since $\mathcal{R}_{s,\varphi}(\Gamma)$ is equal to set (3.41), by virtue of Proposition 3.1 the elliptic boundary-value problem (3.36) has a solution $u\in H^{2q}(\Omega)$. The right-hand sides of equations in this problem are infinitely smooth. Hence [121, Chap. 2, Sec. 5.4], the inclusion $u\in C^{\infty}(\overline{\Omega})$ takes place. Thus, for operator (3.72), we have $u\in K_L^{\infty}(\Omega)$ and $Bu=g$, and this proves (3.74).

Now it is easy to prove the required density. Let us pick any distribution $u\in K_L^{s,\varphi}(\Omega)$ and consider the vector

$$g=Bu\in\mathcal{R}_{s,\varphi}(\Gamma)\subset\mathcal{H}_{s,\varphi}(\Gamma). \tag{3.75}$$

Since the set $C^\infty(\Gamma)$ is dense in $H^{\sigma,\varphi}(\Gamma)$ for every $\sigma \in \mathbb{R}$, there exists a sequence of vectors $g^{(k)}$ such that

$$g^{(k)} \in (C^\infty(\Gamma))^q \quad \text{and} \quad g^{(k)} \to g \ \text{ in } \ \mathcal{H}_{s,\varphi}(\Gamma) \ \text{ as } \ k \to \infty. \tag{3.76}$$

Further, observe that $\mathcal{R}_{s,\varphi}(\Gamma)$ and G are (closed) subspaces in $\mathcal{H}_{s,\varphi}(\Gamma)$ satisfying the conditions $\mathcal{R}_{s,\varphi}(\Gamma) \cap G = \{0\}$ and $\operatorname{codim} \mathcal{R}_{s,\varphi}(\Gamma) = \dim G$. This implies that the space $\mathcal{H}_{s,\varphi}(\Gamma)$ is the direct sum of these subspaces. Due to this sum, we can write $g = g + 0$ and $g^{(k)} = h^{(k)} + \omega^{(k)}$, where $h^{(k)} \in \mathcal{R}_{s,\varphi}(\Gamma)$ and $\omega^{(k)} \in G$. Using this result along with (3.76), we obtain the following two assertions:

$$h^{(k)} = g^{(k)} - \omega^{(k)} \in \mathcal{R}_{s,\varphi}(\Gamma) \cap (C^\infty(\Gamma))^q,$$

$$h^{(k)} \to g \ \text{ in } \ \mathcal{R}_{s,\varphi}(\Gamma) \ \text{ (i.e., in } \mathcal{H}_{s,\varphi}(\Gamma)) \ \text{ as } \ k \to \infty.$$

In view of (3.74), the first assertion implies $B^{-1}h^{(k)} = [\,u_k\,]$ for some $u_k \in K_L^\infty(\Omega)$. In view of (3.73) and (3.75), the second assertion yields

$$[\,u_k\,] = B^{-1}h^{(k)} \to B^{-1}g = [\,u\,],$$

i.e.,

$$[\,u_k - u\,] \to 0 \quad \text{in} \quad K_L^{s,\varphi}(\Omega)/\,N \quad \text{as} \quad k \to \infty.$$

This means that

$$u_k - u + w_k \to 0 \quad \text{in} \quad K_L^{s,\varphi}(\Omega) \quad \text{as} \quad k \to \infty$$

for a sequence of functions $w_k \in N \subset K_L^\infty(\Omega)$. Thus, any distribution $u \in K_L^{s,\varphi}(\Omega)$ can be approximated in the space $K_L^{s,\varphi}(\Omega)$ by a sequence of functions $u_k + w_k \in K_L^\infty(\Omega)$. Hence, the set $K_L^\infty(\Omega)$ is dense in $K_L^{s,\varphi}(\Omega)$.

Theorem 3.11 is proved.

3.3.5 Properties of solutions to the homogeneous elliptic equation

Theorem 3.11 implies that the operator (3.40), which corresponds to problem (3.36), is an isomorphism in the case of trivial kernel N and trivial defect subspace G. In the general case, this operator defines an isomorphism

$$B:\, K_L^{s,\varphi}(\Omega)/\,N \leftrightarrow \mathcal{R}_{s,\varphi}(\Gamma) \quad \text{for all} \quad s \in \mathbb{R} \quad \text{and} \quad \varphi \in \mathcal{M}. \tag{3.77}$$

Recall that $\mathcal{R}_{s,\varphi}(\Gamma)$ is a subspace (3.41). (Note that the operator inverse to (3.77) is bounded by the Banach inverse operator theorem.) The collection of isomorphisms (3.77) gives a solution to problem (3.36) for any distributions $g_1, \ldots, g_q \in \mathcal{D}'(\Gamma)$ that satisfy the condition

$$(g_1,\, C_1^+\, v)_\Gamma + \ldots + (g_q,\, C_q^+\, v)_\Gamma = 0 \quad \text{for any} \quad v \in N^+.$$

In this case, the following *a priori* estimate holds for the solution u.

Theorem 3.13. *Let parameters $s \in \mathbb{R}$, $\varphi \in \mathcal{M}$ and $\varepsilon > 0$ be given. There exists a number $c = c(s, \varphi, \varepsilon) > 0$ such that, for any $u \in K_L^{s,\varphi}(\Omega)$, the following estimate holds:*

$$\|u\|_{H^{s,\varphi}(\Omega)} \le c \left(\|Bu\|_{\mathcal{H}_{s,\varphi}(\Gamma)} + \|u\|_{H^{s-\varepsilon}(\Omega)} \right). \tag{3.78}$$

Proof. Due to isomorphism (3.77), we have

$$\inf\{ \|u + w\|_{H^{s,\varphi}(\Omega)} : w \in N \} \le c_0 \|Bu\|_{\mathcal{H}_{s,\varphi}(\Gamma)} \tag{3.79}$$

for any distribution $u \in K_L^{s,\varphi}(\Omega)$, where c_0 is the norm of the operator, which is inverse to (3.77). Since N is a finite-dimensional subspace for both $H^{s,\varphi}(\Omega)$ and $H^{s-\varepsilon}(\Omega)$, the norms in these two spaces are equivalent on N. In particular, for any $w \in N$, we have

$$\|w\|_{H^{s,\varphi}(\Omega)} \le c_1 \|w\|_{H^{s-\varepsilon}(\Omega)},$$

with a certain number c_1 that does not depend on u and w. In addition, we have

$$\begin{aligned}\|w\|_{H^{s-\varepsilon}(\Omega)} &\le \|u + w\|_{H^{s-\varepsilon}(\Omega)} + \|u\|_{H^{s-\varepsilon}(\Omega)} \\ &\le c_2 \|u + w\|_{H^{s,\varphi}(\Omega)} + \|u\|_{H^{s-\varepsilon}(\Omega)},\end{aligned}$$

where c_2 is the norm of the embedding operator $H^{s,\varphi}(\Omega) \hookrightarrow H^{s-\varepsilon}(\Omega)$. Hence,

$$\begin{aligned}\|u\|_{H^{s,\varphi}(\Omega)} &\le \|u + w\|_{H^{s,\varphi}(\Omega)} + \|w\|_{H^{s,\varphi}(\Omega)} \\ &\le \|u + w\|_{H^{s,\varphi}(\Omega)} + c_1 \|w\|_{H^{s-\varepsilon}(\Omega)} \\ &\le (1 + c_1 c_2) \|u + w\|_{H^{s,\varphi}(\Omega)} + c_1 \|u\|_{H^{s-\varepsilon}(\Omega)}.\end{aligned}$$

By taking the infimum over all $w \in N$ and applying inequality (3.79), we obtain

$$\|u\|_{H^{s,\varphi}(\Omega)} \le (1 + c_1 c_2)\, c_0 \|Bu\|_{\mathcal{H}_{s,\varphi}(\Gamma)} + c_1 \|u\|_{H^{s-\varepsilon}(\Omega)};$$

i.e., estimate (3.78) with $c := \max\{(1 + c_1 c_2) c_0, c_1\}$ is deduced.
Theorem 3.13 is proved.

If the right-hand side of inequality (3.78) is finite, the left-hand side is finite as well.

Theorem 3.14. *Let $s \in \mathbb{R}$, $\varphi \in \mathcal{M}$, and $\varepsilon > 0$. Assume that a distribution $u \in H^{s-\varepsilon}(\Omega)$ is a solution to problem (3.36) where*

$$B_j u = g_j \in H^{s-m_j-1/2,\varphi}(\Gamma) \quad \textit{for each} \quad j \in \{1, \dots, q\}. \tag{3.80}$$

Then $u \in H^{s,\varphi}(\Omega)$.

Proof. By condition of the theorem, we have $u \in K_L^{s-\varepsilon,1}(\Omega)$ and $Bu = g$ where $g = (g_1, \ldots, g_q)$. Hence, due to (3.80) and Theorem 3.11 (the description of the domain of operator (3.40)), we also have

$$g \in B(K_L^{s-\varepsilon,1}(\Omega)) \cap \mathcal{H}_{s,\varphi}(\Gamma) = B(K_L^{s,\varphi}(\Omega)).$$

Therefore, there exists a distribution $u_0 \in K_L^{s,\varphi}(\Omega)$ such that $Bu_0 = g$. From this by using Theorem 3.11 (the description of the kernel of operator (3.40)), we obtain successively:

$$B(u - u_0) = 0,$$

$$w := u - u_0 \in N \subset C^\infty(\overline{\Omega}),$$

$$u = u_0 + w \in H^{s,\varphi}(\Omega).$$

Theorem 3.14 is proved.

Theorem 3.14 states an increase in smoothness of the solution u to problem (3.36) up to the boundary Γ. In this case, we can see that the refined smoothness φ of the right-hand sides of equalities in the problem is inherited by the solution. Note that any solution of the homogeneous elliptic equation $Lu = 0$ in the domain Ω possesses the property $u \in C^\infty(\Omega)$ (see, e.g., [81, Theorem 7.4.1]). Therefore, it is essential in Theorem 3.14 that the smoothness of the solution u increases until it reaches the boundary of the domain Ω.

Corollary 3.2. *Let $\sigma \in \mathbb{R}$. Assume that a distribution $u \in H^\sigma(\Omega)$ is a solution to problem* (3.36) *with*

$$g_j \in H^{m-m_j+(n-1)/2,\varphi}(\Gamma) \quad \text{for each} \quad j \in \{1, \ldots, q\} \tag{3.81}$$

where $m := \max\{m_1, \ldots, m_q\}$ and the function $\varphi \in \mathcal{M}$ satisfies condition (1.37). *Then $u \in C^m(\overline{\Omega})$, and moreover, since $u \in C^\infty(\Omega)$ as well, the distribution u is a classical solution to problem* (3.36).

Proof. Condition (3.81) coincides with (3.80) if we set $s = m + n/2$. Hence, in view of Theorems 3.14 and 3.4, we have

$$u \in H^{m+n/2,\varphi}(\Omega) \hookrightarrow C^m(\overline{\Omega}),$$

Q.E.D.

Note that the left-hand sides of equalities in problem (3.36) for the classical solution u are computed by using classical derivatives. In this case, we have $B_j u \in C(\Gamma)$.

3.4 Elliptic problems with homogeneous boundary conditions

Let us consider the regular elliptic boundary-value problem (3.1), (3.2) in the case where boundary conditions (3.2) are homogeneous:

$$L\,u = f \;\text{ in }\; \Omega, \tag{3.82}$$

$$B_j\,u = 0 \;\text{ on }\; \Gamma \;\text{ for }\; j = 1, \dots, q. \tag{3.83}$$

We study properties of the mapping $u \mapsto Lu$, where u satisfies equalities (3.83), on the refined Sobolev scale.

3.4.1 Theorem on isomorphisms for elliptic operators

We now introduce the required spaces of distributions that satisfy homogeneous boundary conditions. Let $s \in \mathbb{R}$ and $\varphi \in \mathcal{M}$. In this section, for the sake of brevity, we denote the Hilbert space $H^{s,\varphi,(0)}(\Omega)$ by $H^{s,\varphi}$.

Let the abbreviation (b.c.) denote homogeneous boundary conditions (3.83). We set

$$C^\infty(\text{b.c.}) := \{u \in C^\infty(\overline{\Omega}) : \; B_j u = 0 \;\text{ on }\; \Gamma, \;\; j = 1, \dots, q\}$$

and use $H^{s,\varphi}(\text{b.c.})$ in order to denote the closure of the set $C^\infty(\text{b.c.})$ in the topology of the Hilbert space $H^{s,\varphi}$.

Along with problem (3.82), (3.83), we consider the following formally adjoint problem with homogeneous boundary conditions:

$$L^+\,v = g \quad \text{in} \quad \Omega, \tag{3.84}$$

$$B_j^+\,v = 0 \quad \text{on} \quad \Gamma \quad \text{for} \quad j = 1, \dots, q. \tag{3.85}$$

Let $(\text{b.c.})^+$ stand for the homogeneous boundary conditions (3.85). Then we set

$$C^\infty(\text{b.c.})^+ := \{v \in C^\infty(\overline{\Omega}) : \; B_j^+ v = 0 \;\text{ on }\; \Gamma, \;\; j = 1, \dots, q\}$$

and use $H^{s,\varphi}(\text{b.c.})^+$ in order to denote the closure of the set $C^\infty(\text{b.c.})^+$ in the topology of the Hilbert space $H^{s,\varphi}$.

Linear spaces $H^{s,\varphi}(\text{b.c.})$ and $H^{s,\varphi}(\text{b.c.})^+$ are Hilbert spaces with respect to the inner product in $H^{s,\varphi}$.

In view of (3.8), the set $C^\infty(\text{b.c.})^+$ and, consequently, the space $H^{s,\varphi}(\text{b.c.})^+$ are independent of the choice of the system of boundary expressions

$$\{B_1^+, \dots, B_q^+\},$$

which is adjoint to the system $\{B_1, \dots, B_q\}$ with respect to the differential expression L.

Due to Green formula (3.6), we have

$$(Lu, v)_\Omega = (u, L^+ v)_\Omega \tag{3.86}$$

$$\text{for all } u \in C^\infty(\text{b.c.}) \text{ and } v \in C^\infty(\text{b.c.})^+.$$

Therefore, the image Lu of an arbitrary function $u \in C^\infty(\text{b.c.})$ can be naturally treated as the bounded antilinear functional $(Lu, \cdot)_\Omega$ on the space

$$H^{2q-s,1/\varphi}(\text{b.c.})^+.$$

Moreover, since spaces $H^{2q-s,\,1/\varphi}$ and $H^{s-2q,\,\varphi}$ are mutually dual with respect to the form $(\cdot,\cdot)_\Omega$, the subspace $H^{2q-s,\,1/\varphi}(\text{b.c.})^+$ and the factor space $H^{s-2q,\,\varphi}/M_{s-2q,\,\varphi}$, where

$$M_{s-2q,\,\varphi} := \big\{h \in H^{s-2q,\,\varphi} : (h, w)_\Omega = 0 \text{ for all } w \in C^\infty(\text{b.c.})^+\big\},$$

are also mutually dual with respect to this form. Hence, Lu can be treated as the coset

$$\{Lu + h : h \in M_{s-2q,\,\varphi}\},$$

which belongs to the factor space $H^{s-2q,\,\varphi}/M_{s-2q,\,\varphi}$.

We treat the linear mapping $u \mapsto Lu$ as an operator that acts from the space $H^{s,\varphi}(\text{b.c.})$ into the space $H^{s-2q,\,\varphi}/M_{s-2q,\,\varphi}$, which is identified with the dual space $(H^{2q-s,\,1/\varphi}(\text{b.c.})^+)'$ (it consists of antilinear functionals). In order to formulate the theorem on isomorphisms for the operator L, we have to define projectors of the space $H^{s,\varphi}$ onto its subspaces, which are orthogonal, respectively, to N and N^+ regarding to the sesquilinear form $(\,\cdot\,,\,\cdot\,)_\Omega$. Such projectors do exist since spaces N and N^+ are finite-dimensional.

Lemma 3.2. *Let $s \in \mathbb{R}$, and let $\varphi \in \mathcal{M}$. Every element $u \in H^{s,\varphi}$ admits the unique representation in the form $u = u_0 + u_1$, where $u_0 \in N$, and $u_1 \in H^{s,\varphi}$ satisfies the condition $(u_1, w)_\Omega = 0$ for any $w \in N$. In this case, the mapping $P : u \mapsto u_1$ is a projector of the space $H^{s,\varphi}$ onto the subspace*

$$\{u_1 \in H^{s,\varphi} : (u_1, w)_\Omega = 0 \text{ for all } w \in N\}, \tag{3.87}$$

and the image Pu does not depend on s and φ. The restriction of the mapping P on $H^{s,\varphi}(\text{b.c.})$ is a projector of the space $H^{s,\varphi}(\text{b.c.})$ onto the subspace

$$\{u_1 \in H^{s,\varphi}(\text{b.c.}) : (u_1, w)_\Omega = 0 \text{ for all } w \in N\}. \tag{3.88}$$

Furthermore, the statement of the lemma remains true if we replace N by N^+, P by P^+, and (b.c.) *by* (b.c.)$^+$. *In such a case, $M_{s,\varphi}$ is a subspace of $P^+(H^{s,\varphi})$.*

Proof. As it was mentioned above, N is a finite-dimensional subspace of $H^{s,\varphi}$. It is clear that $\dim N$ is equal to the codimension of subspace (3.87) and, moreover, N and (3.87) have the trivial intersection. Hence, the space $H^{s,\varphi}$ can be decomposed into the direct sum of subspaces N and (3.87) with the projector P onto subspace (3.87) such that the projector does not depend on s and φ. In view of the inclusion $N \subset H^{s,\varphi}(\text{b.c.})$, this implies that the space $H^{s,\varphi}(\text{b.c.})$ can be decomposed into the direct sum of subspaces N and (3.88) with the same projector P onto subspace (3.88). Thus, as we can see, the existence of the projector P is implied by two conditions: (1) the space N is finite-dimensional and (2) $N \subset H^{s,\varphi}(\text{b.c.})$. So, since the space N^+ is finite-dimensional and $N^+ \subset H^{s,\varphi}(\text{b.c.})^+$, the lemma remains true for the projector P^+ with the replacements mentioned above in the formulation of the lemma. Finally, we note that the last assertion of the lemma is implied by the inclusion $N^+ \subset C^\infty(\text{b.c.})^+$.

Lemma 3.2 is proved.

We now formulate the main result of Section 3.4, namely, the theorem on isomorphisms, which are generated by the elliptic operator L on the refined Sobolev scale.

Theorem 3.15. *Let $s \in \mathbb{R}$, $\varphi \in \mathcal{M}$, and*

$$s \neq j + 1/2 \quad \textit{for each} \quad j \in \{0, 1, \dots, 2q-1\}. \tag{3.89}$$

The mapping $u \mapsto Lu$, where $u \in C^\infty(\text{b.c.})$ and Lu is treated either as the coset $\{Lu + h : h \in M_{s-2q,\varphi}\}$ or as the functional $(Lu, \cdot)_\Omega$, can be uniquely extended (by continuity) to the bounded linear operator

$$L : H^{s,\varphi}(\text{b.c.}) \to H^{s-2q,\varphi}/M_{s-2q,\varphi} = \big(H^{2q-s,1/\varphi}(\text{b.c.})^+\big)'. \tag{3.90}$$

The restriction of operator (3.90) onto subspace (3.87) generates the isomorphism

$$L : P(H^{s,\varphi}(\text{b.c.})) \leftrightarrow P^+(H^{s-2q,\varphi})/M_{s-2q,\varphi}. \tag{3.91}$$

Remark 3.8. In the Sobolev case ($\varphi \equiv 1$), Theorem 3.15 was proved by Yu. M. Berezansky, S. G. Krein, and Ya. A. Roitberg for integer s (see [22, Theorem 2] and [21, Chap. 3, Theorem 6.12]). For all real s, it was proved in the monograph by Ya. A. Roitberg [209, Theorem 5.5.2] (see also survey [11, Sec. 7.9 c]). For half-integer $s \in \{j + 1/2 : j = 0, 1, \dots, 2q-1\}$, the spaces of action of operator (3.90) were defined with the help of interpolation.

Note that if spaces N and N^+ are trivial, then operator (3.90) becomes an isomorphism

Theorem 3.15 yields the Fredholm property of operator (3.90) and an *a priori* estimate for the solutions of the equation $Lu = f$.

Theorem 3.16. *Let $s \in \mathbb{R}$ and $\varphi \in \mathcal{M}$. Additionally, let condition* (3.89) *be satisfied. Then the following assertions hold:*

(i) *The bounded operator* (3.90) *is a Fredholm operator with the kernel N and range*

$$\{[f] \in H^{s-2q,\varphi}/\,M_{s-2q,\varphi} : (f,w)_\Omega = 0 \text{ for all } w \in N^+\} \tag{3.92}$$

where $[f] = \{f+h : h \in M_{s-2q,\varphi}\}$ is the coset of the element $f \in H^{s-2q,\varphi}$. The index of this operator is equal to $\dim N - \dim N^+$ *and does not depend on s and φ.*

(ii) *For any solution $u \in H^{s,\varphi}(\mathrm{b.c.})$ of the equation $Lu = [f]$, the following* a priori *estimate takes place:*

For any $\varepsilon > 0$, there exists a number $c = c(s,\varphi,\varepsilon) > 0$, which does not depend on u, such that

$$\|u\|_{H^{s,\varphi}} \le c\,\big(\,\|f\|_{H^{s-2q,\varphi}} + \|u\|_{H^{s-\varepsilon}}\big). \tag{3.93}$$

We prove Theorems 3.15 and 3.16 later in Subsection 3.4.3.

Theorem 3.16 implies that N^+ is a defect subspace of operator (3.90). If $N = \{0\}$, then we can omit the term $\|u\|_{H^{s-\varepsilon}}$ in estimate (3.93).

Let us note the following. Since the formally adjoint boundary-value problem (3.84), (3.85) is regular elliptic, Theorems 3.15 and 3.16 remain true if we replace the operator L by the operator L^+ (with other obvious changes in the formulation). Namely, the linear mapping $v \mapsto L^+v$, where $v \in C^\infty(\mathrm{b.c.})^+$, can be uniquely extended (by continuity) to the bounded Fredholm operator

$$L^+ : H^{2q-s,\,1/\varphi}(\mathrm{b.c.})^+ \to H^{-s,\,1/\varphi}/\,M^+_{-s,\,1/\varphi} = (H^{s,\varphi}(\mathrm{b.c.}))' \tag{3.94}$$

with the kernel N^+ and the defect subspace N. Here, we denote

$$M^+_{-s,\,1/\varphi} = \{h \in H^{-s,\,1/\varphi} : (h,w)_\Omega = 0 \text{ for all } w \in C^\infty(\mathrm{b.c.})\},$$

and use the same assumptions on parameters s and φ as in Theorems 3.15 and 3.16. The restriction of operator (3.94) to the subspace

$$P^+(H^{2q-s,\,1/\varphi}(\mathrm{b.c.})^+)$$
$$= \{v \in H^{2q-s,\,1/\varphi}(\mathrm{b.c.})^+ : (v,w)_\Omega = 0 \text{ for all } w \in N^+\}$$

defines the isomorphism

$$L^+ : P^+(H^{2q-s,\,1/\varphi}(\mathrm{b.c.})^+) \leftrightarrow P(H^{-s,\,1/\varphi})/M^+_{-s,\,1/\varphi}.$$

Since operators (3.90) and (3.94) are bounded, equality (3.86) can be extended by continuity to the relation

$$(Lu, v)_\Omega = (u, L^+ v)_\Omega$$

$$\text{for all} \quad u \in H^{s,\varphi}(\text{b.c.}) \quad \text{and} \quad v \in H^{2q-s,\,1/\varphi}(\text{b.c.})^+.$$

This means that operators (3.90) and (3.94) are mutually adjoint with respect to the sesquilinear form $(\cdot, \cdot)_\Omega$ that plays the role of extension by continuity for the inner product in $L_2(\Omega)$.

3.4.2 Interpolation and homogeneous boundary conditions

Let us study interpolation properties of spaces, which are the action spaces for operator (3.90). These properties play an important role in the proof of Theorems 3.15 and 3.16. Recall that we omit the index φ in the notation of spaces introduced in the previous subsection (Subsection 3.4.1) when we consider the Sobolev case $\varphi \equiv 1$.

Theorem 3.17. *Let $s \in \mathbb{R}$ and $\varphi \in \mathcal{M}$. Additionally, let*

$$s \neq m_j + 1/2 \quad \text{for each} \quad j \in \{1, \dots, q\}. \tag{3.95}$$

Then the following assertions hold:

(i) *For s given in such a way, there exists a number $\varrho = \varrho(s) > 0$ (that does not depend on φ) such that, for any $\varepsilon \in (0, \varrho)$, one has*

$$[H^{s-\varepsilon}(\text{b.c.}), H^{s+\varepsilon}(\text{b.c.})]_\psi = H^{s,\varphi}(\text{b.c.})$$

up to equivalence of norms, where ψ does not depend on s and represents the interpolation parameter from Theorem 1.14 *taken for $\varepsilon = \delta$.*

(ii) *If the number s, which satisfies condition* (3.95), *is positive, then $H^{s,\varphi}(\text{b.c.})$ consists of all distributions $u \in H^{s,\varphi}(\Omega)$ that meet the condition $B_j u = 0$ on Γ for every $j \in \{1, \dots, q\}$ with $s > m_j + 1/2$.*

(iii) *If $s < 1/2$, then $H^{s,\varphi}(\text{b.c.}) = H^{s,\varphi}$.*

(iv) *Assertions* (i), (ii), *and* (iii) *remain true even if we replace m_j by m_j^+,* (b.c.) *by* (b.c.)$^+$, *and B_j by B_j^+ in their statements.*

Remark 3.9. Regarding assertion (i) of Theorem 3.17, we note that the interpolation with the power parameter of Sobolev spaces under homogeneous boundary conditions was studied by P. Grisvard [66] and R. T. Seeley [234] (see also [258, Sec. 4.3.3]). Their results imply that condition (3.95) cannot be neglected. In assertion (ii) of Theorem 3.17, we treat the expression $B_j u$ in the sense of Theorem 3.5 on traces. The presence of condition (3.95) in this assertion is justified by arguments represented in Remark 3.5.

Proof of Theorem 3.17. In this proof, we consider cases $s > 0$ and $s < 1/2$ separately.

Case $s > 0$. By changing (if it is necessary) the numbering of operators in the system $\{B_j : j = 1, \ldots, q\}$, we obtain

$$0 \leq m_1 < m_2 < \ldots < m_q \leq 2q - 1.$$

Additionally, we set $m_0 := -1/2$ and $m_{q+1} := +\infty$. Due to condition (3.95), one can find a number $r \in \{0, 1, \ldots, q\}$ such that $m_r + 1/2 < s < m_{r+1} + 1/2$. Let $\varrho = \varrho(s)$ denote the distance from a point s to the set

$$\{j + 1/2 : j = -1/2,\, 0,\, 1, \ldots, 2q - 1\} \setminus \{s\}.$$

Pick any $\varepsilon \in (0, \varrho)$. Then

$$s \mp \varepsilon \neq j + 1/2 \quad \text{for each} \quad j \in \{0,\, 1, \ldots, 2q - 1\}, \tag{3.96}$$

$$0 \leq m_r + 1/2 < s - \varepsilon < s < s + \varepsilon < m_{r+1} + 1/2. \tag{3.97}$$

Let ψ be the interpolation parameter from Theorem 1.14 taken for $\varepsilon = \delta$. This parameter is independent of s. We interpolate the pair of spaces $H^{s\mp\varepsilon}(\text{b.c.})$ with the parameter ψ by using Theorems 3.10 and 1.6 (the interpolation of subspaces). For this purpose, we have to define a projector $\mathcal{P}$ of each space $H^{s\mp\varepsilon}(\Omega) = H^{s\mp\varepsilon}$ onto the subspace $H^{s\mp\varepsilon}(\text{b.c.})$. Such a projector does exist due to the following arguments.

At first, we assume that $r \neq 0$ and consider the collection $\{B_j : j = 1, \ldots, r\}$. Since the system $\{B_j : j = 1, \ldots, q\}$ is normal, this collection, as its part, is a normal system of boundary expressions. We now refer to the monograph by H. Triebel [258, Lemma 5.4.4] where it was shown how to construct a linear mapping $\mathcal{P}$, which is a projector of each of the spaces $H^{\sigma}(\Omega) = H^{\sigma}$ with $\sigma > m_r + 1/2$ onto the subspace

$$\{\, u \in H^{\sigma}(\Omega) : \ B_j\, u = 0 \ \text{ on } \ \Gamma, \ \ j = 1, \ldots, r \,\}. \tag{3.98}$$

Let $\sigma = s \mp \varepsilon$ in this case. Then due to (3.97) we can assert that subspace (3.98) admits the following description:

$$\{u \in H^{\sigma}(\Omega) : \ B_j\, u = 0 \ \text{ on } \ \Gamma$$

$$\text{for all } \ j \in \{1, \ldots, q\} \ \text{ such that } \ \sigma > m_j + 1/2\}. \tag{3.99}$$

At the same time, as it was shown in the monograph by Ya. A. Roitberg [209, Sec. 5.5.2, p. 167], due to condition (3.96) the regular ellipticity of the problem (3.82), (3.83) implies the density of the set $C^\infty(\text{b.c.})$ in subspace (3.99) of the space $H^\sigma(\Omega)$. Hence, the set $C^\infty(\text{b.c.})$ is dense in space (3.98), i.e., this space coincides with $H^\sigma(\text{b.c.})$. Thus, the mapping $\mathcal{P}$ represents the required projector of the space $H^{s\mp\varepsilon}(\Omega)$ onto the subspace $H^{s\mp\varepsilon}(\text{b.c.})$.

If $r = 0$, then $0 < s \mp \varepsilon < m_j + 1/2$ for each $j \in \{1, \ldots, q\}$. Hence [209, Sec. 5.5.2, p. 167], the set $C^\infty(\text{b.c.})$ is dense in the space $H^{s\mp\varepsilon}(\Omega) = H^{s\mp\varepsilon}$ and, in consequence, $H^{s\mp\varepsilon}(\text{b.c.}) = H^{s\mp\varepsilon}(\Omega)$. This means that for $r = 0$ we can use the identity mapping in order to represent $\mathcal{P}$. Thus, the required projector $\mathcal{P}$ is constructed.

Due to Theorems 1.6 (the interpolation of subspaces) and 3.10, this allows us to write down the following equalities for spaces, up to equivalence of their norms:

$$
\begin{aligned}
&[H^{s-\varepsilon}(\text{b.c.}), H^{s+\varepsilon}(\text{b.c.})]_\psi \\
&\qquad = [H^{s-\varepsilon}(\Omega), H^{s+\varepsilon}(\Omega)]_\psi \cap H^{s-\varepsilon}(\text{b.c.}) \\
&\qquad = H^{s,\varphi}(\Omega) \cap H^{s-\varepsilon}(\text{b.c.}) \\
&\qquad = H^{s,\varphi}(\Omega) \cap \{u \in H^{s-\varepsilon}(\Omega) : \; B_j u = 0 \;\text{ on }\; \Gamma, \;\; j = 1, \ldots, r\} \\
&\qquad = \big\{u \in H^{s,\varphi}(\Omega) : \; B_j u = 0 \;\text{ on }\; \Gamma \\
&\qquad\qquad\qquad \text{for all }\; j \in \{1, \ldots, q\} \;\text{ such that }\; s > m_j + 1/2\,\big\}. \qquad (3.100)
\end{aligned}
$$

Note that the last equality in (3.100) is implied by condition (3.97). Thus, the interpolation space

$$[H^{s-\varepsilon}(\text{b.c.}), H^{s+\varepsilon}(\text{b.c.})]_\psi$$

coincides (up to equivalence of norms) with subspace (3.100) of the space $H^{s,\varphi}(\Omega) = H^{s,\varphi}$. Hence, in view of Theorem 1.1, the space $H^{s+\varepsilon}(\text{b.c.})$ is embedded in (3.100) continuously and densely. This implies that the set $C^\infty(\text{b.c.})$ is dense in (3.100), i.e., space (3.100) coincides with $H^{s,\varphi}(\text{b.c.})$. Thus, assertions (i) (for $s > 0$) and (ii) are proved.

Case $s < 1/2$. We set $\varrho = 1/2 - s > 0$ and pick any number $\varepsilon \in (0, \varrho)$. Since $s - \varepsilon < s < s + \varepsilon < 1/2$, in view of 3.9(ii) the set $C_0^\infty(\Omega)$ is dense in spaces $H^{s,\varphi}$ and $H^{s\mp\varepsilon}$, and then, a fortiori, is the wider set $C^\infty(\text{b.c.})$. Therefore, $H^{s,\varphi}(\text{b.c.}) = H^{s,\varphi}$ and $H^{s\mp\varepsilon}(\text{b.c.}) = H^{s\mp\varepsilon}$. In view of Theorem 3.10, this yields assertion (i) of the theorem for $s < 1/2$ immediately.

Assertion (iii) of the theorem is a consequence of the fact that the set $C^\infty(\text{b.c.})$ is dense in the space $H^{s,\varphi}$.

The proof of Theorem 3.17 under the condition (b.c.) is completed. Note that it is based only on the property of regular ellipticity of the boundary-value problem (3.82), (3.83). Since the formally adjoint problem (3.84), (3.85) is also regular elliptic, assertion (iv) is true.

Theorem 3.17 is proved.

Let $\sigma \in \mathbb{R}$ and $\varphi \in \mathcal{M}$. We study the properties of the space $(H^{-\sigma,1/\varphi}(\text{b.c.})^{+})'$, which is antidual to the space $H^{-\sigma,1/\varphi}(\text{b.c.})^{+}$.

Recall that

$$M_{\sigma,\varphi} = \{h \in H^{\sigma,\varphi} : (h,w)_\Omega = 0 \text{ for all } w \in C^\infty(\text{b.c.})^+\}.$$

We claim that the set $M_{\sigma,\varphi}$ is closed in the space $H^{\sigma,\varphi}$. Indeed, due to Theorem 3.9(iii), the function $w \in C^\infty(\text{b.c.})^+ \subset H^{-\sigma,1/\varphi}$ generates the linear continuous functional $(\cdot\,,w)_\Omega$ on the space $H^{\sigma,\varphi}$. Hence, if for a sequence of distributions $h_j \in M_{\sigma,\varphi}$ we have $h_j \to h$ in $H^{\sigma,\varphi}$ as $j \to \infty$, then

$$(h,w)_\Omega = \lim_{j\to\infty} (h_j,w)_\Omega = 0 \quad \text{for every} \quad w \in C^\infty(\text{b.c.})^+,$$

i.e., $h \in M_{\sigma,\varphi}$. Thus, $M_{\sigma,\varphi}$ is a subspace of the Hilbert space $H^{\sigma,\varphi}$. Therefore, the factor space $H^{\sigma,\varphi}/\,M_{\sigma,\varphi}$ is also a Hilbert space.

Theorem 3.18. *Let $\sigma \in \mathbb{R}$ and $\varphi \in \mathcal{M}$. Then the following assertions hold:*

(i) *If $\sigma > -1/2$, then $M_{\sigma,\varphi} = \{0\}$.*

(ii) *The factor space $H^{\sigma,\varphi}/\,M_{\sigma,\varphi}$ and the subspace $H^{-\sigma,1/\varphi}(\text{b.c.})^{+}$ are mutually dual (with equality of norms for $s \neq 0$ and equivalence of norms for $s = 0$) with respect to extension by continuity for the inner product in $L_2(\Omega)$. More precisely, with respect to the bilinear form $\big([u]\,,v\big)_\Omega := \big(u,v\big)_\Omega$, where $[u] = \{u+h : h \in M_{\sigma,\varphi}\}$ is the coset of the element $u \in H^{\sigma,\varphi}$, and $v \in H^{-\sigma,1/\varphi}(\text{b.c.})^{+}$.*

(iii) *Compact dense embeddings*

$$H^{\sigma+\varepsilon}/\,M_{\sigma+\varepsilon} \hookrightarrow H^{\sigma,\varphi}/\,M_{\sigma,\varphi} \hookrightarrow H^{\sigma-\varepsilon}/\,M_{\sigma-\varepsilon}$$

take place for any $\varepsilon > 0$;

(iv) *If the number σ satisfies the condition*

$$-\sigma \neq m_j^+ + 1/2 \quad \text{for all} \quad j \in \{1,\dots,q\}, \tag{3.101}$$

then there exists a number $\varrho = \varrho(\sigma) > 0$ (which does not depend on φ) such that, for any $\varepsilon \in (0,\varrho)$, the following equality of spaces is valid (up to equivalence of norms):

$$[\,H^{\sigma-\varepsilon}/\,M_{\sigma-\varepsilon},\; H^{\sigma+\varepsilon}/\,M_{\sigma+\varepsilon}\,]_\psi = H^{\sigma,\varphi}/\,M_{\sigma,\varphi}.$$

Here, ψ does not depend on σ and represents the interpolation parameter from Theorem 3.10 *taken for* $\varepsilon = \delta$.

Proof. (i) Assume that $\sigma > -1/2$ and $h \in M_{\sigma,\varphi}$. Due to Theorem 3.9(iii), the functional $(h,\,\cdot\,)_\Omega$ is bounded on the space $H^{-\sigma,\,1/\varphi}$. By assumption, it is equal to zero on the set $C^\infty(\text{b.c.})^+$, which is dense in this space by Theorem 3.9(ii). Hence, this functional is equal to zero as an element of the dual space $(H^{-\sigma,\,1/\varphi})'$. Due to Theorem 3.9(iii), this implies the equality $h = 0$. Assertion (i) is proved.

(ii) Since the set $C^\infty(\text{b.c.})^+$ is dense in the space $H^{-\sigma,\,1/\varphi}(\text{b.c.})^+$, we have

$$M_{\sigma,\varphi} = \{h \in H^{\sigma,\varphi} : (h,w)_\Omega = 0 \text{ for all } w \in H^{-\sigma,\,1/\varphi}(\text{b.c.})^+\}$$

and, hence,

$$H^{-\sigma,\,1/\varphi}(\text{b.c.})^+ = \{w \in H^{-\sigma,\,1/\varphi} : (h,w)_\Omega = 0 \text{ for all } h \in M_{\sigma,\varphi}\}.$$

Now we can see that assertion (ii) is a consequence of Theorem 3.9(iii) and the well-known theorem on duals of a subspace and a factor space of a given space (see, e.g., [107, Chap. 1, § 4, Sec. 5]).

(iii) By Theorem 3.9(iv), the following dense compact embeddings take place:

$$H^{\sigma+\varepsilon} \hookrightarrow H^{\sigma,\varphi} \hookrightarrow H^{\sigma-\varepsilon} \quad \text{for any } \varepsilon > 0.$$

This implies that mappings

$$\{u+h : h \in M_{\sigma+\varepsilon}\} \mapsto \{u+h : h \in M_{\sigma,\varphi}\} \mapsto \{u+h : h \in M_{\sigma-\varepsilon}\},$$

where $u \in H^{\sigma+\varepsilon}$ for the first mapping and $u \in H^{\sigma,\varphi}$ for the second one, define dense compact embeddings indicated in assertion (iii).

(iv) Assume that the number $s = -\sigma$ satisfies condition (3.101). Consider the case $s > 0$ first. Let us reuse the proof of Theorem 3.17 where, instead of problem (3.82), (3.83), we consider formally adjoint problem (3.84), (3.85). For our case, in this proof we have to replace m_j, B_j, (b.c.), and P by m_j^+, B_j^+, (b.c.)$^+$, and P^+ respectively. Due to the above-proved result, $\mathcal{P}^+$ is a projector of every space $H^{s\mp\varepsilon}$ onto the subspace $H^{s\mp\varepsilon}(\text{b.c.})^+$. Here we have $\varepsilon \in (0,\varrho)$ for a sufficiently small positive number ϱ, and $s \mp \varepsilon > 0$. Let Π^+ be the operator adjoint to $\mathcal{P}^+$ relative to the sesquilinear form $(\,\cdot\,,\,\cdot\,)_\Omega$. By virtue of Theorem 3.9(iii) and proved assertion (ii) of the present theorem, we have the linear bounded operator

$$\Pi^+ : H^{\sigma\pm\varepsilon}/\,M_{\sigma\pm\varepsilon} \to H^{\sigma\pm\varepsilon}. \tag{3.102}$$

It has the following property:

$$u \in H^{\sigma\pm\varepsilon} \;\Rightarrow\; u - \Pi^+[u] \in M_{\sigma\pm\varepsilon}. \tag{3.103}$$

Here, $[u] = \{u + h : h \in M_{\sigma\pm\varepsilon}\}$ is the coset of u. The property is implied by the fact that for all $u \in H^{\sigma\pm\varepsilon}$ and $w \in C^\infty(\text{b.c.})^+$ we have

$$(\Pi^+[u], w)_\Omega = (\,[u], P^+w)_\Omega = (\,[u], w)_\Omega = (u, w)_\Omega,$$

i.e., $(u - \Pi^+[u], w)_\Omega = 0$.

Properties (3.102) and (3.103) imply that the mapping

$$u \mapsto u - \Pi^+[u], \quad u \in H^{\sigma\pm\varepsilon},$$

is a projector of the space $H^{\sigma\pm\varepsilon} = H^{\sigma\pm\varepsilon}_{\overline{\Omega}}(\mathbb{R}^n)$ onto the subspace $M_{\sigma\pm\varepsilon}$ (the equality follows from the condition $\sigma \pm \varepsilon = -(s \mp \varepsilon) < 0$). This allows us, due to Theorems 1.6 (the interpolation of factor spaces) and 3.10, to write down the following equalities for spaces up to equivalence of norms:

$$\begin{aligned}
&\big[\,H^{\sigma-\varepsilon}/\,M_{\sigma-\varepsilon}\,,\,H^{\sigma+\varepsilon}/\,M_{\sigma+\varepsilon}\,\big]_\psi \\
&\quad= \big[\,H^{\sigma-\varepsilon}_{\overline{\Omega}}(\mathbb{R}^n)/\,M_{\sigma-\varepsilon}\,,\,H^{\sigma+\varepsilon}_{\overline{\Omega}}(\mathbb{R}^n)/\,M_{\sigma+\varepsilon}\,\big]_\psi \\
&\quad= \big[\,H^{\sigma-\varepsilon}_{\overline{\Omega}}(\mathbb{R}^n),\,H^{\sigma+\varepsilon}_{\overline{\Omega}}(\mathbb{R}^n)\,\big]_\psi\,/\big(\,[\,H^{\sigma-\varepsilon}_{\overline{\Omega}}(\mathbb{R}^n),\,H^{\sigma+\varepsilon}_{\overline{\Omega}}(\mathbb{R}^n)\,]_\psi \cap M_{\sigma-\varepsilon}\,\big) \\
&\quad= H^{\sigma,\varphi}_{\overline{\Omega}}(\mathbb{R}^n)\,/\big(\,H^{\sigma,\varphi}_{\overline{\Omega}}(\mathbb{R}^n) \cap M_{\sigma-\varepsilon}\,\big) \\
&\quad= H^{\sigma,\varphi}\,/\big(\,H^{\sigma,\varphi} \cap M_{\sigma-\varepsilon}\,\big) \\
&\quad= H^{\sigma,\varphi}\,/\,M_{\sigma,\varphi}.
\end{aligned}$$

Here, ψ is the interpolation parameter from Theorem 3.10 where we set $\varepsilon = \delta$. Thus, assertion (iv) is proved for $s = -\sigma > 0$.

In the opposite case where $\sigma \geq 0$, the proof is trivial due to already proved assertion (i). Indeed, if we set $\varrho := 1/2$, then by virtue of Theorems 3.9(i) and 3.10, for any $\varepsilon \in (0, 1/2)$ we can write down the following equalities:

$$\begin{aligned}
[\,H^{\sigma-\varepsilon}/\,M_{\sigma-\varepsilon}\,,\,H^{\sigma+\varepsilon}/\,M_{\sigma+\varepsilon}\,]_\psi &= [\,H^{\sigma-\varepsilon},\,H^{\sigma+\varepsilon}\,]_\psi \\
&= [\,H^{\sigma-\varepsilon}(\Omega),\,H^{\sigma+\varepsilon}(\Omega)\,]_\psi \\
&= H^{\sigma,\varphi}(\Omega) = H^{\sigma,\varphi} \\
&= H^{\sigma,\varphi}\,/\,M_{\sigma,\varphi}.
\end{aligned}$$

As above, the equalities for spaces are valid up to equivalence of norms. Assertion (iv) is proved.

Theorem 3.18 is proved.

As it was already mentioned, we identify the factor space $H^{\sigma,\varphi}/\,M_{\sigma,\varphi}$ and the dual space $(H^{-\sigma,\,1/\varphi}(\text{b.c.})^{+})'$ in the sense of Theorem 3.18(ii). In this relation, for each distribution $u \in H^{\sigma,\varphi}$, its coset $[u] = \{u + h : \, h \in M_{\sigma,\varphi}\}$ is identified with the antilinear bounded functional $(u,\,\cdot\,)_{\Omega}$ on the space $H^{-\sigma,\,1/\varphi}(\text{b.c.})^{+}$. The relevant norms of the coset and the functional are equal for $s \neq 0$ and equivalent for $s = 0$. For the sake of brevity, we represent this identification of spaces (in a somewhat conditional manner) in the form of their equality,

$$H^{\sigma,\varphi}/\,M_{\sigma,\varphi} = \big(\,H^{-\sigma,\,1/\varphi}(\text{b.c.})^{+}\,\big)', \tag{3.104}$$

for any $\sigma \in \mathbb{R}$ and $\varphi \in \mathcal{M}$.

3.4.3 Proofs of theorems on isomorphisms and the Fredholm property

Let us prove Theorems 3.15 and 3.16 formulated in Subsection 3.4.1.

Proof of Theorem 3.15. For $\varphi \equiv 1$ (Sobolev spaces), this theorem was proved in the monograph by Ya. A. Roitberg [209, Theorem 5.5.2]. The case where $\varphi \in \mathcal{M}$ is arbitrary can be deduced from that result by using the interpolation with function parameter.

Due to condition (3.89), we have

$$s \neq m_j + 1/2, \quad 2q - s \neq m_j^{+} + 1/2 \quad \text{for each} \quad j \in \{1, \dots, q\}. \tag{3.105}$$

Hence, by Theorems 3.17(i) and 3.18(iv) (we set $\sigma = s - 2q$ in the latter one) there exists a sufficiently small number $\varepsilon > 0$ such that

$$[\,H^{s-\varepsilon}(\text{b.c.}),\, H^{s+\varepsilon}(\text{b.c.})\,]_{\psi} = H^{s,\varphi}(\text{b.c.}), \tag{3.106}$$

$$[\,H^{s-2q-\varepsilon}/\,M_{s-2q-\varepsilon}\,,\; H^{s-2q+\varepsilon}/\,M_{s-2q+\varepsilon}\,]_{\psi} = H^{s-2q,\,\varphi}/\,M_{s-2q,\,\varphi}, \tag{3.107}$$

where $s \mp \varepsilon \neq j + 1/2$ for each $j \in \{0, 1, \dots, 2q-1\}$, and ψ is a certain interpolation parameter. We now refer to Theorem 5.5.2 from the monograph by Ya. A. Roitberg [209, p. 168]. In accordance with that theorem, the linear mapping $u \mapsto Lu$ for $u \in C^{\infty}(\text{b.c.})$ can be continuously extended to bounded operators

$$L : H^{s\mp\varepsilon}(\text{b.c.}) \to H^{s\mp\varepsilon-2q}/M_{s\mp\varepsilon-2q}$$

and related isomorphisms

$$L : P(H^{s\mp\varepsilon}(\text{b.c.})\,) \leftrightarrow P^{+}(H^{s\mp\varepsilon-2q})/M_{s\mp\varepsilon-2q}.$$

Applying the interpolation with the parameter ψ to these extensions, we obtain the bounded operator

$$\begin{aligned} L :\; & \big[\,H^{s-\varepsilon}(\text{b.c.}),\; H^{s+\varepsilon}(\text{b.c.})\,\big]_{\psi} \\ & \to \big[\,H^{s-\varepsilon-2q}/\,M_{s-\varepsilon-2q}\,,\; H^{s+\varepsilon-2q}/\,M_{s+\varepsilon-2q}\,\big]_{\psi} \end{aligned} \tag{3.108}$$

and the isomorphism

$$L:\ \big[\,P(H^{s-\varepsilon}(\text{b.c.})),\ P(H^{s+\varepsilon}(\text{b.c.}))\,\big]_{\psi}$$
$$\leftrightarrow \big[\,P^{+}(H^{s-\varepsilon-2q})/\,M_{s-\varepsilon-2q}\,,\ P^{+}(H^{s+\varepsilon-2q})/\,M_{s+\varepsilon-2q}\,\big]_{\psi}. \tag{3.109}$$

(The pairs of spaces in (3.109) are admissible; this is implied by Theorem 1.6, see the reasoning below.) Due to interpolation formulas (3.106) and (3.107) (and also equality (3.104)), operator (3.108) turns into the bounded operator (3.90) from the theorem under consideration. It remains to show that (3.109) is equal to isomorphism (3.91).

First we prove that the domain of isomorphism (3.109) coincides with the subspace $P(H^{s,\varphi}(\text{b.c.}))$. By Lemma 3.2 the mapping P is a projector of the space $H^{s\mp\varepsilon}(\text{b.c.})$ onto the subspace

$$P(H^{s\mp\varepsilon}(\text{b.c.})) = \{u \in H^{s\mp\varepsilon}(\text{b.c.}) :\ (u,w)_{\Omega} = 0 \ \text{ for all } w \in N\}.$$

Due to Theorem 1.6 (the interpolation of subspaces) and formula (3.106), this allows us to write down the following equalities for spaces up to equivalence of norms:

$$\begin{aligned}
&[\,P(H^{s-\varepsilon}(\text{b.c.})), P(H^{s+\varepsilon}(\text{b.c.}))\,]_{\psi}\\
&\qquad = [\,H^{s-\varepsilon}(\text{b.c.}),\ H^{s+\varepsilon}(\text{b.c.})\,]_{\psi} \cap P(H^{s-\varepsilon}(\text{b.c.}))\\
&\qquad = H^{s,\varphi}(\text{b.c.}) \cap P(H^{s-\varepsilon}(\text{b.c.}))\\
&\qquad = P(H^{s,\varphi}(\text{b.c.})).
\end{aligned}$$

Thus,

$$[\,P(H^{s-\varepsilon}(\text{b.c.})),\ P(H^{s+\varepsilon}(\text{b.c.}))\,]_{\psi} = P(H^{s,\varphi}(\text{b.c.})). \tag{3.110}$$

Let us show that the range of isomorphism (3.109) is equal to the space

$$P^{+}(H^{s-2q,\varphi})/M_{s-2q,\varphi}.$$

By Lemma 3.2 the mapping P^{+} is a projector of the space $H^{s\mp\varepsilon-2q}$ onto the subspace

$$P^{+}(H^{s\mp\varepsilon-2q}) = \{f \in H^{s\mp\varepsilon-2q} :\ (f,w)_{\Omega} = 0 \ \text{ for all } w \in N^{+}\}.$$

Consider the linear mapping

$$\begin{aligned}
[f] &= \{f + h :\ h \in M_{s\mp\varepsilon-2q}\}\\
&\qquad \mapsto [P^{+}f] = \{P^{+}f + h :\ h \in M_{s\mp\varepsilon-2q}\},
\end{aligned} \tag{3.111}$$

where $f \in H^{s\mp\varepsilon-2q}$. This mapping is well defined. Indeed, since $M_{s\mp\varepsilon-2q} \subset P^{+}(H^{s\mp\varepsilon-2q})$, we have $P^{+}h = h$ for any $h \in M_{s\mp\varepsilon-2q}$. Hence, the coset $[P^{+}f]$

does not depend on the choice of a representative f from the coset $[f]$. Properties of the mapping P^{+} imply that mapping (3.111) is a projector of the space $H^{s\mp\varepsilon-2q}/M_{s\mp\varepsilon-2q}$ onto the subspace $P^{+}(H^{s\mp\varepsilon-2q})/M_{s\mp\varepsilon-2q}$. Due to Theorem 1.6 (the interpolation of subspaces) and formula (3.107), this allows us to write down the following equalities for spaces up to equivalence of their norms:

$$
\begin{aligned}
&[\,P^{+}(H^{s-\varepsilon-2q})/\,M_{s-\varepsilon-2q}\,,\,P^{+}(H^{s+\varepsilon-2q})/\,M_{s+\varepsilon-2q}\,]_{\psi}\\
&\qquad=[\,H^{s-\varepsilon-2q}/\,M_{s-\varepsilon-2q}\,,\,H^{s+\varepsilon-2q}/\,M_{s+\varepsilon-2q}\,]_{\psi}\\
&\qquad\qquad\cap\,(\,P^{+}(H^{s-\varepsilon-2q})/\,M_{s-\varepsilon-2q}\,)\\
&\qquad=(\,H^{s-2q,\varphi}/\,M_{s-2q,\varphi})\,\cap\,(\,P^{+}(H^{s-\varepsilon-2q})/\,M_{s-\varepsilon-2q}\,)\\
&\qquad=(\,H^{s-2q,\varphi}\cap P^{+}(H^{s-\varepsilon-2q})\,)/\,M_{s-2q,\varphi}\\
&\qquad=P^{+}(H^{s-2q,\varphi})/M_{s-2q,\varphi}.
\end{aligned}
$$

Thus,

$$
\begin{aligned}
&\big[\,P^{+}(H^{s-\varepsilon-2q})/\,M_{s-\varepsilon-2q}\,,\,P^{+}(H^{s+\varepsilon-2q})/\,M_{s+\varepsilon-2q}\,\big]_{\psi}\\
&\qquad\qquad=P^{+}(H^{s-2q,\varphi})/M_{s-2q,\varphi}.
\end{aligned}
\tag{3.112}
$$

Interpolation formulas (3.110) and (3.112) imply that mapping (3.109) generates isomorphism (3.91).

Theorem 3.15 is proved.

Proof of Theorem 3.16. Recall that spaces N and N^{+} are finite-dimensional. First we show that N is the kernel of operator (3.90). Since $N\subset C^{\infty}(\text{b.c.})$, the image of any element $u\in N$ under mapping (3.90) is the coset

$$\{Lu+h:\,h\in M_{s-2q,\varphi}\}=\{0+h:\,h\in M_{s-2q,\varphi}\},$$

i.e., it is zero element of the factor space $H^{s-2q,\varphi}/M_{s-2q,\varphi}$. The inverse assertion is also true:

If an element $u\in H^{s,\varphi}$ satisfies the condition $Lu=0$, then due to the expansion $u=u_0+Pu$ stated by Lemma 3.2, where $u_0\in N$, we obtain the equality

$$0=Lu=Lu_0+LPu=LPu.$$

Due to isomorphism (3.91), this yields $Pu=0$, i.e., $u=u_0\in N$. Thus, N is the finite-dimensional kernel of operator (3.90).

Isomorphism (3.91) also implies that the range of operator (3.90) coincides with the factor space $P^{+}(H^{s-2q,\varphi})/M_{s-2q,\varphi}$, i.e., with (3.92). Therefore, the range is closed in the space $H^{s-2q,\varphi}/M_{s-2q,\varphi}$. The codimension of the range is equal to the dimension of the factor space

$$(H^{s-2q,\varphi}/M_{s-2q,\varphi})/(P^{+}(H^{s-2q,\varphi})/M_{s-2q,\varphi}) = H^{s-2q,\varphi}/P^{+}(H^{s-2q,\varphi}).$$

By virtue of Lemma 3.2, this dimension coincides with $\dim N^{+}$, and therefore, it is finite. Thus, operator (3.90) is a Fredholm operator, and its index is equal to $\dim N - \dim N^{+}$.

It remains to prove *a priori* estimate (3.93). Consider an arbitrary distribution $u \in H^{s,\varphi}(\text{b.c.})$ and fix a number $\varepsilon > 0$. By Lemma 3.2 the inclusion $u - Pu \in N$ takes place. At the same time, N is a finite-dimensional subspace of each of the spaces $H^{s,\varphi}$ and $H^{s-\varepsilon}$. Hence, norms in these spaces are equivalent on N. In particular,

$$\|u - Pu\|_{H^{s,\varphi}} \le c_1 \|u - Pu\|_{H^{s-\varepsilon}},$$

where $c_1 > 0$ is a constant that does not depend on u. This implies that

$$\begin{aligned}\|u\|_{H^{s,\varphi}} &\le \|u - Pu\|_{H^{s,\varphi}} + \|Pu\|_{H^{s,\varphi}} \\ &\le c_1 \|u - Pu\|_{H^{s-\varepsilon}} + \|Pu\|_{H^{s,\varphi}} \\ &\le c_1 \|u\|_{H^{s-\varepsilon}} + c_1 \|Pu\|_{H^{s-\varepsilon}} + \|Pu\|_{H^{s,\varphi}} \\ &\le c_1 \|u\|_{H^{s-\varepsilon}} + (c_1 c_2 + 1) \|Pu\|_{H^{s,\varphi}},\end{aligned}$$

where c_2 is the norm of the embedding operator $H^{s,\varphi} \hookrightarrow H^{s-\varepsilon}$ (see Theorem 3.9(iv)). Thus,

$$\|u\|_{H^{s,\varphi}} \le c_1 \|u\|_{H^{s-\varepsilon}} + (c_1 c_2 + 1) \|Pu\|_{H^{s,\varphi}}. \tag{3.113}$$

Now, let $Lu = [f]$. Since it was proved above that N is the kernel of operator (3.90) and $u - Pu \in N$, we have $LPu = [f]$. Thus, Pu is the preimage of the coset $[f]$ under isomorphism (3.91). Hence,

$$\|Pu\|_{H^{s,\varphi}} \le c_0 \|f\|_{H^{s-2q,\varphi}},$$

where c_0 is the norm of the operator inverse to (3.91). This result and inequality (3.113) yield *a priori* estimate (3.93) immediately.

Theorem 3.16 is proved.

3.4.4 Local increase in smoothness of solutions up to the boundary

Let $H^{-\infty}$ denote the union of all spaces $\mathrm{H}^{s,\varphi}$, where $s \in \mathbb{R}$ and $\varphi \in \mathcal{M}$. We set

$$M_{-\infty} := \{h \in H^{-\infty} : \ (h, w)_\Omega = 0 \ \text{for all} \ w \in C(\mathrm{b.c.})^{+}\}.$$

For $s \in \mathbb{R}$ and $\varphi \in \mathcal{M}$, operators (3.90) define a linear mapping

$$L : \ H^{-\infty} \to H^{-\infty}/M_{-\infty}$$

and an isomorphism

$$L : \ P(H^{-\infty}) \leftrightarrow P^{+}(H^{-\infty})/M_{-\infty}. \tag{3.114}$$

Consider the following problem:

Suppose that a distribution $u \in H^{-\infty}$ satisfies the equation

$$Lu = \{f + h : h \in M_{-\infty}\}, \tag{3.115}$$

where f has a given smoothness in the refined Sobolev scale on some set U, which is open in $\overline{\Omega}$. Then what can be asserted on the smoothness of solutions on this set?

We now answer this question.

Let U be an open set in the space $\mathbb{R}^n$, and let $\Omega_0 := \overline{\Omega} \cap U \neq \varnothing$. We set $\Gamma_0 := \Gamma \cap U$ (the case where $\Gamma_0 = \varnothing$ is also possible). Let us introduce the following spaces, which are characterized by a refined smoothness on Ω_0.

Let $s \in \mathbb{R}$ and $\varphi \in \mathcal{M}$. We define

$$H^{s,\varphi,(0)}_{\mathrm{loc}}(\Omega_0) := \{u \in H^{-\infty} : \ \chi u \in H^{s,\varphi} \ \text{for all} \ \chi \in C^\infty(\overline{\Omega}), \ \operatorname{supp}\chi \subset \Omega_0\}.$$

Note that the operator of multiplication by a function $\chi \in C^\infty(\overline{\Omega})$ is bounded on the space $H^{s,\varphi}$. This fact is well known in the Sobolev case where $\varphi \equiv 1$ (see, e.g., [209, Sec 1.12, p. 57]); due to interpolation theorem 3.10, this is also true for any $\varphi \in \mathcal{M}$.

In view of Theorem 3.17(ii), we set

$$\begin{aligned} H^{s,\varphi}_{\mathrm{loc}}(\Omega_0, \mathrm{b.c.}, \Gamma_0) := \{u \in H^{s,\varphi,(0)}_{\mathrm{loc}}(\Omega_0) : \ B_j\, u = 0 \ \text{ on } \Gamma_0 \\ \text{for each } \ j \in \{1, \dots, q\} \ \text{ such that } \ s > m_j + 1/2\}, \end{aligned}$$

where we assume that $s \neq m_j + 1/2$ for any $j \in \{1, \dots, q\}$.

Theorem 3.19. *Let $u \in H^{-\infty}$ be a solution of equation* (3.115), *where $f \in H^{s-2q,\varphi,(0)}_{\mathrm{loc}}(\Omega_0)$ for some $s \in \mathbb{R}$ satisfying* (3.89), *and some $\varphi \in \mathcal{M}$. Then $u \in H^{s,\varphi}_{\mathrm{loc}}(\Omega_0, \mathrm{b.c.}, \Gamma_0)$.*

Theorem 3.19 is a statement on a local increase in the refined smoothness of the solution u to the boundary-value problem (3.82), (3.83) up to the boundary of the domain Ω. Note that in the case where $\overline{\Omega} \subset U$, i.e., in the case where $\Omega_0 = \overline{\Omega}$ and $\Gamma_0 = \Gamma$, the "local" spaces coincide with "global" ones:

$$H^{s,\varphi,(0)}_{\mathrm{loc}}(\Omega_0) = H^{s,\varphi} \quad \text{and} \quad H^{s,\varphi}_{\mathrm{loc}}(\Omega_0, \mathrm{b.c.}, \Gamma_0) = H^{s,\varphi}(\mathrm{b.c.}).$$

Therefore, Theorem 3.19 also states the global increase in smoothness, i.e., it states that the smoothness increases globally in the whole closed domain $\overline{\Omega}$. At last, in addition, we emphasize the case where $U \subset \Omega$, i.e., $\Gamma_0 = \varnothing$, that leads to a statement on increase in smoothness inside the domain Ω.

Remark 3.10. For the Sobolev case $\varphi \equiv 1$, Theorem 3.19 was proved by Yu. M. Berezansky, S. G. Krein, and Ya. A. Roitberg (see papers [22, 201] and the monograph [209, Theorem 7.3.1]).

Proof of Theorem 3.19. As it was mentioned in Remark 3.9, this theorem is known for the case where $\varphi \equiv 1$. By using this known result, we prove it for any $\varphi \in \mathcal{M}$ with the help of Theorem 3.15 on isomorphism.

For any number $\varepsilon > 0$, we set

$$U_\varepsilon := \{x \in U : \mathrm{dist}(x, \partial U) > \varepsilon\}, \quad \Omega_\varepsilon := \overline{\Omega} \cap U_\varepsilon, \quad \Gamma_\varepsilon := \Gamma \cap U_\varepsilon.$$

Here, as usual, ∂U is the boundary of the set U. Then we consider a function $\chi_\varepsilon \in C^\infty(\overline{\Omega})$ such that $\mathrm{supp}\,\chi_\varepsilon \subset \Omega_0$ and $\chi_\varepsilon \equiv 1$ on Ω_ε.

Let us represent the distribution f in the form $f = \chi_\varepsilon f + (1 - \chi_\varepsilon) f$. Since $f \in H^{s-2q,\varphi,(0)}_{\mathrm{loc}}(\Omega_0)$ and $\mathrm{supp}\,\chi_\varepsilon \subset \Omega_0$, we have $\chi_\varepsilon f \in H^{s-2q,\varphi}$. Hence, due to isomorphism (3.91) from Theorem 3.15, we conclude that there exists a distribution $u_\varepsilon \in P(H^{s,\varphi}(\mathrm{b.c.}))$ such that

$$Lu_\varepsilon = [\, P^+ \chi_\varepsilon f \,]. \tag{3.116}$$

Since $1 - \chi_\varepsilon = 0$ on the set Ω_ε, we have $\chi(1 - \chi_\varepsilon) f \equiv 0$ for any function $\chi \in C^\infty(\overline{\Omega})$ with $\mathrm{supp}\,\chi \subset \Omega_\varepsilon$. In particular, this yields $(1-\chi_\varepsilon) f \in H^{s+1-2q,(0)}_{\mathrm{loc}}(\Omega_\varepsilon)$. By Lemma 3.2 we write

$$(1 - \chi_\varepsilon) f = g_\varepsilon + P^+(1 - \chi_\varepsilon) f,$$

where $g_\varepsilon \in N^+ \subset C^\infty(\mathrm{b.c.})$. Hence,

$$P^+(1 - \chi_\varepsilon) f \in H^{s+1-2q,(0)}_{\mathrm{loc}}(\Omega_\varepsilon).$$

In addition, due to (3.114), there exists $v_\varepsilon \in P(H^{-\infty})$ such that

$$Lv_\varepsilon = [\, P^+(1 - \chi_\varepsilon) f \,]. \tag{3.117}$$

Hence, by [209, Theorem 7.3.1], the inclusion $v_\varepsilon \in H^{s+1}_{\mathrm{loc}}(\Omega_\varepsilon, \text{b.c.}, \Gamma_\varepsilon)$ holds. Due to (3.116), (3.117), and the condition $u_\varepsilon \in H^{s,\varphi}(\text{b.c.})$, this implies the equality

$$L(u_\varepsilon + v_\varepsilon) = [\,P^+\chi_\varepsilon f\,] + [\,P^+(1-\chi_\varepsilon)f\,] = [\,P^+ f\,],$$

where

$$u_\varepsilon + v_\varepsilon \in H^{s,\varphi}(\text{b.c.}) \cup H^{s+1}_{\mathrm{loc}}(\Omega_\varepsilon, \text{b.c.}, \Gamma_\varepsilon) \subset H^{s,\varphi}_{\mathrm{loc}}(\Omega_\varepsilon, \text{b.c.}, \Gamma_\varepsilon).$$

To prove this, we used the embedding $H^{s+1} \hookrightarrow H^{s,\varphi}$ and the definition of the space $H^{s,\varphi}_{\mathrm{loc}}(\Omega_\varepsilon, \text{b.c.}, \Gamma_\varepsilon)$.

At the same time, $Lu = [f]$. Hence, due to (3.114), the relation $f = P^+ f$ holds. Therefore,

$$L(u - u_\varepsilon - v_\varepsilon) = [f] - [\,P^+ f\,] = 0,$$

where $u_\varepsilon + v_\varepsilon \in H^{s,\varphi}_{\mathrm{loc}}(\Omega_\varepsilon, \text{b.c.}, \Gamma_\varepsilon)$. Due to Theorem 3.15, this implies

$$v := u - u_\varepsilon - v_\varepsilon \in N \subset C^\infty(\text{b.c.}).$$

Thus,

$$u = u_\varepsilon + v_\varepsilon + v \in H^{s,\varphi}_{\mathrm{loc}}(\Omega_\varepsilon, \text{b.c.}, \Gamma_\varepsilon).$$

Recall that the number $\varepsilon > 0$ in the proof is taken arbitraryly. It is clear that

$$H^{s,\varphi}_{\mathrm{loc}}(\Omega_0, \text{b.c.}, \Gamma_0) = \bigcap_{\varepsilon>0} H^{s,\varphi}_{\mathrm{loc}}(\Omega_\varepsilon, \text{b.c.}, \Gamma_\varepsilon).$$

Hence, $u \in H^{s,\varphi}_{\mathrm{loc}}(\Omega_0, \text{b.c.}, \Gamma_0)$.

Theorem 3.19 is proved.

3.5 Some properties of Hörmander spaces

In the present section, we study some subspaces of the space $H^{s,\varphi}(\Omega)$ and give an equivalent description of this space. These results are used, in particular, in Section 4.5.

3.5.1 Space $H^{s,\varphi}_0(\Omega)$ and its properties

Let $s \in \mathbb{R}$ and $\varphi \in \mathcal{M}$. By $H^{s,\varphi}_0(\Omega)$ we denote the closure of the set $C^\infty_0(\Omega)$ in the space $H^{s,\varphi}(\Omega)$. We treat $H^{s,\varphi}_0(\Omega)$ as a Hilbert space with respect to the inner product in $H^{s,\varphi}(\Omega)$. In this subsection, we study the properties of the space $H^{s,\varphi}_0(\Omega)$ that lead to various equivalent descriptions of the space $H^{s,\varphi}(\Omega)$ for negative nonhalf-integer indices s. These important results are used in what follows. We delayed the representation of the proof for these results to the present subsection because our proof is based on a result obtained in Subsection 3.4.2.

For $k \in \mathbb{N}$, we set

$$C^{\infty}_{\nu,k}(\overline{\Omega}) := \{u \in C^{\infty}(\overline{\Omega}) : D^{j-1}_{\nu} u = 0 \text{ on } \Gamma, \ j = 1, \dots, k\}. \tag{3.118}$$

Here and in what follows, for the sake of convenience we use the notation $D_\nu := i\,\partial/\partial\nu$. Let $H^{s,\varphi}_{\nu,k}(\Omega)$ denote the closure of the set $C^{\infty}_{\nu,k}(\overline{\Omega})$ in the space $H^{s,\varphi}(\Omega)$. We treat $H^{s,\varphi}_{\nu,k}(\Omega)$ as a Hilbert space with respect to the inner product in $H^{s,\varphi}(\Omega)$.

If $s \geq 0$, then $H^{s,\varphi}_{\nu,k}(\Omega)$ is equal to the space $H^{s,\varphi}(\text{b.c.})$ introduced in Subsection 3.4.1, where (b.c.) stands for the Dirichlet homogeneous boundary conditions represented in (3.118). Therefore, by Theorem 3.17(ii),

$$H^{s,\varphi}_{\nu,k}(\Omega) = \{u \in H^{s,\varphi}(\Omega) : \ D^{j-1}_{\nu} u = 0 \ \text{ on } \ \Gamma$$

$$\text{for all } \ j = 1, \dots, k\} \quad \text{if} \quad s > k - 1/2. \tag{3.119}$$

By virtue of Theorem 3.3(ii), we have

$$H^{s,\varphi}_{0}(\Omega) = H^{s,\varphi}(\Omega) \quad \text{for } \ s < 1/2. \tag{3.120}$$

If $s > 1/2$, then the inclusion $H^{s,\varphi}_{0}(\Omega) \subset H^{s,\varphi}(\Omega)$ is proper. This is implied by Theorem 3.5 on traces.

Theorem 3.20. *Let $s > 1/2$ and $s - 1/2 \notin \mathbb{Z}$, and let $\varphi \in \mathcal{M}$. Then the following assertions hold:*

(i) *$H^{s,\varphi}_{0}(\Omega)$ coincides with space (3.119), where $k := [s + 1/2]$.*

(ii) *The norms in spaces $H^{s,\varphi}_{0}(\Omega)$ and $H^{s,\varphi}_{\overline{\Omega}}(\mathbb{R}^n)$ are equivalent on the dense subset $C^{\infty}_{0}(\Omega)$, and therefore, $H^{s,\varphi}_{0}(\Omega) = H^{s,\varphi}_{\overline{\Omega}}(\mathbb{R}^n)$ up to equivalence of norms.*

(iii) *The spaces $H^{s,\varphi}_{0}(\Omega)$ and $H^{-s,1/\varphi}(\Omega)$ are mutually dual (up to equivalence of norms) with respect to the inner product in $L_2(\Omega)$.*

Proof. In the Sobolev case where $\varphi \equiv 1$, this theorem is well known. Its proof is represented, e.g, in the monograph by H. Triebel [258, Chap. 4]. Namely, assertions (i), (ii), and (iii) are contained, respectively, in Theorems 4.7.1(a), 4.3.2/1(c), and 4.8.2(a) of this monograph. From results for this case, we deduce assertions (i) – (iii) for arbitrary $\varphi \in \mathcal{M}$.

(i) Let $k := [s+1/2]$. Then $k - 1/2 < s < k + 1/2$. Pick a number $\varepsilon > 0$ such that

$$k - 1/2 < s \mp \varepsilon < k + 1/2. \tag{3.121}$$

Then we have the following dense continuous embedding:

$$H^{s+\varepsilon}_{0}(\Omega) = H^{s+\varepsilon}_{\nu,k}(\Omega) \hookrightarrow H^{s,\varphi}_{\nu,k}(\Omega).$$

As it is indicated above, this equality for Sobolev spaces is true since $[s+\varepsilon+1/2]=k$. Therefore, the set $C_0^\infty(\Omega)$ is dense in $H^{s,\varphi}_{\nu,k}(\Omega)$. Hence, $H^{s,\varphi}_0(\Omega)=H^{s,\varphi}_{\nu,k}(\Omega)$, and assertion (i) is proved.

(ii) We deduce assertion (ii) from the Sobolev case ($\varphi\equiv 1$) by using the interpolation. Let numbers k and ε be like those that considered in the previous paragraph. By $\mathcal{O}$ we denote the operator of extension by zero over the space $\mathbb{R}^n$ for functions from the domain $\overline{\Omega}$. The mapping $u\mapsto \mathcal{O}u$, where $u\in C_0^\infty(\Omega)$, can be extended by continuity to isomorphisms

$$\mathcal{O}:\,H_0^{s\mp\varepsilon}(\Omega)\leftrightarrow H_{\overline{\Omega}}^{s\mp\varepsilon}(\mathbb{R}^n).$$

Let us apply the interpolation with the functional parameter ψ from Theorem 1.14 taken for $\delta=\varepsilon$. Due to Theorem 3.7, additionally we can obtain the isomorphism

$$\mathcal{O}:\,\bigl[H_0^{s-\varepsilon}(\Omega),H_0^{s+\varepsilon}(\Omega)\bigr]_\psi\leftrightarrow H_{\overline{\Omega}}^{s,\varphi}(\mathbb{R}^n). \tag{3.122}$$

Let us describe the domain of operator (3.122). Due to inequality (3.121) and assertion (i), we have

$$H_0^{s\mp\varepsilon}(\Omega)=H_{\nu,k}^{s\mp\varepsilon}(\Omega)\quad\text{and}\quad H_{\nu,k}^{s,\varphi}(\Omega)=H_0^{s,\varphi}(\Omega).$$

Hence, on the base of Theorem 3.17(i), we obtain

$$\begin{aligned}[H_0^{s-\varepsilon}(\Omega),H_0^{s+\varepsilon}(\Omega)]_\psi &=\bigl[H_{\nu,k}^{s-\varepsilon}(\Omega),H_{\nu,k}^{s+\varepsilon}(\Omega)\bigr]_\psi\\ &=H_{\nu,k}^{s,\varphi}(\Omega)\\ &=H_0^{s,\varphi}(\Omega).\end{aligned}$$

Here, the second equality for spaces is satisfied up to equivalence of norms. Hence, (3.122) means the isomorphism

$$\mathcal{O}:\,H_0^{s,\varphi}(\Omega)\leftrightarrow H_{\overline{\Omega}}^{s,\varphi}(\mathbb{R}^n).$$

Thus, assertion (ii) is proved.

Assertion (iii) is immediately implied by Theorem 3.8(iii) and assertion (ii). Theorem 3.20 is proved.

3.5.2 Equivalent description of $H^{s,\varphi}(\Omega)$

By Theorem 3.20(iii) the space $H^{s,\varphi}(\Omega)$ with negative nonhalf-integer indices s can be defined as a space, which is dual to $H_0^{-s,1/\varphi}(\Omega)$ with respect to the inner product in $L_2(\Omega)$. In the Sobolev case, such a definition was used by J.-L. Lions and E. Magenes [119, 120, 121, 126] in their study of boundary-value problems (see Remark 3.6).

This yields a distinct (equivalent) description of the space $H^{s,\varphi}(\Omega)$. Let us recall that by definition we have

$$H^{s,\varphi}(\Omega) = \{w \restriction \Omega : \, w \in H^{s,\varphi}(\mathbb{R}^n)\}.$$

It turns out that here we can restrict the study to the set of distributions w with supports in $\overline{\Omega}$.

Theorem 3.21. *Let $s < 1/2$ and $s - 1/2 \notin \mathbb{Z}$, and let $\varphi \in \mathcal{M}$. Then*

$$H^{s,\varphi}(\Omega) = H^{s,\varphi}_{\overline{\Omega}}(\mathbb{R}^n)/H^{s,\varphi}_{\Gamma}(\mathbb{R}^n) = \big\{w \restriction \Omega : \, w \in H^{s,\varphi}_{\overline{\Omega}}(\mathbb{R}^n)\big\}, \tag{3.123}$$

$$\|u\|_{H^{s,\varphi}(\Omega)} \asymp \inf\big\{ \|w\|_{H^{s,\varphi}(\mathbb{R}^n)} : \, w \in H^{s,\varphi}_{\overline{\Omega}}(\mathbb{R}^n), \; w = u \textit{ in } \Omega \big\}. \tag{3.124}$$

Proof. By virtue of Theorem 3.20(iii), for each nonhalf-integer $s < -1/2$ we have the equality

$$H^{s,\varphi}(\Omega) = (H^{-s,1/\varphi}_0(\Omega))' \quad \text{with equivalence of norms.} \tag{3.125}$$

Consider the duality of spaces with respect to the inner product in $L_2(\Omega)$. If $-1/2 < s < 1/2$, then (3.125) is also true due to Theorem 3.8(ii), (iii) and equality (3.120). Namely,

$$\begin{aligned} H^{s,\varphi}(\Omega) &= \big(H^{-s,1/\varphi}_{\overline{\Omega}}(\mathbb{R}^n)\big)' \\ &= \big(H^{-s,1/\varphi}(\Omega)\big)' \\ &= \big(H^{-s,1/\varphi}_0(\Omega)\big)'. \end{aligned}$$

Let us describe $(H^{-s,1/\varphi}_0(\Omega))'$ as a space, which is dual to the subspace $H^{-s,1/\varphi}_0(\Omega)$ of the space $H^{-s,1/\varphi}(\Omega)$. In view of Theorem 3.8(iii), we have

$$\begin{aligned} (H^{-s,1/\varphi}_0(\Omega))' &= (H^{-s,1/\varphi}(\Omega))'/G_{s,\varphi} \\ &= H^{s,\varphi}_{\overline{\Omega}}(\mathbb{R}^n)/G_{s,\varphi}. \end{aligned}$$

Here,

$$\begin{aligned} G_{s,\varphi} &:= \big\{w \in H^{s,\varphi}_{\overline{\Omega}}(\mathbb{R}^n) : \, (w,v)_\Omega = 0 \text{ for all } v \in H^{-s,1/\varphi}_0(\Omega)\big\} \\ &= \big\{w \in H^{s,\varphi}_{\overline{\Omega}}(\mathbb{R}^n) : \, (w,v)_\Omega = 0 \text{ for all } v \in C^\infty_0(\Omega)\big\} \\ &= H^{s,\varphi}_{\Gamma}(\mathbb{R}^n). \end{aligned}$$

Thus, by virtue of (3.125),

$$H^{s,\varphi}(\Omega) = H^{s,\varphi}_{\overline{\Omega}}(\mathbb{R}^n)/H^{s,\varphi}_{\Gamma}(\mathbb{R}^n)$$

up to equivalence of norms. This implies relations (3.123) and (3.124) immediately.

Theorem 3.21 is proved.

3.6 Remarks and comments

Section 3.1. The notion of general elliptic boundary-value problem was first formulated by Ya. B. Lopatinskii [123, 124]. In special cases, Z. Ya. Shapiro [230], N. Aronszajn and A. N. Milgram [16], and independently, M. Schechter [222] introduced the important condition of normality for a system of boundary expressions. This condition ensures the existence of a formally adjoint boundary-value problem in the class of differential operators.

Systematic presentations of the theory of general elliptic boundary-value problems can be found, e.g., in the monographs by S. Agmon [3], Yu. M. Berezansky [21], L. Hörmander [81, 86], J.-L. Lions and E. Magenes [121], O. I. Panich [184], Ya. A. Roitberg [209], M. Schechter [227], H. Triebel [258], and in the survey by M. S. Agranovich [11].

Section 3.2. For open or closed Euclidean domains, the Hörmander inner product spaces were introduced and studied by L. R. Volevich and B. P. Paneah [269]. They used standard definitions from the theory of function spaces (see, e.g., the monograph by H. Triebel [258, Sec. 4.2.1 and 4.3.2]).

The refined scales over Euclidean domains were introduced in [143]. Unlike L. R. Volevich and B. P. Paneah, we established the properties of Hörmander spaces over the domains by using interpolation formulas that relate these spaces with the Sobolev scale. Theorems 3.1–3.5 on the properties of the refined scale over Euclidean domains were proved in [145, Sec. 3] in a more general situation of the refined scale over a manifold with boundary. Theorems 3.6–3.8 on the properties of the refined scale over closed Euclidean domains were established in [149, Sec. 4]. Theorem 3.9 was also proved in the same work. The proof of the interpolation theorem 3.10 was not published earlier. The last two theorems describe the properties of the rigging of the space $L_2(\Omega)$ with Hörmander spaces.

The definition of Hilbert rigging and related notions can be found, e.g., in the monographs by Yu. M. Berezansky [21, Chap. 1, Sec. 1.1] and Yu. M. Berezansky, G. F. Us, and Z. G. Sheftel [23, Chap. 14, Sec. 1.1]. The rigging with Sobolev spaces was introduced and studied by Yu. M. Berezansky (see [21, Chap. 1, Sec. 3] and [23, Chap. 14, Sec. 3 and 4]).

In [231, Sec. 2], G. Shlenzak proved an interpolation formula connecting the Sobolev scale with some Hörmander spaces given over a Euclidean domain with smooth boundary. They were applied by G. Shlenzak to the theory of general elliptic boundary-value problems.

Section 3.3. All theorems of this section were proved by the authors in [148]. The main result (Theorem 3.11) is new even in the Sobolev case for half integer $s < 0$. For the remaining values of the parameter s, it is contained in the Lions–Magenes theorem [121, Sec. 7.3] on solvability of regular inhomogeneous elliptic boundary-value problems in the two-sided scale of Sobolev spaces (see also [119, 120]).

Section 3.4. The theorem on isomorphisms generated by an elliptic operator in the two-sided Sobolev scale under homogeneous normal boundary conditions was proved by Yu. M. Berezansky, S. G. Krein, and Ya. A. Roitberg [22, Theorem 2]. In the same paper, they also established the theorem on local increase in smoothness of solutions of elliptic equations up to the boundary of the domain. Independently, M. Schechter [225] established an appropriate *a priori* estimate for the solutions of elliptic equations. The proofs of these results can be found in the monographs by Yu. M. Berezansky [21, Chap. 3, Sec. 6.6 and 6.12] and Ya. A. Roitberg [209, Sec. 5.5 and 7.3]. See also the survey by M. S. Agranovich [11, Sec. 7.9 c] and the book by Yu. M. Berezansky, G. F. Us, and Z. G. Sheftel [23, Chap. 16, Sec. 1.1 and 2.3] (in the last book, the authors consider the case of second-order elliptic expressions with real coefficients).

A generalization of these results to the case of boundary conditions that are not normal was obtained by Yu. V. Kostarchuk and Ya. A. Roitberg [97, Sec. 4]. The case of elliptic systems was studied by I. Ya. Roitberg and Ya. A. Roitberg [200, Sec. 3.7]; see also the monograph by Ya. A. Roitberg [210, Sec. 1.3.7].

In connection with Subsection 3.4.2, we note that the interpolation with number parameters between the Sobolev spaces satisfying homogeneous boundary conditions was studied by P. Grisvard [66], R. T. Seeley [234], and J. Löfström [125] (the results of the first two authors can also be found in Triebel's monograph [258, Sec. 4.3.3]).

All theorems in Section 3.4 (except Theorem 3.19) were proved in [149]. Theorem 3.19 was announced, together with Theorems 3.15 and 3.16, in [143].

Section 3.5. The results obtained in this section give other equivalent definitions of the Hörmander spaces $H^{s,\varphi}(\Omega)$ over Euclidean domains for any nonhalf-integer $s < 0$. In the Sobolev case, these definitions were used by J.-L. Lions and E. Magenes [119, 120, 121, 126] in their investigation of boundary-value problems. The theorems presented in this section were not published earlier.

Chapter 4

Inhomogeneous elliptic boundary-value problems

4.1 Elliptic boundary-value problems in the positive one-sided scale

In this section, we investigate inhomogeneous elliptic boundary-value problems in Hörmander spaces that form the positive one-sided refined Sobolev scale. For these spaces, the number index defining the main smoothness is positive. In Subsections 4.1.1 and 4.1.2, a regular elliptic boundary-value problem is investigated. In the rest subsections, other important classes of elliptic boundary-value problems are considered.

4.1.1 Theorems on Fredholm property and isomorphisms

Consider the inhomogeneous regular elliptic boundary-value problem (3.1), (3.2), namely

$$Lu = f \quad \text{in} \quad \Omega, \qquad B_j\, u = g_j \quad \text{on} \quad \Gamma \quad \text{for} \quad j = 1, \ldots, q. \tag{4.1}$$

With it, we associate the linear mapping

$$u \mapsto (Lu, Bu) := (Lu, B_1 u, \ldots, B_q u), \quad u \in C^\infty(\overline{\Omega}). \tag{4.2}$$

We study properties of the operator (L, B), which is an extension of this mapping by continuity in the corresponding pairs of positive Hörmander spaces. Recall that the finite-dimensional spaces $N, N^+ \subset C^\infty(\overline{\Omega})$ are defined in Section 3.1

Theorem 4.1. *For arbitrary parameters $s > 2q$ and $\varphi \in \mathcal{M}$, mapping* (4.2) *extends uniquely (by continuity) to the bounded linear operator*

$$(L, B) : H^{s,\varphi}(\Omega) \to H^{s-2q,\varphi}(\Omega) \oplus \bigoplus_{j=1}^{q} H^{s-m_j-1/2,\varphi}(\Gamma) =: \mathcal{H}_{s,\varphi}(\Omega, \Gamma). \tag{4.3}$$

This operator is Fredholm. Its kernel is equal to N and the range consists of all vectors $(f, g_1, \ldots, g_q) \in \mathcal{H}_{s,\varphi}(\Omega, \Gamma)$ such that

$$(f, v)_\Omega + \sum_{j=1}^{q} (g_j, C_j^+ v)_\Gamma = 0 \quad \textit{for all} \quad v \in N^+. \tag{4.4}$$

The index of operator (4.3) *is equal to* $\dim N - \dim N^+$ *and does not depend on* s *and* φ.

Proof. In the Sobolev case where $\varphi \equiv 1$ and $s \geq 2q$, this theorem is the classical result on solvability of regular elliptic boundary-value problems. It is proved, e.g., in the monographs by Yu. M. Berezansky [21, Chap. 3, Sec. 6] and J.-L. Lions and E. Magenes [121, Part. 2, Sec. 5.2]. The general case of $\varphi \in \mathcal{M}$ is obtained from the Sobolev case with the use of interpolation with a function parameter.

Namely, let $s > 2q$ and let $\varepsilon := s - 2q$. Mapping (4.2) is extended by continuity to the bounded and Fredholm operators

$$(L, B) : H^{s\mp\varepsilon}(\Omega) \to H^{s\mp\varepsilon-2q}(\Omega) \oplus \bigoplus_{j=1}^{q} H^{s\mp\varepsilon-m_j-1/2}(\Gamma)$$
$$=: \mathcal{H}_{s\mp\varepsilon}(\Omega, \Gamma). \tag{4.5}$$

They have the common kernel N, the same index $\varkappa := \dim N - \dim N^+$, and the range

$$(L, B)(H^{s\mp\varepsilon}(\Omega)) = \{\, (f, g_1, \ldots, g_q) \in \mathcal{H}_{s\mp\varepsilon}(\Omega, \Gamma) : \ (4.4) \text{ is true}\,\}. \tag{4.6}$$

We apply the interpolation with the function parameter ψ in Theorem 1.14, where $\varepsilon = \delta$, to (4.5). We obtain the operator

$$(L, B) : \ [H^{s-\varepsilon}(\Omega), H^{s+\varepsilon}(\Omega)]_\psi \to [\mathcal{H}_{s-\varepsilon}(\Omega, \Gamma), \mathcal{H}_{s+\varepsilon}(\Omega, \Gamma)]_\psi, \tag{4.7}$$

which extends mapping (4.2) by continuity. By virtue of the interpolation theorems 1.5, 2.2, and 3.2, this implies that mapping (4.2) extends by continuity to the bounded operator (4.3) equal to (4.7). By Theorem 1.7, the Fredholm property of operators (4.5) yields the Fredholm property of operator (4.3), which inherits their kernel N and index $\varkappa = \dim N - \dim N^+$. In addition, the range of operator (4.3) is equal to $\mathcal{H}_{s,\varphi}(\Omega, \Gamma) \cap (L, B)(H^{s-\varepsilon}(\Omega))$. By virtue of (4.6), this implies that the range is the same as that in the statement of the theorem being demonstrated.

Theorem 4.1 is proved.

By virtue of Theorem 4.1, for an arbitrary function $u \in H^{s,\varphi}(\Omega)$, where $s > 2q$ and $\varphi \in \mathcal{M}$, the right-hand sides $f \in H^{s-2q,\varphi}(\Omega)$ and $g_j \in H^{s-m_j-1/2,\varphi}(\Gamma)$ of the boundary-value problem (4.1) are determined, as images of u at the mapping (4.3).

If $N = N^+ = \{0\}$ (the defect of the boundary-value problem is absent), then operator (4.3) is an isomorphism of $H^{s,\varphi}(\Omega)$ onto $\mathcal{H}_{s,\varphi}(\Omega, \Gamma)$. This follows from Theorem 4.1 and the Banach theorem on inverse operator. In the general case, it is convenient to define the corresponding isomorphism with the use of the following projectors.

Let $s > 2q$ and $\varphi \in \mathcal{M}$. We represent spaces in which operator (4.3) acts in the form of the direct sums of (closed) subspaces

$$H^{s,\varphi}(\Omega) = N \dotplus \{u \in H^{s,\varphi}(\Omega) : (u,w)_\Omega = 0 \text{ for all } w \in N\}, \tag{4.8}$$

$$\mathcal{H}_{s,\varphi}(\Omega,\Gamma) = \{(v,0,\dots,0) : v \in N^+\} \dotplus (L,B)(H^{s,\varphi}(\Omega)). \tag{4.9}$$

These decomposition in direct sums exist. Indeed, equality (4.8) is the restriction of the decomposition of the space $L_2(\Omega)$ in the orthogonal sum of the subspace N and its complement. Equality (4.9) follows from Theorem 4.1 according to which the subspaces on its right-hand side have a trivial intersection, while the (finite) dimension of the first of the subspaces coincides with the codimension of the second.

Let P and Q^+ denote the oblique projectors of the spaces $H^{s,\varphi}(\Omega)$ and $\mathcal{H}_{s,\varphi}(\Omega,\Gamma)$, respectively, onto the second summands in sums (4.8) and (4.9) parallel to the first ones. These projectors are independent of s and φ.

Theorem 4.2. *For arbitrary parameters $s > 2q$ and $\varphi \in \mathcal{M}$, the restriction of mapping* (4.3) *onto the subspace $P(H^{s,\varphi}(\Omega))$ is the isomorphism*

$$(L,B) : P(H^{s,\varphi}(\Omega)) \leftrightarrow Q^+(\mathcal{H}_{s,\varphi}(\Omega,\Gamma)). \tag{4.10}$$

Proof. By Theorem 4.1, N is the kernel and $Q^+(\mathcal{H}_{s,\varphi}(\Omega,\Gamma))$ is the range of operator (4.3). Therefore, the bounded operator (4.10) is a bijection. Thus, by virtue of the Banach theorem on inverse operator, it is an isomorphism.

Theorem 4.2 is proved.

The following *a priori* estimate for the solutions to the elliptic boundary-value problem (4.1) results from Theorem 4.2.

Theorem 4.3. *Let $s > 2q$ and $\varphi \in \mathcal{M}$. Suppose that a function $u \in H^{s,\varphi}(\Omega)$ is a solution to the boundary-value problem* (4.1) *with $(f,g_1,\dots,g_q) \in \mathcal{H}_{s,\varphi}(\Omega,\Gamma)$. Then the following estimate is true:*

$$\|u\|_{H^{s,\varphi}(\Omega)} \le c\,\big(\,\|(f,g_1,\dots,g_q)\|_{\mathcal{H}_{s,\varphi}(\Omega,\Gamma)} + \|u\|_{L_2(\Omega)}\,\big), \tag{4.11}$$

where the number $c = c(s,\varphi) > 0$ is independent of u and $(f,g_1,\dots,g_q)$.

Proof. We use decomposition (4.8) and rewrite the function $u \in H^{s,\varphi}(\Omega)$ in the form $u = u_0 + u_1$, where $u_0 := (1-P)u \in N$ and $u_1 := Pu \in P(H^{s,\varphi}(\Omega))$. By virtue of Theorem 4.2, we have

$$\|u_1\|_{H^{s,\varphi}(\Omega)} \le c_1\|(L,B)u_1\|_{\mathcal{H}_{s,\varphi}(\Omega,\Gamma)}$$

$$= c_1\|(L,B)u\|_{\mathcal{H}_{s,\varphi}(\Omega,\Gamma)}$$

$$= c_1\|(f,g_1,\dots,g_q)\|_{\mathcal{H}_{s,\varphi}(\Omega,\Gamma)}.$$

Here, c_1 is the norm of the inverse to (4.10). In addition, since N is finite-dimensional and since $1-P$ is the orthoprojector of $L_2(\Omega)$ onto N, we have

$$\|u_0\|_{H^{s,\varphi}(\Omega)} \le c_0\|u_0\|_{L_2(\Omega)} \le c_0\|u\|_{L_2(\Omega)}.$$

Here, the number $c_0 > 0$ is independent of u and $(f,g_1,\dots,g_q)$. Summing these inequalities, we obtain (4.11).

Theorem 4.3 is proved.

If $N=\{0\}$, i.e., the boundary-value problem (4.1) has at most one solution, then the term $\|u\|_{L_2(\Omega)}$ on the right-hand side of the *a priori* estimate (4.11) can be omitted.

At the end of this subsection, we discuss the connection between inhomogeneous and semihomogeneous elliptic boundary-value problems. For the sake of simplicity, we suppose that $N = N^+ = \{0\}$. Let $s > 2q$ and $\varphi \in \mathcal{M}$. It follows from Theorems 3.11 and 3.17(ii) that the space $H^{s,\varphi}(\Omega)$ is the direct sum of the subspaces $K_L^{s,\varphi}(\Omega)$ and $H^{s,\varphi}(\text{b.c.})$. Therefore, Theorems 3.11 and 3.15 (on solvability of inhomogeneous problems) give the isomorphism

$$(L,B) : H^{s,\varphi}(\Omega) \leftrightarrow \mathcal{H}_{s,\varphi}(\Omega,\Gamma).$$

(Note that the antidual space $(H^{2q-s,1/\varphi}(\text{b.c.})^+)'$ coincides with $H^{s-2q,\varphi}(\Omega)$ in view of Theorem 3.18(i) and (ii)). Thus, the inhomogeneous problem (4.1) can be immediately reduced to the semihomogeneous problems provided that $s > 2q$. The same conclusion is also true in more general case of $s > m + 1/2$, we denoting $m := \max\{m_1,\dots,m_q\}$.

This reduction fails for $s < m+1/2$. Indeed, if $0 \le s < m+1/2$, then the operator (L,B) cannot be reasonably defined on $K_L^{s,\varphi}(\Omega) \cup H^{s,\varphi}(\text{b.c.})$ because $K_L^{s,\varphi}(\Omega) \cap H^{s,\varphi}(\text{b.c.}) \neq \varnothing$. This inequality follows from Theorems 3.11 and 3.17(ii) if we note that the boundary-value problem (3.36), with $g_q \equiv 1$ and $g_j \equiv 0$ for $j < q$, has a nonzero solution $u \in K_L^{\infty}(\Omega)$ belonging to $H^{s,\varphi}(\text{b.c.})$. Here we may suppose that $m_q = m$.

So much the more, the above reduction is impossible for negative s. Note, if $s < -1/2$, then the solutions to the different semihomogeneous problems pertain to the spaces of distributions of different nature. Namely, the solutions to the problem (3.36) belong to $K_L^{s,\varphi}(\Omega) \subset H^{s,\varphi}(\Omega)$ and are distributions given in the open domain Ω, whereas the solutions to the problem (3.82) and (3.83) belong to $H^{s,\varphi}(\text{b.c.}) \subset H^{s,\varphi}_{\overline{\Omega}}(\mathbb{R}^n)$ and are distributions supported on the closed domain $\overline{\Omega}$.

The same conclusions remain valid for nontrivial N and/or N^+.

4.1.2 Smoothness of the solutions up to the boundary

Assume that the right-hand side of the elliptic boundary-value problem (4.1) has, in a set open in $\overline{\Omega}$, a certain smoothness on the positive refined Sobolev scale We study the smoothness of the solution u on this set. First, consider the case of the global smoothness on the whole closed domain $\overline{\Omega}$.

Theorem 4.4. *Suppose that a function $u \in H^{2q}(\Omega)$ is a solution to the boundary-value problem* (4.1) *in which*

$$f \in H^{s-2q,\varphi}(\Omega) \quad \text{and} \quad g_j \in H^{s-m_j-1/2,\varphi}(\Gamma), \ j = 1, \dots, q,$$

for certain parameters $s > 2q$ and $\varphi \in \mathcal{M}$. Then $u \in H^{s,\varphi}(\Omega)$.

Proof. According to Theorem 4.1, which is true in the Sobolev case of $s = 2q$ and $\varphi \equiv 1$, the vector $F := (f, g_1, \dots, g_q) \in \mathcal{H}_{s,\varphi}(\Omega, \Gamma)$ satisfies condition (4.4). Therefore, by virtue of the same theorem, $F \in (L, B)(H^{s,\varphi}(\Omega))$. Thus, parallel with condition $(L, B)u = F$, the equality $(L, B)v = F$ is true for a certain $v \in H^{s,\varphi}(\Omega)$. This implies that $(L, B)(u - v) = 0$, which, by Theorem 4.1, yields the inclusion

$$w := u - v \in N \subset C^{\infty}(\overline{\Omega}) \subset H^{s,\varphi}(\Omega).$$

Thus, $u = v + w \in H^{s,\varphi}(\Omega)$.

Theorem 4.4 is proved.

Now consider the case of local smoothness. Let U be an open set in $\mathbb{R}^n$ that has a nonempty intersection with domain Ω. We set $\Omega_0 := U \cap \Omega$ and $\Gamma_0 := U \cap \Gamma$ (the case where $\Gamma_0 = \varnothing$ is possible). For arbitrary parameters $\sigma \in \mathbb{R}$ and $\varphi \in \mathcal{M}$, we introduce a local analog of the Hörmander space over the Euclidean domain. Namely, we set

$$H^{\sigma,\varphi}_{\mathrm{loc}}(\Omega_0, \Gamma_0) := \{u \in \mathcal{D}'(\Omega) : \chi u \in H^{\sigma,\varphi}(\Omega) \quad \text{for all } \chi \in C^{\infty}(\overline{\Omega}) \text{ with } \operatorname{supp} \chi \subset \Omega_0 \cup \Gamma_0\}.$$

Recall that $\mathcal{D}'(\Omega)$ is the topological linear space of all distributions defined in the domain Ω. We also need the local space $H^{\sigma,\varphi}_{\mathrm{loc}}(\Gamma_0)$ introduced in Sec. 2.2.3. As above, in the case of $\varphi \equiv 1$, we omit the index φ in the notation of these spaces.

We note the inclusions

$$H^{\sigma,\varphi}(\Omega) \subset H^{\sigma,\varphi}_{\mathrm{loc}}(\Omega_0, \Gamma_0) \quad \text{and} \quad H^{\sigma,\varphi}(\Gamma) \subset H^{\sigma,\varphi}_{\mathrm{loc}}(\Gamma_0). \tag{4.12}$$

They follow from the fact that the multiplication by every function from the class $C^{\infty}(\overline{\Omega})$ (respectively, from $C^{\infty}(\Gamma)$) is a bounded operator on the space

$H^{\sigma,\varphi}(\Omega)$ (on $H^{\sigma,\varphi}(\Gamma)$). In the Sobolev case of $\varphi \equiv 1$, this fact is known [11, Sec. 2.1 c, p. 13]. The general case of an arbitrary $\varphi \in \mathcal{M}$ follows from the Sobolev case by virtue of the interpolation theorems 2.2 and 3.2.

Theorem 4.5. *Suppose that a function $u \in H^{2q}(\Omega)$ is a solution to the boundary-value problem* (4.1) *in which*

$$f \in H^{s-2q,\varphi}_{\mathrm{loc}}(\Omega_0, \Gamma_0), \tag{4.13}$$

$$g_j \in H^{s-m_j-1/2,\varphi}_{\mathrm{loc}}(\Gamma_0), \quad j = 1, \dots, q, \tag{4.14}$$

for certain parameters $s > 2q$ and $\varphi \in \mathcal{M}$. Then $u \in H^{s,\varphi}_{\mathrm{loc}}(\Omega_0, \Gamma_0)$.

Proof. We follow the scheme of the proof given in [15]. Put

$$\Upsilon := \{\chi \in C^\infty(\overline{\Omega}) : \operatorname{supp} \chi \subset \Omega_0 \cup \Gamma_0\}.$$

Beforehand, we prove that, by virtue of the condition of this theorem, the following implication holds for each $r \geq 0$:

$$\begin{aligned} &\bigl(\chi u \in H^{s,\varphi}(\Omega) + H^{r+2q}(\Omega) \ \text{ for all } \ \chi \in \Upsilon\bigr) \\ &\qquad \Rightarrow \bigl(\chi u \in H^{s,\varphi}(\Omega) + H^{r+1+2q}(\Omega) \ \text{ for all } \ \chi \in \Upsilon\bigr). \end{aligned} \tag{4.15}$$

Here and below in this proof, we use algebraic sums of sets.

Let us choose $r \geq 0$ arbitrarily and assume that the premise of implication (4.15) is true. Consider an arbitrary function $\chi \in \Upsilon$; let another function $\eta \in \Upsilon$ be such that $\eta = 1$ in a neighborhood of $\operatorname{supp} \chi$. By the condition of this theorem, we have $\chi F \in \mathcal{H}_{s,\varphi}(\Omega, \Gamma)$, where $F := (L, B)u = (f, g_1, \dots, g_q)$.

Interchanging the operator of multiplication by the function χ and the differential operators L and B_j, $j = 1, \dots, q$, we may write

$$\chi F = \chi(L, B)(\eta u) = (L, B)(\chi\eta u) - (L', B')(\eta u)$$

and, hence,

$$(L, B)(\chi u) = \chi F + (L', B')(\eta u). \tag{4.16}$$

Here, L' is a certain linear differential expression on $\overline{\Omega}$, and $B' := (B'_1, \dots, B'_g)$, where each B'_j is a certain boundary linear differential expression on Γ. The coefficients of these expressions are infinitely smooth, while the orders satisfy the conditions $\operatorname{ord} L' \leq 2q - 1$ and $\operatorname{ord} B'_j \leq m_j - 1$.

By the premise of implication (4.15), we have $\eta u = u_1 + u_2$ for certain functions

$$u_1 \in H^{s,\varphi}(\Omega) \quad \text{and} \quad u_2 \in H^{r+2q}(\Omega).$$

Hence and by (4.16), we may write $(L,B)(\chi u) = F_1 + F_2$, with

$$F_1 := \chi F + (L',B')u_1 \in \mathcal{H}_{s,\varphi}(\Omega,\Gamma), \tag{4.17}$$

$$F_2 := (L',B')u_2 \in \mathcal{H}_{r+2q+1}(\Omega,\Gamma). \tag{4.18}$$

Here, similar to (4.3), we denote

$$\mathcal{H}_\sigma(\Omega,\Gamma) := H^{\sigma-2q}(\Omega) \oplus \bigoplus_{j=1}^{q} H^{\sigma-m_j-1/2}(\Gamma) \quad \text{for every} \quad \sigma \in \mathbb{R}.$$

Let us argue the inclusions appearing in (4.17) and (4.18). Since $\operatorname{ord} L' \le 2q-1$ and $\operatorname{ord} B_j' \le m_j - 1$, the mapping $v \mapsto (L'v, B'v)$, with $v \in C^\infty(\overline{\Omega})$, extends by continuity to the bounded operator

$$(L',B') :\, H^\sigma(\Omega) \to \mathcal{H}_{\sigma+1}(\Omega,\Gamma) \quad \text{for every} \quad \sigma \ge 2q-1$$

(see, e.g., [11, Sec. 2.2, p. 16]). Hence, $u_2 \in H^{r+2q}(\Omega) \Rightarrow$ (4.18).

Moreover, using the interpolation with the function parameter ψ, we obtain the bounded operator

$$(L',B') : H^{s,\varphi}(\Omega) = [H^{s-\varepsilon}(\Omega), H^{s+\varepsilon}(\Omega)]_\psi$$
$$\to [\mathcal{H}_{s-\varepsilon+1}(\Omega,\Gamma), \mathcal{H}_{s+\varepsilon+1}(\Omega,\Gamma)]_\psi = \mathcal{H}_{s+1,\varphi}(\Omega,\Gamma).$$

Here, $\varepsilon := s - 2q$, while ψ is the same as that in Theorem 1.14, where $\varepsilon = \delta$. Therefore, (4.17) is a consequence of the inclusions $\chi F \in \mathcal{H}_{s,\varphi}(\Omega,\Gamma)$ and $u_1 \in H^{s,\varphi}(\Omega)$.

Now, we use the projector Q^+ and apply Theorem 4.2 (in the Sobolev case as well). It follows from the equality $(L,B)(\chi u) = F_1 + F_2$ and inclusions (4.17) and (4.18) that

$$(L,B)(\chi u) = Q^+(L,B)(\chi u) = Q^+F_1 + Q^+F_2 = (L,B)v_1 + (L,B)v_2.$$

Here, the functions

$$v_1 \in P\big(H^{s,\varphi}(\Omega)\big) \quad \text{and} \quad v_2 \in P\big(H^{r+2q+1}(\Omega)\big) \tag{4.19}$$

are the (unique) solutions to the boundary value-problems

$$(L,B)v_1 = Q^+F_1 \in Q^+\big(\mathcal{H}_{s,\varphi}(\Omega,\Gamma)\big)$$

and

$$(L,B)v_2 = Q^+F_2 \in Q^+\big(\mathcal{H}_{r+2q+1}(\Omega,\Gamma)\big).$$

Hence, it follows from the equality

$$(L,B)(\chi u) = (L,B)(v_1 + v_2)$$

that

$$\chi u = v_1 + (v_2 + w) \quad \text{for a certain} \quad w \in N \subset C^{\infty}(\overline{\Omega}).$$

Therefore, by virtue of (4.19), we arrive at the inference of implication (4.15). This implication is proved for each $r \geq 0$.

Now, we may complete the proof in the following way. The premise of implication (4.15) is true for $r = 0$ because of the condition $u \in H^{2q}(\Omega)$. Using this implication in succession for values $r = 0$, $r = 1$, ..., and $r = [s - 2q]$, we conclude that

$$\chi u \in H^{s,\varphi}(\Omega) + H^{[s]+1}(\Omega) = H^{s,\varphi}(\Omega) \quad \text{for all} \quad \chi \in \Upsilon.$$

Thus, $u \in H^{s,\varphi}_{\mathrm{loc}}(\Omega_0, \Gamma_0)$.

Theorem 4.5 is proved.

In Theorem 4.5, we note the case $\Gamma_0 = \varnothing$, which leads to the statement on the increase in the local smoothness of the solution in neighborhoods of interior points in the domain Ω.

As an application of Theorems 4.4 and 4.5, we establish a sufficient condition for the solution u of the elliptic boundary-value problem (4.1) to be *classical*, i.e., to belong to the class $C^{2q}(\Omega) \cap C^{m}(\overline{\Omega})$, where $m := \max\{m_1, \ldots, m_q\}$. If u is a solution in this class, then the left-hand sides of the equalities in (4.1) are calculated with the use of classical derivatives, and these equalities are fulfilled at every point in the set Ω or Γ, respectively. Moreover, their right-hand sides have the following smoothness:

$$f \in C(\Omega) \quad \text{and} \quad g_j \in C^{m-m_j}(\Gamma) \quad \text{for each} \quad j \in \{1, \ldots, q\}. \tag{4.20}$$

The converse is not true; namely, the condition (4.20) does not imply that the solution u is classical [64, Chap. 4, Notes]. With the use of the refined Sobolev scale, we strengthen this condition so that it becomes sufficient for the solution to be classical.

Theorem 4.6. *Suppose that a function $u \in H^{2q}(\Omega) \cap H^{2q,\varphi}(\Omega)$ is a solution to problem* (4.1), *where*

$$f \in H^{n/2,\varphi}_{\mathrm{loc}}(\Omega, \varnothing) \cap H^{m-2q+n/2,\varphi}(\Omega), \tag{4.21}$$

$$g_j \in H^{m-m_j+(n-1)/2,\varphi}(\Gamma) \quad \textit{for all} \quad j = 1, \ldots, q, \tag{4.22}$$

and the function parameter $\varphi \in \mathcal{M}$ satisfies condition (1.37). *Then the solution u is classical, i.e., $u \in C^{2q}(\Omega) \cap C^{m}(\overline{\Omega})$.*

Remark 4.1. Conditions (4.21) and (4.22) imply property (4.20). This follows from Theorems 2.8 and 3.4. We also note that, in Theorem 4.6, condition (1.37) not only is sufficient for the solution u to be classical but also is necessary on the class of all the considered solutions to the elliptic boundary-value problem; see also Remark 1.9.

Proof of Theorem 4.6. In view of condition (4.21), for $s := 2q + n/2$, we have the inclusions

$$f \in H^{s-2q,\varphi}_{\mathrm{loc}}(\Omega, \varnothing),$$

$$g_j \in \mathcal{D}'(\Gamma) = H^{s-m_j-1/2,\varphi}_{\mathrm{loc}}(\varnothing) \quad \text{for each} \quad j \in \{1, \dots, q\}.$$

Based on Theorems 4.5 and 3.4, we obtain

$$u \in H^{s,\varphi}_{\mathrm{loc}}(\Omega, \varnothing) = H^{2q+n/2,\varphi}_{\mathrm{loc}}(\Omega, \varnothing) \subset C^{2q}(\Omega).$$

Now let us prove that $u \in C^m(\overline{\Omega})$. If $s := m+n/2 > 2q$, then, in view of (4.21) and (4.22), the condition of Theorem 4.4 is satisfied. Therefore, by virtue of Theorems 4.4 and 3.4,

$$u \in H^{s,\varphi}(\Omega) = H^{m+n/2,\varphi}(\Omega) \subset C^m(\overline{\Omega}).$$

If $m + n/2 \leq 2q$, then, by the condition,

$$u \in H^{2q,\varphi}(\Omega) \subseteq H^{m+n/2,\varphi}(\Omega) \subset C^m(\overline{\Omega}).$$

Thus, $u \in C^{2q}(\Omega) \cap C^m(\overline{\Omega})$. Theorem 4.6 is proved.

Remark 4.2. If we restricted ourselves to the Sobolev scale in Theorem 4.6, we would fix small $\varepsilon > 0$ and replace (4.21) and (4.22) with the stronger conditions

$$f \in H^{n/2+\varepsilon}_{\mathrm{loc}}(\Omega, \varnothing) \cap H^{m-2q+n/2+\varepsilon}(\Omega),$$

$$g_j \in H^{m-m_j+(n-1)/2+\varepsilon}(\Gamma) \quad \text{for each} \quad j \in \{1, \dots, q\}.$$

These conditions set the main smoothness of the right-hand sides of the boundary-value problem (4.1) too high, which coarsens the result.

4.1.3 Nonregular elliptic boundary-value problems

In this section, we assume that the boundary-value problem (4.1) is elliptic in the domain Ω but not regular. This means that it satisfies conditions (i) and (ii) of Definition 3.1 but does not satisfy condition (iii) of this definition. For this problem, all results in Subsections 4.1.1 and 4.1.2 remain true (with certain changes). Changes refer only to Theorems 4.1 and 4.2 and are caused by the fact that a nonregular elliptic boundary-value problem does not have a formally conjugate boundary-value problem in a class of differential ones. We give analogs of these theorems.

Theorem 4.7. *For arbitrary parameters $s > 2q$ and $\varphi \in \mathcal{M}$, mapping* (4.2) *extends uniquely (by continuity) to the Fredholm bounded operator* (4.3). *The kernel of this operator is equal to N and the range consists of all vectors $(f, g_1, \dots, g_q) \in \mathcal{H}_{s,\varphi}(\Omega, \Gamma)$ such that*

$$(f, v)_\Omega + \sum_{j=1}^{q} (g_j, v_j)_\Gamma = 0 \quad \textit{for all} \quad (v, v_1, \dots, v_q) \in W. \tag{4.23}$$

Here, W is a certain finite-dimensional space that lies in $C^\infty(\overline{\Omega}) \times (C^\infty(\Gamma))^q$ and is independent of s and φ. The index of operator (4.3) *is equal to* $\dim N - \dim W$ *and does not depend on s and φ.*

In the Sobolev case of $\varphi \equiv 1$, this theorem is known (see, e.g., [86, Theorems 20.1.2 and 20.1.8] or [11, Sec. 2.4 a]). The general case of $\varphi \in \mathcal{M}$ is deduced from the latter with the use of interpolation by analogy with the proof of Theorem 4.1.

By virtue of Theorem 4.7, for arbitrary $s > 2q$ and $\varphi \in \mathcal{M}$, we can write

$$\mathcal{H}_{s,\varphi}(\Omega, \Gamma) = W \dotplus (L, B)(H^{s,\varphi}(\Omega)).$$

Let Q denote the oblique projector of the space $\mathcal{H}_{s,\varphi}(\Omega, \Gamma)$ onto the subspace $(L, B)(H^{s,\varphi}(\Omega))$ parallel to W. This projector is independent of s and φ.

Theorem 4.8. *For arbitrary parameters $s > 2q$ and $\varphi \in \mathcal{M}$, the restriction of mapping* (4.3) *to the subspace $P(H^{s,\varphi}(\Omega))$ is the isomorphism*

$$(L, B) : P(H^{s,\varphi}(\Omega)) \leftrightarrow Q(\mathcal{H}_{s,\varphi}(\Omega, \Gamma)).$$

Theorem 4.8 directly follows from Theorem 4.7 and the Banach theorem on inverse operator.

Example 4.1. The oblique derivative problem for the Laplace equation

$$\Delta u = f \ \text{ in } \ \Omega, \quad \frac{\partial u}{\partial \tau} = g \ \text{ on } \ \Gamma.$$

Here, τ is an infinitely smooth field of unit vectors tangent to the boundary Γ. If $\dim \Omega = 2$, then this problem is elliptic but not regular. If $\dim \Omega \geq 3$, then it is not elliptic at all [11, Sec. 1.4].

Example 4.2. Let $\tau_1, \dots, \tau_{n-1}$ be a linearly independent system of infinitely smooth fields of nonzero vectors tangent to the boundary Γ. (Recall that $n = \dim \Omega$.) The following boundary-value problem is elliptic but not regular:

$$\Delta^{n-1} u = f \text{ in } \Omega, \qquad \frac{\partial u}{\partial \tau_j} = g_j \text{ on } \Gamma \text{ for } j = 1, \dots, n-1.$$

Example 4.3. Let $\dim \Omega = 2$. Consider the boundary-value problem

$$\Delta^2 u = f \text{ in } \Omega, \qquad \frac{\partial u}{\partial \nu} + \frac{\partial u}{\partial \tau} = g_1, \quad \frac{\partial u}{\partial \nu} - \frac{\partial u}{\partial \tau} = g_2 \text{ on } \Gamma.$$

Here, τ is an infinitely smooth field of unit vectors tangent to the boundary Γ. This problem is elliptic but not regular.

Other examples of elliptic but nonregular boundary-value problems are given, e.g., in the paper by Ya. A. Roitberg [205].

In conclusion of this section, we recall the following important fact [11, Sec. 2.4 a]: if, for the boundary-value problem (4.1), the bounded operator (4.3) is Fredholm for a certain $s > 2q$ under the condition that $\varphi \equiv 1$, then this problem is elliptic in the domain Ω, i.e., satisfies conditions (i) and (ii) of Definition 3.1.

4.1.4 Parameter-elliptic boundary-value problems

S. Agmon, L. Nirenberg [2, 6] and M. S. Agranovich, M. I. Vishik [13] selected an important subclass of elliptic boundary-value problems, called parameter-elliptic (see also review [11, Sec. 4]). These problems depend on a complex-valued parameter and characterized by the following important property. For sufficiently large absolute values of the parameter, the corresponding operator sets isomorphisms between appropriate Sobolev spaces, and, furthermore, the norm of the operator admits a two-sided estimate with constants independent of the parameter. In this section, we prove that this property is preserved for the refined Sobolev scale.

We give the definition of a parameter-elliptic boundary-value problem. Consider the inhomogeneous boundary-value problem

$$L(\lambda)\, u = f \quad \text{in} \quad \Omega, \tag{4.24}$$

$$B_j(\lambda)\, u = g_j \quad \text{on} \quad \Gamma \quad \text{for} \quad j = 1, \dots, q \tag{4.25}$$

that depends on the complex-valued parameter λ as follows:

$$L(\lambda) := \sum_{r=0}^{2q} \lambda^{2q-r} L_r \quad \text{and} \quad B_j(\lambda) := \sum_{r=0}^{m_j} \lambda^{m_j - r} B_{j,r}. \tag{4.26}$$

Here, L_r is a linear differential expression on $\overline{\Omega}$, and $B_{j,r}$ is a boundary linear differential expression on Γ; the coefficients of these expressions are infinitely smooth complex-valued functions, and the orders do not exceed the number r. As above, the fixed integers q and m_j satisfy the conditions $q \geq 1$ and $0 \leq m_j \leq 2q-1$. Note that $L(0) = L_{2q}$ and $B_j(0) = B_{j,m_j}$.

We associate the differential expressions (4.26) with certain homogeneous polynomials in $(\xi, \lambda) \in \mathbb{C}^{n+1}$. We set

$$L^{(0)}(x;\xi,\lambda) := \sum_{r=0}^{2q} \lambda^{2q-r} L_r^{(0)}(x,\xi) \quad \text{for} \quad x \in \overline{\Omega},\ \xi \in \mathbb{C}^n,\ \lambda \in \mathbb{C}.$$

Here, $L_r^{(0)}(x,\xi)$ is the principal symbol of the expression L_r in the case where $\operatorname{ord} L_r = r$, otherwise $L_r^{(0)}(x,\xi) \equiv 0$. By analogy, for each $j \in \{1,\ldots,q\}$, we set

$$B_j^{(0)}(x;\xi,\lambda) := \sum_{r=0}^{m_j} \lambda^{m_j-r} B_{j,r}^{(0)}(x,\xi) \quad \text{for} \quad x \in \Gamma,\ \xi \in \mathbb{C}^n,\ \lambda \in \mathbb{C}.$$

Here, $B_{j,r}^{(0)}(x,\xi)$ is the principal symbol of the boundary differential expression $B_{j,r}$ in the case where $\operatorname{ord} B_{j,r} = r$, otherwise $B_{j,r}^{(0)}(x,\xi) \equiv 0$. Note that $L^{(0)}(x;\xi,\lambda)$ and $B_j^{(0)}(x;\xi,\lambda)$ are homogeneous polynomials in $(\xi,\lambda) \in \mathbb{C}^{n+1}$ of degrees $2q$ and m_j, respectively.

Let K be a fixed closed angle on the complex plane with vertex at the origin (the case where K degenerates into a ray is not excluded).

Definition 4.1. The boundary-value problem (4.24), (4.25) is called *parameter-elliptic* in the angle K if the following conditions are satisfied:

(i) For all $x \in \overline{\Omega}$, $\xi \in \mathbb{R}^n$, and $\lambda \in K$ such that $|\xi| + |\lambda| \neq 0$, the inequality $L^{(0)}(x;\xi,\lambda) \neq 0$ is true.

(ii) For arbitrarily fixed point $x \in \Gamma$, vector $\xi \in \mathbb{R}^n$ tangent to the boundary Γ at the point x, and the parameter $\lambda \in K$ such that $|\xi| + |\lambda| \neq 0$, the polynomials $B_j^{(0)}(x;\xi+\tau\nu(x),\lambda)$, $j = 1,\ldots,q$, in τ are linearly independent modulo the polynomial

$$\prod_{j=1}^{q}(\tau - \tau_j^+(x;\xi;\lambda)).$$

Here, $\tau_1^+(x;\xi;\lambda),\ldots,\tau_q^+(x;\xi;\lambda)$ are all τ-roots of the polynomial $L^{(0)}(x;\xi+\tau\nu(x),\lambda)$ that have the positive imaginary part and are written with regard for their multiplicity.

Remark 4.3. Condition (ii) of Definition 4.1 is correctly defined in the sense that the polynomial $L^{(0)}(x;\xi+\tau\nu(x),\lambda)$ has exactly q τ-roots with positive imaginary part and the same number of roots with negative imaginary part (with regard for their multiplicity). Indeed, it follows from condition (i) that, for every point $x \in \overline{\Omega}$, the differential expression

$$L(x; D, D_t) := \sum_{r=0}^{2q} D_t^{2q-r} L_r(x, D)$$

is elliptic. Since it contains the operators of differentiation with respect to $n+1 \geq 3$ real variables $x_1, \dots, x_n, t$, its ellipticity leads to the proper ellipticity [11, Sec. 1.2, p. 7]. Therefore, the τ-roots of the polynomial $L^{(0)}(x; \xi+\tau\nu(x), \lambda)$ have the indicated property.

We give some examples of parameter-elliptic boundary-value problems [11, Sec. 3.1 b].

Example 4.4. Let the differential expression $L(\lambda)$ satisfy condition (i) of Definition 4.1. Then the Dirichlet boundary-value problem for the equation $L(\lambda) = f$ is parameter-elliptic in the angle K. Here, the boundary conditions are independent of the parameter λ.

Example 4.5. The boundary-value problem

$$\Delta u + \lambda^2 u = f \quad \text{in} \quad \Omega, \qquad \frac{\partial u}{\partial \nu} - \lambda u = g \quad \text{on} \quad \Gamma$$

is parameter-elliptic in each angle

$$K_\varepsilon := \{\lambda \in \mathbb{C} : \varepsilon \leq |\arg \lambda| \leq \pi - \varepsilon\},$$

where $0 < \varepsilon < \pi/2$ and the complex plane is slitted along the negative ray.

Further, in this section, we assume that the boundary-value problem (4.24), (4.25) is parameter-elliptic in the angle K.

For $\lambda = 0$, condition (i) of Definition 4.1 implies that the differential expression $L(0)$ is properly elliptic on $\overline{\Omega}$ (see Remark 4.3). Condition (ii) means that the collection $\{B_1(0), \dots, B_q(0)\}$ of boundary differential expressions satisfies the complementing condition with respect to $L(0)$ on Γ. Therefore, for $\lambda = 0$, the boundary-value problem (4.24), (4.25) is elliptic (not necessarily regular) in the domain $\overline{\Omega}$. Since the parameter λ affects only the lower terms of the differential expressions $L(\lambda)$ and $B_j(\lambda)$, this problem is elliptic for all $\lambda \in \mathbb{C}$. According to Theorem 4.7, the mapping

$$u \mapsto (L(\lambda)u, B(\lambda)u) := (L(\lambda)u, B_1(\lambda)u, \dots, B_q(\lambda)u), \quad u \in C^\infty(\overline{\Omega}),$$

extends by continuity to the Fredholm bounded operator

$$(L(\lambda), B(\lambda)) : H^{s,\varphi}(\Omega) \to \mathcal{H}_{s,\varphi}(\Omega, \Gamma) \tag{4.27}$$

for arbitrary parameters $s > 2q$, $\varphi \in \mathcal{M}$, and $\lambda \in \mathbb{C}$. Its index is independent of both s, φ, and λ because λ affects only the lower terms [86, Theorem 20.1.8].

Since the boundary-value problem (4.24), (4.25) is parameter-elliptic in the angle K, operator (4.27) has the following important properties.

Theorem 4.9. *The following assertions are true:*

(i) *There exists a number $\lambda_0 > 0$ such that, for each $\lambda \in K$ with $|\lambda| \geq \lambda_0$ and for arbitrary $s > 2q$ and $\varphi \in \mathcal{M}$, operator (4.27) is an isomorphism of $H^{s,\varphi}(\Omega)$ onto $\mathcal{H}_{s,\varphi}(\Omega, \Gamma)$.*

(ii) *For arbitrarily fixed parameters $s > 2q$ and $\varphi \in \mathcal{M}$, there exists a number $c = c(s, \varphi) \geq 1$ such that, for each $\lambda \in K$ with $|\lambda| \geq \max\{\lambda_0, 1\}$ and for every function $u \in H^{s,\varphi}(\Omega)$, we have the two-sided estimate*

$$\begin{aligned}
c^{-1}\big(\, \|u\|_{H^{s,\varphi}(\Omega)} + |\lambda|^{s}\varphi(|\lambda|)\, \|u\|_{L_2(\Omega)} \big)& \\
\leq \|L(\lambda)u\|_{H^{s-2q,\varphi}(\Omega)} &+ |\lambda|^{s-2q}\varphi(|\lambda|)\, \|L(\lambda)u\|_{L_2(\Omega)} \\
&+ \sum_{j=1}^{q} \big(\, \|B_j(\lambda)u\|_{H^{s-m_j-1/2,\varphi}(\Gamma)} \\
&+ |\lambda|^{s-m_j-1/2}\varphi(|\lambda|)\, \|B_j(\lambda)u\|_{L_2(\Gamma)} \big) \\
\leq c\big(\, \|u\|_{H^{s,\varphi}(\Omega)} &+ |\lambda|^{s}\varphi(|\lambda|)\, \|u\|_{L_2(\Omega)} \big). \qquad (4.28)
\end{aligned}$$

Here, the number c is independent of u and λ.

Remark 4.4. Assertion (ii) of Theorem 4.9 needs commenting. For a fixed λ, estimate (4.28) is written for norms equivalent to the norms $\|u\|_{H^{s,\varphi}(\Omega)}$ and $\|(L(\lambda), B(\lambda))u\|_{\mathcal{H}_{s,\varphi}(\Omega,\Gamma)}$. To avoid awkward expressions, we write this estimate for non-Hilbert norms. It is also true for the corresponding Hilbert norms (generating inner products in $H^{s,\varphi}(\Omega)$ and $\mathcal{H}_{s,\varphi}(\Omega, \Gamma)$) because they are estimated via the used norms with constants independent of s, φ, and λ. The additional condition $|\lambda| \geq 1$ is caused by the fact that the function $\varphi(t)$ is defined only for $t \geq 1$. Note that estimate (4.28) is of interest only for $|\lambda| \gg 1$.

In the Sobolev case of $\varphi \equiv 1$ and $s \geq 2q$, Theorem 4.9 is proved by M. S. Agranovich and M. I. Vishik in [13, Sec. 4 and 5] (see also [11, Sec. 3.2]). The two-sided *a priori* estimate (4.28) has a number of applications, specifically, in the theory of parabolic problems [13]. Note [13, Proposition 4.1] that the right-hand side of estimate (4.28) is true without assumption that problem (4.24), (4.25) is parameter-elliptic.

We separately prove assertions (i) and (ii) of Theorem 4.9.

Proof of assertion (i) of Theorem 4.9. We take the number $\lambda_0 > 0$ from the statement of this theorem in the case of $\varphi \equiv 1$. Let $\lambda \in K$, $|\lambda| \geq \lambda_0$, and let $s > 2q$, $\varphi \in \mathcal{M}$. We set $\varepsilon := s - 2q$. We have the isomorphisms

$$(L(\lambda), B(\lambda)) : H^{s\mp\varepsilon}(\Omega) \leftrightarrow \mathcal{H}_{s\mp\varepsilon}(\Omega, \Gamma).$$

Using the interpolation with the function parameter ψ from Theorem 1.14, where $\varepsilon = \delta$, and applying the interpolation theorems 1.5, 2.2, and 3.2, we obtain the isomorphism

$$(L(\lambda), B(\lambda)) : H^{s,\varphi}(\Omega) \leftrightarrow \mathcal{H}_{s,\varphi}(\Omega, \Gamma).$$

Assertion (i) is proved.

Prior to the proof of assertion (ii), it is useful to introduce several Hörmander spaces whose norms depend on the additional parameter ϱ. We need one interpolation property of these spaces.

Let $\sigma > 0$, $\varphi \in \mathcal{M}$, and $\varrho \geq 1$. Assume that $G \in \{\mathbb{R}^k, \Omega, \Gamma\}$, where $k \in \mathbb{N}$. Let $H^{\sigma,\varphi}(G, \varrho)$ denote the space $H^{\sigma,\varphi}(G)$ endowed with the norm that depends on the parameter ϱ as follows:

$$\|u\|_{H^{\sigma,\varphi}(G,\varrho)} := \left(\|u\|^2_{H^{\sigma,\varphi}(G)} + \varrho^{2\sigma}\varphi^2(\varrho)\|u\|^2_{L_2(G)} \right)^{1/2}. \tag{4.29}$$

This norm is equivalent to the norm in the space $H^{\sigma,\varphi}(G)$ for every value of the parameter ϱ. Therefore, the space $H^{\sigma,\varphi}(G, \varrho)$ is complete. It is a Hilbert space because norm (4.29) is generated by the inner product

$$(u_1, u_2)_{H^{\sigma,\varphi}(G,\varrho)} := (u_1, u_2)_{H^{\sigma,\varphi}(G)} + \varrho^{2\sigma}\varphi^2(\varrho)(u_1, u_2)_{L_2(\Omega)}.$$

As usual, in the Sobolev case of $\varphi \equiv 1$, we omit the index φ in notation.

By virtue of Theorems 1.14, 2.2, and 3.2, the spaces

$$[H^{\sigma-\varepsilon}(G, \varrho), H^{\sigma+\delta}(G, \varrho)]_\psi \quad \text{and} \quad H^{\sigma,\varphi}(G, \varrho)$$

are equal up to equivalence of norms. It turns out that the constants in estimates for norms of these spaces can be chosen so that they are independent of the parameter ϱ.

Lemma 4.1. *Let a function $\varphi \in \mathcal{M}$ and positive numbers σ, ε, and δ be given and, furthermore, let $\sigma - \varepsilon > 0$. Then there exists a number $c \geq 1$ such that, for arbitrary $\varrho \geq 1$ and $u \in H^{\sigma,\varphi}(G)$, the following two-sided estimate for norms is true:*

$$c^{-1} \|u\|_{H^{\sigma,\varphi}(G,\varrho)} \leq \|u\|_{[H^{\sigma-\varepsilon}(G,\varrho), H^{\sigma+\delta}(G,\varrho)]_\psi} \leq c \|u\|_{H^{\sigma,\varphi}(G,\varrho)}. \tag{4.30}$$

Here, $G \in \{\mathbb{R}^k, \Omega, \Gamma\}$, ψ is the interpolation parameter from Theorem 1.14, and the number c is independent of both ϱ and u.

Proof. First, we prove Lemma 4.1 for $G = \mathbb{R}^k$, where $k \in \mathbb{N}$. Using this result, we prove the lemma in the required cases $G = \Omega$ and $G = \Gamma$.

The case $G = \mathbb{R}^k$. Let $\varrho \geq 1$ and let $u \in H^{\sigma,\varphi}(\mathbb{R}^k)$. Using the definition of norms in the spaces $H^{\sigma,\varphi}(\mathbb{R}^k, \varrho)$ and $H^{\sigma,\varphi}(\mathbb{R}^k)$, we write

$$\|u\|_{H^{\sigma,\varphi}(\mathbb{R}^k,\varrho)} = \Bigl(\int\limits_{\mathbb{R}^k} \bigl(\langle\xi\rangle^{2\sigma}\varphi^2(\langle\xi\rangle) + (2\pi)^{-k}\varrho^{2\sigma}\varphi^2(\varrho)\bigr)\, |\widehat{u}(\xi)|^2 \, d\xi \Bigr)^{1/2}. \tag{4.31}$$

Parallel with (4.31), consider one more Hilbert norm of the function u:

$$\Bigl(\int\limits_{\mathbb{R}^k} (\langle\xi\rangle + \varrho)^{2\sigma} \, \varphi^2(\langle\xi\rangle + \varrho) \, |\widehat{u}(\xi)|^2 \, d\xi \Bigr)^{1/2}. \tag{4.32}$$

Norms (4.31) and (4.32) are equivalent and, furthermore, the constants whereby one norm is estimated in terms of the second depend only on σ and φ and, hence, do not depend on the parameter ϱ. This follows from Lemma 4.2, which will be established after the proof of this lemma. Let $H^{\sigma,\varphi}(\mathbb{R}^k, \varrho, 1)$ denote the Hilbert space $H^{\sigma,\varphi}(\mathbb{R}^k)$ endowed with norm (4.32) and the corresponding inner product

$$(u_1, u_2)_{H^{\sigma,\varphi}(\mathbb{R}^k,\varrho,1)} := \int\limits_{\mathbb{R}^k} (\langle\xi\rangle + \varrho)^{2\sigma} \, \varphi^2(\langle\xi\rangle + \varrho) \, \widehat{u}_1(\xi) \, \overline{\widehat{u}_2(\xi)} \, d\xi.$$

We have the equivalence of the norms

$$c_0^{-1} \, \|u\|_{H^{\sigma,\varphi}(\mathbb{R}^k,\varrho)} \leq \|u\|_{H^{\sigma,\varphi}(\mathbb{R}^k,\varrho,1)} \leq c_0 \, \|u\|_{H^{\sigma,\varphi}(\mathbb{R}^k,\varrho)}. \tag{4.33}$$

Here, the number $c_0 = c_0(\sigma, \varphi) \geq 1$ does not depend on u and ϱ.

We interpolate the pair $[H^{\sigma-\varepsilon}(\mathbb{R}^k, \varrho, 1), H^{\sigma+\delta}(\mathbb{R}^k, \varrho, 1)]$ with the parameter ψ from Theorem 1.14. The PsDO J_ϱ with symbol $(\langle\xi\rangle + \varrho)^{\varepsilon+\delta}$ is the generating operator for this pair. With the use of the Fourier transform

$$\mathcal{F} : H^{\sigma-\varepsilon}(\mathbb{R}^k, \varrho, 1) \leftrightarrow L_2\bigl(\mathbb{R}^k, (\langle\xi\rangle + \varrho)^{2(\sigma-\varepsilon)} d\xi\bigr),$$

the operator $\psi(J_\varrho)$ is reduced to the form of multiplication by the function $\psi((\langle\xi\rangle + \varrho)^{(\varepsilon+\delta)}) = (\langle\xi\rangle + \varrho)^{\varepsilon}\varphi(\langle\xi\rangle + \varrho)$ of argument $\xi \in \mathbb{R}^k$. Therefore,

$$\begin{aligned}
\|u\|^2_{[H^{\sigma-\varepsilon}(\mathbb{R}^k,\varrho,1),H^{\sigma+\delta}(\mathbb{R}^k,\varrho,1)]_\psi} &= \|\psi(J_\varrho)u\|^2_{H^{\sigma-\varepsilon}(\mathbb{R}^k,\varrho,1)} \\
&= \int\limits_{\mathbb{R}^k} (\langle\xi\rangle + \varrho)^{2(\sigma-\varepsilon)} \, \psi^2((\langle\xi\rangle + \varrho)^{\varepsilon+\delta}) \, |\widehat{u}(\xi)|^2 \, d\xi \\
&= \int\limits_{\mathbb{R}^k} (\langle\xi\rangle + \varrho)^{2\sigma} \, \varphi^2(\langle\xi\rangle + \varrho) \, |\widehat{u}(\xi)|^2 \, d\xi \\
&= \|u\|^2_{H^{\sigma,\varphi}(\mathbb{R}^k,\varrho,1)}.
\end{aligned}$$

Thus,

$$\|u\|_{[H^{\sigma-\varepsilon}(\mathbb{R}^k,\varrho,1),H^{\sigma+\delta}(\mathbb{R}^k,\varrho,1)]_\psi} = \|u\|_{H^{\sigma,\varphi}(\mathbb{R}^k,\varrho,1)}. \tag{4.34}$$

Note that

$$c_1^{-1}\,\|u\|_{[H^{\sigma-\varepsilon}(\mathbb{R}^k,\varrho),H^{\sigma+\delta}(\mathbb{R}^k,\varrho)]_\psi} \le \|u\|_{[H^{\sigma-\varepsilon}(\mathbb{R}^k,\varrho,1),H^{\sigma+\delta}(\mathbb{R}^k,\varrho,1)]_\psi}$$

$$\le c_1\,\|u\|_{[H^{\sigma-\varepsilon}(\mathbb{R}^k,\varrho),H^{\sigma+\delta}(\mathbb{R}^k,\varrho)]_\psi}, \tag{4.35}$$

where the number $c_1 \geq 1$ does not depend on u and ϱ. Indeed, the identity operator I defines the isomorphisms

$$I : H^{\sigma-\varepsilon}(\mathbb{R}^k,\varrho) \leftrightarrow H^{\sigma-\varepsilon}(\mathbb{R}^k,\varrho,1),$$

$$I : H^{\sigma+\delta}(\mathbb{R}^k,\varrho) \leftrightarrow H^{\sigma+\delta}(\mathbb{R}^k,\varrho,1).$$

Here, the norms of direct and inverse operators are uniformly bounded in the parameter ϱ. By virtue of Theorem 1.8, this yields the isomorphism

$$I : [H^{\sigma-\varepsilon}(\mathbb{R}^k,\varrho),H^{\sigma+\delta}(\mathbb{R}^k,\varrho)]_\psi \leftrightarrow [H^{\sigma-\varepsilon}(\mathbb{R}^k,\varrho,1),H^{\sigma+\delta}(\mathbb{R}^k,\varrho,1)]_\psi$$

such that the norms of the direct and inverse operators are uniformly bounded in the parameter ϱ. (Note that the pairs of spaces written above are normal.) This means the two-sided estimate for norms (4.35).

Now, relations (4.33)–(4.35) yield the required estimate (4.30) for $G = \mathbb{R}^k$, where the number $c := c_0 c_1$ does not depend on the parameter ϱ.

The case $G = \Omega$. We derive it from the previous case, in which $k := n$. Let $\varrho \geq 1$. Let R_Ω denote the linear operator that restricts a distribution from the space $\mathbb{R}^n$ to the domain Ω. We have the bounded operators

$$R_\Omega : H^{\sigma,\varphi}(\mathbb{R}^n,\varrho) \to H^{\sigma,\varphi}(\Omega,\varrho), \tag{4.36}$$

$$R_\Omega : H^{\alpha}(\mathbb{R}^n,\varrho) \to H^{\alpha}(\Omega,\varrho), \quad \alpha > 0. \tag{4.37}$$

It is obvious that their norms does not exceed number 1. We apply the interpolation with parameter ψ to spaces in which operators (4.37), where $\alpha \in \{\sigma-\varepsilon, \sigma+\delta\}$, act. By virtue of Theorem 1.8, we conclude that the norm of the operator

$$R_\Omega : [H^{\sigma-\varepsilon}(\mathbb{R}^n,\varrho),H^{\sigma+\delta}(\mathbb{R}^n,\varrho)]_\psi \to [H^{\sigma-\varepsilon}(\Omega,\varrho),H^{\sigma+\delta}(\Omega,\varrho)]_\psi$$

is uniformly bounded in the parameter ϱ. (Since, here, the left pair of spaces is normal, the right pair is also normal.) Using inequality (4.30) for the case $G = \mathbb{R}^n$, we conclude that the norm of the operator

$$R_\Omega : H^{\sigma,\varphi}(\mathbb{R}^n,\varrho) \to [H^{\sigma-\varepsilon}(\Omega,\varrho),H^{\sigma+\delta}(\Omega,\varrho)]_\psi \tag{4.38}$$

is uniformly bounded in the parameter ϱ.

We need a linear bounded operator which is a right inverse to (4.38). In monograph [258, Sec. 4.2.2], for each $l \in \mathbb{N}$, it is constructed a linear map T_l that extends an arbitrary distribution $u \in H^{-l}(\Omega)$ to the space $\mathbb{R}^n$ and is a bounded operator

$$T_l : H^{\alpha}(\Omega) \to H^{\alpha}(\mathbb{R}^n) \quad \text{for each} \quad \alpha \in \mathbb{R}, \ |\alpha| < l. \tag{4.39}$$

(This operator has been used in Section 3.2.1.) We choose the integer $l > \sigma + \delta$. Applying the interpolation with the parameter ψ to (4.39), where $\alpha \in \{\sigma - \varepsilon, \sigma + \delta\}$, and using Theorems 1.14 and 3.2, we obtain the bounded operator

$$T_l : H^{\sigma,\varphi}(\Omega) \to H^{\sigma,\varphi}(\mathbb{R}^n). \tag{4.40}$$

It follows from the boundedness of operators (4.39) and (4.40) that the norms of the operators

$$T_l : H^{\alpha}(\Omega, \varrho) \to H^{\alpha}(\mathbb{R}^n, \varrho), \quad 0 < \alpha < l, \tag{4.41}$$

$$T_l : H^{\sigma,\varphi}(\Omega, \varrho) \to H^{\sigma,\varphi}(\mathbb{R}^n, \varrho) \tag{4.42}$$

are uniformly bounded in the parameter ϱ. We apply the interpolation with the parameter ψ to the spaces in which operators (4.41), where $\alpha \in \{\sigma - \varepsilon, \sigma + \delta\}$, act. Using Theorem 1.8 and inequality (4.30) for $G = \mathbb{R}^n$, we conclude that the norm of the operator

$$T_l : [H^{\sigma-\varepsilon}(\Omega, \varrho), H^{\sigma+\delta}(\Omega, \varrho)]_{\psi} \to H^{\sigma,\varphi}(\mathbb{R}^n, \varrho) \tag{4.43}$$

is uniformly bounded in ϱ.

The operator $R_{\Omega} T_l = I$ is the identity mapping It follows from the uniform boundedness in the parameter ϱ of the norms of operators (4.43), (4.36) and (4.42), (4.38) that the norms of the embedding operators

$$I = R_{\Omega} T_l : [H^{\sigma-\varepsilon}(\Omega, \varrho), H^{\sigma+\delta}(\Omega, \varrho)]_{\psi} \to H^{\sigma,\varphi}(\Omega, \varrho),$$

$$I = R_{\Omega} T_l : H^{\sigma,\varphi}(\Omega, \varrho) \to [H^{\sigma-\varepsilon}(\Omega, \varrho), H^{\sigma+\delta}(\Omega, \varrho)]_{\psi}$$

are uniformly bounded in ϱ. This immediately gives the two-sided estimate (4.30) for $G = \Omega$.

The case $G = \Gamma$. We derive it from the first case $G = \mathbb{R}^k$ with $k := n - 1$. Let $\varrho \geq 1$. We reason by analogy with the proof of Lemma 2.6. We use the local definition 2.1 of the spaces $H^{s,\varphi}(\Gamma)$, $s \in \mathbb{R}$, $\varphi \in \mathcal{M}$, for fixed finite atlas and partition of unity on Γ. Consider the linear mapping of "flattening" of the manifold Γ:

$$T : u \mapsto ((\chi_1 u) \circ \alpha_1, \ldots, (\chi_r u) \circ \alpha_r), \quad u \in \mathcal{D}'(\Gamma).$$

It is directly verified that this mapping defines the isometric operators

$$T:\ H^{\sigma,\varphi}(\Gamma,\varrho)\to(H^{\sigma,\varphi}(\mathbb{R}^{n-1},\varrho))^{r},\tag{4.44}$$

$$T:\ H^{\alpha}(\Gamma,\varrho)\to(H^{\alpha}(\mathbb{R}^{n-1},\varrho))^{r},\quad \alpha>0.\tag{4.45}$$

We apply the interpolation with the parameter ψ to the spaces in which operators (4.45), where $\alpha\in\{\sigma-\varepsilon,\sigma+\delta\}$, act. Using Theorem 1.8, we conclude that the norm of the operator

$$T:\ [H^{\sigma-\varepsilon}(\Gamma,\varrho),H^{\sigma+\delta}(\Gamma,\varrho)]_{\psi}\to\big[(H^{\sigma-\varepsilon}(\mathbb{R}^{n-1},\varrho))^{r},(H^{\sigma+\delta}(\mathbb{R}^{n-1},\varrho))^{r}\big]_{\psi}$$

is uniformly bounded in the parameter ϱ. (It is obvious that the pairs of spaces written here are normal.) By virtue of Theorem 1.5 and relation (4.30) proved for $G=\mathbb{R}^{n-1}$, we conclude that the norm of the operator

$$T:\ [H^{\sigma-\varepsilon}(\Gamma,\varrho),H^{\sigma+\delta}(\Gamma,\varrho)]_{\psi}\to(H^{\sigma,\varphi}(\mathbb{R}^{n-1},\varrho))^{r}\tag{4.46}$$

is also uniformly bounded in ϱ.

Parallel with T, consider the linear mapping of "sewing"

$$K:\ (w_1,\dots,w_r)\mapsto\sum_{j=1}^{r}\Theta_j((\eta_j w_j)\circ\alpha_j^{-1}),$$

where $w_1,\dots,w_r$ are distributions in $\mathbb{R}^{n-1}$. Here, the function $\eta_j\in C^{\infty}(\mathbb{R}^{n-1})$ is finite and equal to 1 on the set $\alpha_j^{-1}(\operatorname{supp}\chi_j)$, and Θ_j is the operator of extension by zero to Γ. By virtue of (2.8), we have the bounded operators

$$K:\ (H^{\sigma,\varphi}(\mathbb{R}^{n-1}))^{r}\to H^{\sigma,\varphi}(\Gamma),$$

$$K:\ (H^{\alpha}(\mathbb{R}^{n-1}))^{r}\to H^{\alpha}(\Gamma),\quad \alpha\in\mathbb{R}.$$

It is directly verified that the norm of each of the operators

$$K:\ (H^{\sigma,\varphi}(\mathbb{R}^{n-1},\varrho))^{r}\to H^{\sigma,\varphi}(\Gamma,\varrho),\tag{4.47}$$

$$K:\ (H^{\alpha}(\mathbb{R}^{n-1},\varrho))^{r}\to H^{\alpha}(\Gamma,\varrho),\quad \alpha>0,\tag{4.48}$$

is uniformly bounded in the parameter ϱ.

We apply the interpolation with the parameter ψ to the spaces in which operators (4.48), where $\alpha\in\{\sigma-\varepsilon,\sigma+\delta\}$, act. Based on Theorem 1.8, we conclude that the norm of the operator

$$K:\ [(H^{\sigma-\varepsilon}(\mathbb{R}^{n-1},\varrho))^{r},(H^{\sigma+\delta}(\mathbb{R}^{n-1},\varrho))^{r}]_{\psi}\to[H^{\sigma-\varepsilon}(\Gamma,\varrho),H^{\sigma+\delta}(\Gamma,\varrho)]_{\psi}$$

is uniformly bounded in ϱ. By virtue of Theorem 1.5 and relation (4.30) proved for $G=\mathbb{R}^{n-1}$, we conclude that the norm of the operator

$$K:\,(H^{\sigma,\varphi}(\mathbb{R}^{n-1},\varrho))^{r}\to[H^{\sigma-\varepsilon}(\Gamma,\varrho),H^{\sigma+\delta}(\Gamma,\varrho)]_{\psi} \tag{4.49}$$

is also uniformly bounded in ϱ.

By virtue of (2.7), the product $KT=I$ is the identity operator. Therefore, the uniform boundedness in the parameter ϱ of the norms of operators (4.44), (4.49) and (4.46), (4.47) leads to the uniform boundedness in ϱ of the norms of the embedding operators

$$I=KT:\,H^{\sigma,\varphi}(\Gamma,\varrho)\to[H^{\sigma-\varepsilon}(\Gamma,\varrho),H^{\sigma+\delta}(\Gamma,\varrho)]_{\psi},$$

$$I=KT:\,[H^{\sigma-\varepsilon}(\Gamma,\varrho),H^{\sigma+\delta}(\Gamma,\varrho)]_{\psi}\to H^{\sigma,\varphi}(\Gamma,\varrho).$$

This immediately yields the two-sided estimate (4.30) for $G=\Gamma$.

Lemma 4.1 is proved.

In the proof of Lemma 4.1, we used the following result.

Lemma 4.2. *Let $\sigma>0$, $\varphi\in\mathcal{M}$, and $\varphi_\sigma(t):=t^{\sigma}\varphi(t)$ for $t\geq1$. Then there exists a number $c=c(\sigma,\varphi)\geq1$ such that*

$$c^{-1}\varphi_\sigma(t_1+t_2)\leq\varphi_\sigma(t_1)+\varphi_\sigma(t_2)\leq c\,\varphi_\sigma(t_1+t_2) \tag{4.50}$$

for all $t_1,t_2\geq1$.

Proof. Since $\varphi_\sigma\in\mathrm{RO}$ (see Subsection 2.8.1), we have

$$\varphi_\sigma(2t)\asymp\varphi_\sigma(t)\quad\text{for}\quad t\geq1. \tag{4.51}$$

In addition, since the function φ_σ has the order of variation $\sigma>0$, it is (weakly) equivalent to a certain increasing positive function ψ:

$$\varphi_\sigma(t)\asymp\psi(t)\quad\text{for}\quad t\geq1. \tag{4.52}$$

Now (4.50) follows from relations (4.51) and (4.52) because the function ψ increases. Indeed, for each number $j=1,2$, we can write

$$\varphi_\sigma(t_j)\asymp\psi(t_j)\leq\psi(t_1+t_2)\asymp\varphi_\sigma(t_1+t_2)\quad\text{for}\quad t_1,t_2\geq1.$$

Therefore, there exists a number $c_1>0$ such that

$$\varphi_\sigma(t_1)+\varphi_\sigma(t_2)\leq c_1\varphi_\sigma(t_1+t_2)\quad\text{for}\quad t_1,t_2\geq1. \tag{4.53}$$

Conversely, assuming without loss of generality that $t_1 \le t_2$, we have

$$\varphi_\sigma(t_1+t_2) \asymp \psi(t_1+t_2) \le \psi(2t_2) \asymp \varphi_\sigma(2t_2) \asymp \varphi_\sigma(t_2)$$

$$\le \varphi_\sigma(t_1)+\varphi_\sigma(t_2) \quad \text{for} \quad t_1,t_2 \ge 1.$$

Therefore, there exists a number $c_2 > 0$ such that

$$\varphi_\sigma(t_1+t_2) \le c_2\,(\varphi_\sigma(t_1)+\varphi_\sigma(t_2)) \quad \text{for} \quad t_1,t_2 \ge 1. \tag{4.54}$$

Relations (4.53) and (4.54) mean the two-sided inequality (4.50).

Lemma 4.2 is proved.

Using Lemma 4.1, we can give the following:

Proof of assertion (ii) of Theorem 4.9. Let $s > 2q$ and $\varphi \in \mathcal{M}$, and let the parameter $\lambda \in K$ be such that $|\lambda| \ge \max\{\lambda_0, 1\}$, where the number $\lambda_0 > 0$ is taken from assertion (i) of this theorem. We set $\varepsilon = \delta = (s-2q)/2 > 0$. As indicated above, Theorem 4.9 is true in the Sobolev case of $\varphi \equiv 1$. Therefore, we have the isomorphisms

$$(L(\lambda), B(\lambda)) : \; H^{s\mp\varepsilon}(\Omega,|\lambda|) \leftrightarrow H^{s\mp\varepsilon-2q}(\Omega,|\lambda|) \oplus \bigoplus_{j=1}^{q} H^{s\mp\varepsilon-m_j-1/2}(\Gamma,|\lambda|)$$

$$=: \; \mathcal{H}_{s\mp\varepsilon}(\Omega,\Gamma,|\lambda|) \tag{4.55}$$

such that the norms of the direct and inverse operators are uniformly bounded in the parameter λ. [Note that we passed to Hilbert norms in estimate (4.28).] Let ψ be the interpolation parameter from Theorem 1.14. Applying the interpolation with this parameter to (4.55), we obtain one more isomorphism

$$(L(\lambda), B(\lambda)) : \; [H^{s-\varepsilon}(\Omega,|\lambda|), H^{s+\varepsilon}(\Omega,|\lambda|)]_\psi$$

$$\leftrightarrow [\mathcal{H}_{s-\varepsilon}(\Omega,\Gamma,|\lambda|), \mathcal{H}_{s+\varepsilon}(\Omega,\Gamma,|\lambda|)]_\psi. \tag{4.56}$$

Here, the norms of the direct and inverse operators are uniformly bounded in λ by virtue of Theorem 1.8. [Note that, the pairs of spaces written in (4.56) are normal.]

Further, according to Theorem 1.5 on interpolation of orthogonal sums of spaces, we can write

$$[\mathcal{H}_{s-\varepsilon}(\Omega,\Gamma,|\lambda|), \mathcal{H}_{s+\varepsilon}(\Omega,\Gamma,|\lambda|)]_\psi$$

$$= [H^{s-\varepsilon-2q}(\Omega,|\lambda|), H^{s+\varepsilon-2q}(\Omega,|\lambda|)]_\psi$$

$$\oplus \bigoplus_{j=1}^{q} [H^{s-\varepsilon-m_j-1/2}(\Gamma,|\lambda|), H^{s+\varepsilon-m_j-1/2}(\Gamma,|\lambda|)]_\psi$$

with equality of norms. By virtue of Lemma 4.1 and (4.56), this yields the isomorphism

$$(L(\lambda), B(\lambda)) : H^{s,\varphi}(\Omega, |\lambda|) \leftrightarrow H^{s-2q,\varphi}(\Omega, |\lambda|) \oplus \bigoplus_{j=1}^{q} H^{s-m_j-1/2,\varphi}(\Gamma, |\lambda|)$$

such that the norms of the direct and inverse operators are uniformly bounded in the parameters λ. This immediately gives the two-sided estimate (4.28).

Assertion (ii) is proved.

Thus, Theorem 4.9 is proved.

Corollary 4.1. *Let the boundary-value problem* (4.24), (4.25) *be parameter-elliptic on a certain closed beam* $K := \{\lambda \in \mathbb{C} : \arg \lambda = \text{const}\}$. *Then bounded operator* (4.27) *has the zero index for arbitrary* $s > 2q$, $\varphi \in \mathcal{M}$, *and* $\lambda \in \mathbb{C}$.

Proof. As indicated above, operator (4.27) is Fredholm, and its index is independent of the mentioned parameters s, φ, and λ. By virtue of Theorem 4.9, the index of operator (4.27) is equal to 0 for $|\lambda| \gg 1$. Therefore, it is equal to 0 for every $\lambda \in \mathbb{C}$. Corollary 4.1 is proved.

4.1.5 Formally mixed elliptic boundary-value problem

In this section, we consider the elliptic boundary-value problem for the linear differential equation $Lu = f$ in the multiply connected domain Ω. Unlike the previous sections, it is assumed that orders of boundary expressions are different on different connected components of the boundary Γ. For example, for the Laplace equation in a ring, one can define the Dirichlet boundary condition on one component of the boundary and the Neumann boundary condition on the other. The considered problem relates to the class of mixed elliptic boundary-value problems. The theory of such problems is more complicated than that of unmixed problems (see, e.g., papers [223, 186, 265, 236, 76, 43], monograph [77] and references therein). In the considered problem, parts of the boundary on which the orders of the boundary expression are different do not border on each other. We formally call this problem mixed. With the use of local constructions, this problem can be reduced to an elliptic model problem in the half-space.

In this subsection we assume that the boundary Γ of the domain Ω consists of $r \geq 2$ nonempty connected components $\Gamma_1, \dots, \Gamma_r$. Consider the formally mixed boundary-value problem in the domain Ω

$$Lu = f \quad \text{in} \quad \Omega \tag{4.57}$$

$$B_{k,j}u = g_{k,j} \quad \text{on} \quad \Gamma_k \quad \text{for} \quad j = 1, \dots, q \quad \text{and} \quad k = 1, \dots, r. \tag{4.58}$$

Here, as above, L is a linear differential expression on $\overline{\Omega}$ of even order $2q$ and $\{B_{k,j} : j = 1, \ldots, q\}$ is a system of linear differential expressions defined on the component Γ_k. It is assumed that all orders $m_{k,j} := \operatorname{ord} B_{k,j} \leq 2q - 1$. The coefficients of the differential expressions L and $B_{k,j}$ are infinitely smooth complex-valued functions. We set

$$\Lambda := (L, B_{1,1}, \ldots, B_{1,q}, \ldots, B_{r,1}, \ldots, B_{r,q}),$$

$$N_\Lambda := \{u \in C^\infty(\overline{\Omega}) : \Lambda u = 0\}.$$

Definition 4.2. The formally mixed boundary-value problem (4.57), (4.58) is called *elliptic* in the multiply connected domain Ω if the following conditions are satisfied:

(i) The differential expression L is properly elliptic on $\overline{\Omega}$.

(ii) For each $k \in \{1, \ldots, r\}$, the system of boundary expressions

$$\{B_{k,j} : j = 1, \ldots, q\}$$

satisfies the complementing condition with respect to L on Γ_k.

Theorem 4.10. *Suppose that the boundary-value problem* (4.57), (4.58) *is elliptic in the domain* Ω. *Let* $s > 2q$ *and* $\varphi \in \mathcal{M}$. *Then the mapping* $u \mapsto \Lambda u$, *where* $u \in C^\infty(\overline{\Omega})$, *extends uniquely (by continuity) to the Fredholm bounded operator*

$$\Lambda : H^{s,\varphi}(\Omega) \to H^{s-2q,\,\varphi}(\Omega) \oplus \bigoplus_{k=1}^{r} \bigoplus_{j=1}^{q} H^{s-m_{k,j}-1/2,\varphi}(\Gamma_k) \qquad (4.59)$$

$$=: \mathcal{H}_{s,\varphi}.$$

The kernel of this operator coincides with N_Λ. *Its range consists of all vectors*

$$(f, g_{1,1}, \ldots, g_{1,q}, \ldots, g_{r,1}, \ldots, g_{r,q}) \in \mathcal{H}_{s,\varphi}$$

such that

$$(f, w_0)_\Omega + \sum_{k=1}^{r} \sum_{j=1}^{q} (g_{k,j}, w_{k,j})_{\Gamma_k} = 0$$

for every vector-valued function

$$(w_0, w_{1,1}, \ldots, w_{1,q}, \ldots, w_{r,1}, \ldots, w_{r,q}) \in W_\Lambda.$$

Here, W_Λ *is a certain finite-dimensional subspace of*

$$C^\infty(\overline{\Omega}) \times \prod_{j=1}^{r} (C^\infty(\Gamma_j))^q,$$

this subspace not depending on s and φ. The index of operator (4.59) *is equal to* $\dim N_\Lambda - \dim W_\Lambda$ *and does not depend on s and φ as well.*

In the Sobolev case of $\varphi \equiv 1$, this theorem is a special case of Theorem 1 in [163], where the boundedness and Fredholm property of the operator Λ are established for the scale of Lizorkin–Triebel spaces, which contains the Sobolev scale. The general case of $\varphi \in \mathcal{M}$ is derived with the use of interpolation by analogy with the proof of Theorem 4.1.

4.2 Elliptic boundary-value problems in the two-sided scale

In this section, we study the regular elliptic boundary-value problem (4.1) in the modified two-sided scale of Hörmander inner product spaces. Now, unlike the previous Section 4.1, the numerical index defining the main smoothness passes through the entire real axis.

4.2.1 Preliminary remarks

Theorems on solvability of elliptic boundary-value problems proved in Section 4.1 are, generally speaking, not true for an arbitrary real parameter s that defines the main smoothness of a solution to the problem. This is caused by the fact that the mapping $u \mapsto u \restriction \Gamma$, where $u \in C^\infty(\overline{\Omega})$, cannot be extended to the continuous trace operator $R_\Gamma : H^{s,\varphi}(\Omega) \to \mathcal{D}'(\Gamma)$ for $s < 1/2$ (see Remark 3.5). Therefore, operator (4.3) corresponding to the boundary-value problem (4.1) is not defined for $s < m + 1/2$, where m is the maximum of the orders of the boundary expressions $B_1, \dots, B_q$. For the other boundary-value problems considered in Section 4.1, the situation is analogous.

To obtain the bounded operator (L, B) for any $s < 2q$, it is necessary to take another space, somewhat different from $H^{s,\varphi}(\Omega)$, as a domain of this operator. There are two essentially different methods for the construction of the domain proposed by Ya. A. Roitberg [202, 203, 209] and J.-L. Lions and E. Magenes [121, 126, 119, 120] in the Sobolev case. These methods lead to different types of the theorems on solvability of elliptic boundary-value problems (the general theorem and the individual theorems). In the general theorem, the domain of the operator (L, B) does not depend on coefficients of the elliptic expression L and is unique for all boundary-value problems of the same order. In individual theorems, the domain depends on coefficients of the expression L (even on the coefficients in lower order derivatives).

Note that the theorems (about solvability of elliptic boundary-value problems) proved in Section 4.1 are general (for the corresponding classes of problems).

In Section 4.2, we extend Ya. A. Roitberg's approach over the two-sided refined Sobolev scale. Namely, we modify this scale in the sense of Roitberg and then prove the general theorem on the solvability of the regular elliptic boundary-value problem (4.1) in this modified scale. (The theory developed below can be easily extend over the other elliptic boundary-value problems considered in Section 4.1. We do not dwell on it.)

J.-L. Lions and E. Magenes' approach leading to the individual theorems on solvability will be considered in Sections 4.4 and 4.5.

Note again (see the end of Subsection 4.1.1) that, in the case of $s < m + 1/2$, the properties of the inhomogeneous elliptic boundary-value problem (4.1) can not be directly derived from the properties of the two semihomogeneous boundary-value problems studied in Sections 3.3 and 3.4.

4.2.2 The refined scale modified in the sense of Roitberg

First, following Ya. A. Roitberg [202, 203, 209], we introduce the notion of the generalized solution to the boundary-value problem (4.1).

In a neighborhood of the boundary Γ, we write the differential expressions L and B_j in the form

$$L = \sum_{k=0}^{2q} L_k\, D_\nu^k \quad \text{and} \quad B_j = \sum_{k=0}^{m_j} B_{j,k}\, D_\nu^k. \tag{4.60}$$

Here, $D_\nu := i\,\partial/\partial\nu$ as above, and L_k and $B_{j,k}$ are certain tangential differential expressions (with respect to the boundary Γ). Integrating by parts, we write the following Green formula:

$$(Lu, v)_\Omega = (u, L^+ v)_\Omega - i \sum_{k=1}^{2q} (D_\nu^{k-1} u, L^{(k)} v)_\Gamma \tag{4.61}$$

for arbitrary functions $u, v \in C^\infty(\overline{\Omega})$. Here,

$$L^{(k)} := \sum_{r=k}^{2q} D_\nu^{r-k} L_r^+,$$

where L_r^+ denotes the differential expression which is formally conjugate to L_r. Passing to the limit, we establish that relation (4.61) is true for each function $u \in H^{2q}(\Omega)$. Denote

$$u_0 := u \quad \text{and} \quad u_k := (D_\nu^{k-1} u) \restriction \Gamma \quad \text{for} \quad k = 1, \ldots 2q. \tag{4.62}$$

By virtue of (4.60) and (4.61), the boundary-value problem (4.1) for the unknown function $u \in H^{2q}(\Omega)$ is equivalent to the system of conditions

$$(u_0, L^+ v)_\Omega - i \sum_{k=1}^{2q} (u_k, L^{(k)} v)_\Gamma = (f, v)_\Omega \quad \text{for all} \quad v \in C^\infty(\overline{\Omega}), \tag{4.63}$$

$$\sum_{k=0}^{m_j} B_{j,k}\, u_{k+1} = g_j \quad \text{on } \Gamma \quad \text{for} \quad j = 1, \ldots, q. \tag{4.64}$$

Note that these conditions make sense in the case of arbitrary distributions

$$u_0 \in \mathcal{D}'(\mathbb{R}^n), \quad \operatorname{supp} u_0 \subseteq \overline{\Omega}, \quad u_1, \ldots, u_{2q} \in \mathcal{D}'(\Gamma), \tag{4.65}$$

$$f \in \mathcal{D}'(\mathbb{R}^n), \quad \operatorname{supp} f \subseteq \overline{\Omega}, \quad g_1, \ldots, g_q \in \mathcal{D}'(\Gamma).$$

For this reason, it is useful to introduce the following definition.

Definition 4.3. A vector $u = (u_0, u_1, \ldots, u_{2q})$ satisfying (4.65) is said to be a *Roitberg generalized solution* to the boundary-value problem (4.1) if conditions (4.63) and (4.64) are fulfilled.

We introduce Hilbert spaces whose elements can be considered as Roitberg generalized solutions.

Let $r \in \mathbb{N}$, $s \in \mathbb{R}$, and $\varphi \in \mathcal{M}$. We set

$$E_r := \{k - 1/2 : k = 1, \ldots, r\}.$$

Definition 4.4. In the case of $s \in \mathbb{R} \setminus E_r$, the linear space $H^{s,\varphi,(r)}(\Omega)$ is, by definition, the completion of $C^\infty(\overline{\Omega})$ with respect to the Hilbert norm

$$\|u\|_{H^{s,\varphi,(r)}(\Omega)} := \biggl(\|u\|^2_{H^{s,\varphi,(0)}(\Omega)} + \sum_{k=1}^{r} \|(D_\nu^{k-1} u) \restriction \Gamma\|^2_{H^{s-k+1/2,\varphi}(\Gamma)} \biggr)^{1/2}. \tag{4.66}$$

In the case $s \in E_r$, we define the space $H^{s,\varphi,(r)}(\Omega)$ by the interpolation

$$H^{s,\varphi,(r)}(\Omega) := [H^{s-\varepsilon,\varphi,(r)}(\Omega),\, H^{s+\varepsilon,\varphi,(r)}(\Omega)]_{t^{1/2}} \quad \text{for} \quad 0 < \varepsilon < 1. \tag{4.67}$$

Remark 4.5. In relation (4.67), we use the interpolation of Hilbert spaces with the power parameter $\psi(t) = t^{1/2}$. In what follows, we will show that the pair of spaces on the right-hand side of this relation is admissible [Theorem 4.12(i), (iv)] and that the result of this interpolation is independent of ε up to equivalence of norms (Section 4.2.5, Theorem 4.21).

In the Sobolev case of $\varphi \equiv 1$, the space $H^{s,\varphi,(r)}(\Omega)$ was introduced by Ya. A. Roitberg in [202, 203] (see also his monograph [209, Sec. 2.1]). As usual, we set $H^{s,(r)}(\Omega) := H^{s,1,(r)}(\Omega)$.

Definition 4.5. The family of Hilbert spaces

$$\{H^{s,\varphi,(r)}(\Omega) : s \in \mathbb{R}, \varphi \in \mathcal{M}\} \tag{4.68}$$

is said to be the refined Sobolev scale modified in the sense of Roitberg, or, briefly, the modified refined scale. The number r is called the order of modification.

From the point of view of the application to the boundary-value problem (4.1), we are interested in the case of even order of modification $r = 2q$. By virtue of Definition 4.4, the mapping $u \mapsto (D_\nu^{k-1}u) \restriction \Gamma$, where $u \in C^\infty(\overline{\Omega})$, extends by continuity to a bounded operator from $H^{s,\varphi,(r)}(\Omega)$ to $H^{s-k+1/2,\varphi}(\Gamma)$ for arbitrary $s \in \mathbb{R}$ and $\varphi \in \mathcal{M}$ provided that $k \in \{0, 1, \dots, 2q\}$. Therefore, for every element $u \in H^{s,\varphi,(2q)}(\Omega)$, the vector

$$(u_0, u_1, \dots, u_{2q}) \in H^{s,\varphi,(0)}(\Omega) \oplus \bigoplus_{k=1}^{2q} H^{s-k+1/2,\varphi}(\Gamma) \tag{4.69}$$

is well-defined by relations (4.62) whereby the closure. Thus, the element u can be regarded as the Roitberg generalized solution (4.69) of the boundary-value problem (4.1).

We study properties of the modified refined scale (4.68). For arbitrary $s \in \mathbb{R}$ and $\varphi \in \mathcal{M}$, we set

$$\Pi_{s,\varphi,(r)}(\Omega, \Gamma) := H^{s,\varphi,(0)}(\Omega) \oplus \bigoplus_{k=1}^{r} H^{s-k+1/2,\varphi}(\Gamma).$$

In addition, denote

$$K_{s,\varphi,(r)}(\Omega, \Gamma) := \{(u_0, u_1, \dots, u_r) \in \Pi_{s,\varphi,(r)}(\Omega, \Gamma) : \; u_k = (D_\nu^{k-1}u_0) \restriction \Gamma$$
$$\text{for all } k \in \{1, \dots r\} \text{ such that } s > k - 1/2\}.$$

By virtue of Theorem 3.5, $K_{s,\varphi,(r)}(\Omega, \Gamma)$ is closed in $\Pi_{s,\varphi,(r)}(\Omega, \Gamma)$. We consider $K_{s,\varphi,(r)}(\Omega, \Gamma)$ as a Hilbert space with respect to the inner product in $\Pi_{s,\varphi,(r)}(\Omega, \Gamma)$. In the Sobolev case $\varphi \equiv 1$, we omit the index φ in notation of the spaces introduced in this chapter.

Theorem 4.11. *Let $r \in \mathbb{N}$, $s \in \mathbb{R} \setminus E_r$, and $\varphi \in \mathcal{M}$. The following assertions are true:*

(i) *The linear mapping*

$$T_r : u \mapsto (u, u \restriction \Gamma, \dots, (D_\nu^{r-1}u) \restriction \Gamma), \quad u \in C^\infty(\overline{\Omega}), \tag{4.70}$$

extends uniquely (by continuity) to the isometric isomorphism

$$T_r : H^{s,\varphi,(r)}(\Omega) \leftrightarrow K_{s,\varphi,(r)}(\Omega, \Gamma). \tag{4.71}$$

(ii) *For arbitrary positive numbers ε and δ such that all the numbers s, $s - \varepsilon$, and $s + \delta$ belong to one of the intervals*

$$\begin{aligned} &\alpha_0 := (-\infty, 1/2), \\ &\alpha_k := (k - 1/2,\, k + 1/2), \;\; k = 1, \dots, r - 1, \\ &\alpha_r := (r - 1/2, +\infty), \end{aligned} \tag{4.72}$$

we have

$$[H^{s-\varepsilon,(r)}(\Omega), H^{s+\delta,(r)}(\Omega)]_\psi = H^{s,\varphi,(r)}(\Omega) \tag{4.73}$$

up to equivalence of norms. Here, ψ *is the interpolation parameter from Theorem* 1.14.

Proof. In the case of $\varphi \equiv 1$, assertion (i) is established by Ya. A. Roitberg [209, Lemma 2.2.1]. Using this result and interpolation with a function parameter, we first derive assertion (ii) and then prove assertion (i) for an arbitrary $\varphi \in \mathcal{M}$.

Let X_ψ denote the left-hand side of equality (4.73). (The pair of spaces in (4.73) is obviously admissible.) Consider the isometric operators

$$T_r : H^{\sigma,(r)}(\Omega) \to \Pi_{\sigma,(r)}(\Omega,\Gamma), \quad \sigma \in \{s-\varepsilon, s+\delta\}.$$

Using the interpolation with the parameter ψ, we obtain the bounded operator

$$T_r : X_\psi \to [\Pi_{s-\varepsilon,(r)}(\Omega,\Gamma), \Pi_{s+\delta,(r)}(\Omega,\Gamma)]_\psi.$$

By virtue of the interpolation theorems 1.5, 2.2, and 3.10, we obtain the following equalities of spaces with equivalence of norms:

$$\begin{aligned}
&[\Pi_{s-\varepsilon,(r)}(\Omega,\Gamma), \Pi_{s+\delta,(r)}(\Omega,\Gamma)]_\psi \\
&\quad = [H^{s-\varepsilon,(0)}(\Omega), H^{s+\delta,(0)}(\Omega)]_\psi \oplus \bigoplus_{k=1}^{r} [H^{s-\varepsilon-k+1/2}(\Gamma), H^{s+\delta-k+1/2}(\Gamma)]_\psi \\
&\quad = H^{s,\varphi,(0)}(\Omega) \oplus \bigoplus_{k=1}^{r} H^{s-k+1/2,\varphi}(\Gamma) = \Pi_{s,\varphi,(r)}(\Omega,\Gamma).
\end{aligned}$$

Therefore, the operator

$$T_r : X_\psi \to \Pi_{s,\varphi,(r)}(\Omega,\Gamma) \tag{4.74}$$

is bounded. This yields the estimate

$$\|u\|_{H^{s,\varphi,(r)}(\Omega)} = \|T_r\, u\|_{\Pi_{s,\varphi,(r)}(\Omega,\Gamma)} \le c_1 \, \|u\|_{X_\psi} \tag{4.75}$$

for any $u \in C^\infty(\overline{\Omega})$. Here, c_1 is the norm of operator (4.74).

We prove the inequality inverse to (4.75). By condition, $s, s-\varepsilon, s+\delta \in \alpha_p$ for a certain number $p \in \{0, 1, \ldots, r\}$. Consider the linear mapping

$$T_{r,p} : u \mapsto \big(u,\, \{(D_\nu^{k-1} u) \restriction \Gamma : p+1 \le k \le r\}\big), \quad u \in C^\infty(\overline{\Omega}).$$

(As above, the index k is integer.)

This mapping extends by continuity to the isomorphism

$$T_{r,p} : H^{\sigma,(r)}(\Omega) \leftrightarrow H^{\sigma,(0)}(\Omega) \oplus \bigoplus_{p+1\le k\le r} H^{\sigma-k+1/2}(\Gamma) \tag{4.76}$$

$$\text{for each} \quad \sigma \in \{s-\varepsilon, s+\delta\}.$$

Indeed, the boundedness of operator (4.76) follows from the definition of the space $H^{\sigma,(r)}(\Omega)$. We show that this operator is bijective. Let $u \in H^{\sigma,(r)}(\Omega)$ and let

$$\big(u_0,\, \{u_k : p+1 \le k \le r\}\big) \in H^{\sigma,(0)}(\Omega) \oplus \bigoplus_{p+1\le k\le r} H^{\sigma-k+1/2}(\Gamma).$$

We set $u_k := (D_\nu^{k-1}u_0)\upharpoonright\Gamma$ for $1 \le k \le p$. By virtue of Theorem 3.5, the distribution u_k is well-defined because $\sigma > k-1/2$ for the indicated numbers k. Note that $\sigma < k-1/2$ for $p+1 \le k \le r$. Therefore, $(u_0, u_1, \dots, u_r) \in K_{\sigma,(r)}(\Omega,\Gamma)$. As was mentioned above, assertion (i) of the theorem is true in the case of $\varphi \equiv 1$. Therefore, we have the isomorphisms

$$T_r : H^{\sigma,(r)}(\Omega) \leftrightarrow K_{\sigma,(r)}(\Omega,\Gamma), \quad \sigma \in \{s-\varepsilon, s+\delta\}.$$

Since

$$T_r\, u = (u_0, u_1, \dots, u_r) \;\Leftrightarrow\; T_{r,p}\, u = (\, u_0,\, \{u_k : p+1 \le k \le r\}\,),$$

we establish that the bounded operator (4.76) is bijective. Therefore, it is an isomorphism (by the Banach theorem on inverse operator).

We apply the interpolation with the parameter ψ to (4.76). By virtue of Theorems 1.5, 2.2, and 3.10, we obtain the isomorphism

$$T_{r,p} : X_\psi \leftrightarrow H^{s,\varphi,(0)}(\Omega) \oplus \bigoplus_{p+1\le k\le r} H^{s-k+1/2,\varphi}(\Gamma). \tag{4.77}$$

This yields the following inequality, which is inverse to (4.75):

$$\|u\|_{X_\psi} \le c_2\Big(\|u\|^2_{H^{s,\varphi,(0)}(\Omega)} + \sum_{p+1\le k\le r} \|(D_\nu^{k-1}u)\upharpoonright\Gamma\|^2_{H^{s-k+1/2,\varphi}(\Gamma)}\Big)^{1/2}$$

$$\le c_2\, \|u\|_{H^{s,\varphi,(r)}(\Omega)}$$

for all $u \in C^\infty(\overline{\Omega})$. Here, c_2 is the norm of the inverse to (4.77). Thus, the norms in the spaces X_ψ and $H^{s,\varphi,(r)}(\Omega)$ are equivalent on the set $C^\infty(\overline{\Omega})$. It is dense in $H^{s,\varphi,(r)}(\Omega)$ by definition and in X_ψ by virtue of Theorem 1.1. Therefore, $X_\psi = H^{s,\varphi,(r)}(\Omega)$ up to equivalent norms. Assertion (ii) is proved.

Let us prove assertion (i). According to the definition of the space $H^{s,\varphi,(r)}(\Omega)$, mapping (4.70) extends by continuity to the isometric operator

$$T_r : H^{s,\varphi,(r)}(\Omega) \to \Pi_{s,\varphi,(r)}(\Omega,\Gamma). \tag{4.78}$$

On the basis of Theorem 3.5, we have the inclusion

$$T_r(H^{s,\varphi,(r)}(\Omega)) \subseteq K_{s,\varphi,(r)}(\Omega,\Gamma).$$

We prove the inverse inclusion. Let

$$(u_0, u_1, \ldots, u_r) \in K_{s,\varphi,(r)}(\Omega,\Gamma).$$

By virtue of (4.77) and the equality $X_\psi = H^{s,\varphi,(r)}(\Omega)$, we obtain the isomorphism

$$T_{r,p} : H^{s,\varphi,(r)}(\Omega) \leftrightarrow H^{s,\varphi,(0)}(\Omega) \oplus \bigoplus_{p+1\le k\le r} H^{s-k+1/2,\varphi}(\Gamma).$$

Therefore, there exists $u \in H^{s,\varphi,(r)}(\Omega)$ such that

$$T_{r,p}\, u = (u_0,\ \{u_k : p+1 \le k \le r\}\,).$$

In view of Theorem 3.5, this yields the equality $T_r\, u = (u_0, u_1, \ldots, u_r)$. Therefore, the inclusion

$$K_{s,\varphi,(r)}(\Omega,\Gamma) \subseteq T_r(H^{s,\varphi,(r)}(\Omega))$$

is proved. Thus,

$$T_r(H^{s,\varphi,(r)}(\Omega)) = K_{s,\varphi,(r)}(\Omega,\Gamma),$$

which, together with the isometric operator (4.78), implies the isometric isomorphism (4.71). Assertion (i) is proved.

Theorem 4.11 is proved.

Assertion (i) of Theorem 4.11 gives the useful isometric representation of the space $H^{s,\varphi,(r)}(\Omega)$ for $s \notin E_r$, namely

$$T_r(H^{s,\varphi,(r)}(\Omega)) = K_{s,\varphi,(r)}(\Omega,\Gamma).$$

Remark 4.6. If $s \in E_r$, then mapping (4.70) extends by continuity to the bounded operator (4.78). However, we can only state that

$$T_r(H^{s,\varphi,(r)}(\Omega)) \subseteq K_{s,\varphi,(r)}(\Omega,\Gamma).$$

This follows from relation (4.67), Theorem 4.11(i), and the interpolation lemma given below.

Lemma 4.3. *For arbitrary* $\sigma \in \mathbb{R}$, $\varepsilon > 0$, *and* $\varphi \in \mathcal{M}$, *we have*

$$[H^{\sigma-\varepsilon,\varphi,(0)}(\Omega), H^{\sigma+\varepsilon,\varphi,(0)}(\Omega)]_{t^{1/2}} = H^{\sigma,\varphi,(0)}(\Omega) \tag{4.79}$$

$$[H^{\sigma-\varepsilon,\varphi}(\Gamma), H^{\sigma+\varepsilon,\varphi}(\Gamma)]_{t^{1/2}} = H^{\sigma,\varphi}(\Gamma) \tag{4.80}$$

up to equivalence of norms.

Proof. We derive equality (4.79) from the interpolation theorems 3.10 and 1.3. According to the first of them,

$$H^{\sigma\mp\varepsilon,\varphi,(0)}(\Omega) = [H^{\sigma-2\varepsilon,(0)}(\Omega), H^{\sigma+2\varepsilon,(0)}(\Omega)]_{\psi_\mp}.$$

Here, the interpolation parameters $\psi_\mp$ are defined by the relations

$$\psi_-(t) := t^{1/4}\varphi(t^{1/(4\varepsilon)}), \quad \psi_+(t) := t^{3/4}\varphi(t^{1/(4\varepsilon)}) \quad \text{for} \quad t \geq 1,$$

and $\psi_\mp(t) := 1$ for $0 < t < 1$. By virtue of Theorem 1.3 on reiterated interpolation, we obtain

$$\begin{aligned}
&[H^{\sigma-\varepsilon,\varphi,(0)}(\Omega), H^{\sigma+\varepsilon,\varphi,(0)}(\Omega)]_{t^{1/2}} \\
&= \big[\,[H^{\sigma-2\varepsilon,(0)}(\Omega), H^{\sigma+2\varepsilon,(0)}(\Omega)]_{\psi_-}, [H^{\sigma-2\varepsilon,(0)}(\Omega), H^{\sigma+2\varepsilon,(0)}(\Omega)]_{\psi_+}\,\big]_{t^{1/2}} \\
&= [H^{\sigma-2\varepsilon,(0)}(\Omega), H^{\sigma+2\varepsilon,(0)}(\Omega)]_{\psi}.
\end{aligned} \tag{4.81}$$

Here, the interpolation parameter ψ is defined by the formulas

$$\psi(t) := \psi_-(t)\,(\psi_+(t)/\psi_-(t))^{1/2} = t^{1/2}\varphi(t^{1/(4\varepsilon)}) \quad \text{for} \quad t \geq 1$$

and $\psi(t) = 1$ for $0 < t < 1$. Therefore, on the basis of Theorem 3.10, we have

$$[H^{\sigma-2\varepsilon,(0)}(\Omega), H^{\sigma+2\varepsilon,(0)}(\Omega)]_{\psi} = H^{\sigma,\varphi,(0)}(\Omega). \tag{4.82}$$

Now equalities (4.81) and (4.82) yield (4.79). Since the obtained equalities of spaces hold up to equivalence of norms, so does equality (4.79). Equality (4.80) is a special case of Theorem 2.4.

Lemma 4.3 is proved.

The proved lemma will also be useful in what follows (in the cases where we will refer to the interpolation formula (4.67)).

We continue the study of properties of the modified refined scale.

Theorem 4.12. *Let* $r \in \mathbb{N}$, $s \in \mathbb{R}$, *and* $\varphi, \varphi_1 \in \mathcal{M}$. *The following assertions are true:*

(i) *The Hilbert space* $H^{s,\varphi,(r)}(\Omega)$ *is separable.*

(ii) *The set* $C^\infty(\overline{\Omega})$ *is dense in the space* $H^{s,\varphi,(r)}(\Omega)$.

(iii) *If* $s > r - 1/2$, *then the norms in the spaces* $H^{s,\varphi,(r)}(\Omega)$ *and* $H^{s,\varphi}(\Omega)$ *are equivalent on the dense set* $C^\infty(\overline{\Omega})$ *and, hence, these spaces are equal up to equivalence of norms.*

(iv) *For an arbitrary number* $\varepsilon > 0$, *we have the continuous and dense embedding* $H^{s+\varepsilon,\varphi_1,(r)}(\Omega) \hookrightarrow H^{s,\varphi,(r)}(\Omega)$. *This embedding is compact.*

(v) *If the function* φ/φ_1 *is bounded in the neighborhood of* ∞, *then we have the continuous and dense embedding* $H^{s,\varphi_1,(r)}(\Omega) \hookrightarrow H^{s,\varphi,(r)}(\Omega)$. *This embedding is compact if* $\varphi(t)/\varphi_1(t) \to 0$ *as* $t \to \infty$.

Remark 4.7. The continuous embedding $H^{s+\varepsilon,\varphi_1,(r)}(\Omega) \hookrightarrow H^{s,\varphi,(r)}(\Omega)$, appeared in Theorem 4.12(iv), (v) for $\varepsilon \geq 0$, is understood in the following sense [23, Chap. 14, Sec. 7]:

(i) there exists a number $c > 0$ such that

$$\|u\|_{H^{s+\varepsilon,\varphi_1,(r)}(\Omega)} \leq c\,\|u\|_{H^{s,\varphi,(r)}(\Omega)} \quad \text{for all} \quad u \in C^\infty(\overline{\Omega});$$

(ii) the identity mapping given on the functions $u \in C^\infty(\overline{\Omega})$ extends by continuity to an *injective* operator from $H^{s+\varepsilon,\varphi_1,(r)}(\Omega)$ to $H^{s,\varphi,(r)}(\Omega)$.

Proof of Theorem 4.12. (i) For the parameter $s \notin E_r$, the separability of the space $H^{s,\varphi,(r)}(\Omega)$ follows from Theorem 4.11(i) and the separability of the space $K_{s,\varphi,(r)}(\Omega,\Gamma)$. If $s \in E_r$, then the space $H^{s,\varphi,(r)}(\Omega)$ is separable by virtue of (4.67) as the result of interpolation of separable Hilbert spaces.

(ii) Assertion (ii) for $s \notin E_r$ is contained in the definition of the space $H^{s,\varphi,(r)}(\Omega)$. If $s \in E_r$, then, by virtue of (4.67) and Theorem 1.1, the dense continuous embedding $H^{s+\varepsilon,\varphi,(r)}(\Omega) \hookrightarrow H^{s,\varphi,(r)}(\Omega)$ is true for a sufficiently small $\varepsilon > 0$. Since $s + \varepsilon \notin E_r$, the set $C^\infty(\overline{\Omega})$ is dense in the space $H^{s+\varepsilon,\varphi,(r)}(\Omega)$. Therefore, this set is also dense in the space $H^{s,\varphi,(r)}(\Omega)$.

(iii) If $s > r - 1/2$ and $k \in \{1,\dots,r\}$, then, by virtue of Theorem 3.5, we have

$$\|(D_\nu^{k-1}u)\restriction\Gamma\|_{H^{s-k+1/2,\varphi}(\Gamma)} \leq c\,\|u\|_{H^{s,\varphi}(\Omega)}$$

for any $u \in C^\infty(\overline{\Omega})$, where the number $c > 0$ is independent of u. Therefore, norms in the spaces $H^{s,\varphi,(r)}(\Omega)$ and $H^{s,\varphi,(0)}(\Omega) = H^{s,\varphi}(\Omega)$ are equivalent on the dense linear manifold $C^\infty(\overline{\Omega})$. Therefore, these spaces are equal.

(iv) By virtue of Theorems 3.9(iv) and 2.3(iii), the compact embedding

$$\Pi_{s+\varepsilon,\varphi_1,(r)}(\Omega,\Gamma) \hookrightarrow \Pi_{s,\varphi,(r)}(\Omega,\Gamma)$$

is true. This implies the compact embedding of subspaces

$$K_{s+\varepsilon,\varphi_1,(r)}(\Omega,\Gamma) \hookrightarrow K_{s,\varphi,(r)}(\Omega,\Gamma).$$

Based on Theorem 4.11(i), we obtain, in the case of $s, s+\varepsilon \notin E_r$, the compact embedding of the space

$$H^{s+\varepsilon,\varphi_1,(r)}(\Omega) = T_r^{-1}(K_{s+\varepsilon,\varphi_1,(r)}(\Omega,\Gamma))$$

in the space

$$H^{s,\varphi,(r)}(\Omega) = T_r^{-1}(K_{s,\varphi,(r)}(\Omega,\Gamma)).$$

If $\{s, s+\varepsilon\} \cap E_r \neq \varnothing$, then, by virtue of (4.67) and Theorem 1.1, the following embeddings are true for a sufficiently small number $\varepsilon_0 > 0$:

$$H^{s+\varepsilon,\varphi_1,(r)}(\Omega) \hookrightarrow H^{s+\varepsilon-\varepsilon_0,\varphi_1,(r)}(\Omega) \hookrightarrow H^{s+\varepsilon_0,\varphi,(r)}(\Omega) \hookrightarrow H^{s,\varphi,(r)}(\Omega).$$

Here, the number ε_0 must satisfy the conditions

$$0 < \varepsilon_0 < 1, \quad \varepsilon_0 < \varepsilon/2, \quad s+\varepsilon-\varepsilon_0 \notin E_r, \quad \text{and} \quad s+\varepsilon_0 \notin E_r.$$

By the proved result, the mean embedding is compact. Therefore, the embedding of the flanked spaces is also compact. This embedding is dense by virtue of assertion (ii).

(v) Assume that the function φ/φ_1 is bounded in a neighborhood of ∞. Then, by virtue of assertion (iv) of Theorems 2.3, 3.3, and 3.8, we obtain the continuous embedding

$$K_{s,\varphi_1,(r)}(\Omega,\Gamma) \hookrightarrow K_{s,\varphi,(r)}(\Omega,\Gamma).$$

Based on Theorem 4.11(i), we obtain the required continuous embedding in the case $s \notin E_r$, namely,

$$\begin{aligned} H^{s,\varphi_1,(r)}(\Omega) &= T_r^{-1}(K_{s,\varphi_1,(r)}(\Omega,\Gamma)) \\ &\hookrightarrow T_r^{-1}(K_{s,\varphi,(r)}(\Omega,\Gamma)) = H^{s,\varphi,(r)}(\Omega). \end{aligned} \tag{4.83}$$

If $s \in E_r$, then, by virtue of the interpolation formula (4.67), we have

$$H^{s\mp 1/2,\varphi_1,(r)}(\Omega) \hookrightarrow H^{s\mp 1/2,\varphi,(r)}(\Omega) \;\Rightarrow\; H^{s,\varphi_1,(r)}(\Omega) \hookrightarrow H^{s,\varphi,(r)}(\Omega). \tag{4.84}$$

Here, the left continuous embeddings has already been proved because

$$s \mp 1/2 \notin E_r.$$

Therefore, the right continuous embedding is true. It is dense for any $s \in \mathbb{R}$ according to assertion (ii).

Further, if $\varphi(t)/\varphi_1(t) \to 0$ as $t \to \infty$, then, by virtue of assertion (iv) of Theorems 2.3, 3.3, and 3.8, the embedding

$$\Pi_{s,\varphi_1,(r)}(\Omega,\Gamma) \hookrightarrow \Pi_{s,\varphi,(r)}(\Omega,\Gamma)$$

is compact. This yields the compactness of the embedding of subspaces

$$K_{s,\varphi_1,(r)}(\Omega,\Gamma) \hookrightarrow K_{s,\varphi,(r)}(\Omega,\Gamma).$$

Therefore, embedding (4.83) is compact in the case of $s \notin E_r$. Now if $s \in E_r$, then the left embeddings in implication (4.84) are compact. This yields the compactness of the right embedding by virtue of (4.67) and the theorem which states that the compactness of operators is preserved in the case of interpolation with power parameter [258, Sec. 1.16.4]. Assertion (v) is proved.

Theorem (4.12) is proved.

In conclusion of this section, we study properties of differential expressions as operators on the space $H^{s,\varphi,(r)}(\Omega)$. Let $K = K(x,D)$ be a linear differential expression defined on $\overline{\Omega}$, and let $R = R(x,D)$ be a boundary linear differential expression defined on Γ. The coefficients of these expressions are infinitely smooth complex-valued functions, and orders are arbitrary.

Theorem 4.13. *Let $r \in \mathbb{N}$, $s \in \mathbb{R}$, and $\varphi \in \mathcal{M}$. The following assertions are true:*

(i) *If $\varkappa := \operatorname{ord} K \le r$, then*

$$\|Ku\|_{H^{s-\varkappa,\varphi,(0)}(\Omega)} \le c_1 \, \|u\|_{H^{s,\varphi,(r)}(\Omega)} \quad \text{for all} \quad u \in C^\infty(\overline{\Omega}),$$

where the number $c_1 > 0$ is independent of u. Therefore, the mapping

$$u \mapsto Ku, \quad \text{where} \quad u \in C^\infty(\overline{\Omega}),$$

extends uniquely (by continuity) to the bounded linear operator

$$K : \, H^{s,\varphi,(r)}(\Omega) \to H^{s-\varkappa,\varphi,(0)}(\Omega).$$

(ii) *If $\varrho := \operatorname{ord} R \le r-1$, then*

$$\|Ru\|_{H^{s-\varrho-1/2,\varphi}(\Gamma)} \le c_2 \, \|u\|_{H^{s,\varphi,(r)}(\Omega)} \quad \text{for all} \quad u \in C^\infty(\overline{\Omega}),$$

where the number $c_2 > 0$ is independent of u. Therefore, the mapping

$$u \mapsto Ru, \quad \text{where} \quad u \in C^\infty(\overline{\Omega}),$$

extends uniquely (by continuity) to the bounded linear operator

$$R : \, H^{s,\varphi,(r)}(\Omega) \to H^{s-\varrho-1/2,\varphi}(\Gamma).$$

Proof. In the Sobolev case, where $\varphi \equiv 1$, this theorem is proved by Ya. A. Roitberg [209, Lemma 2.3.1]. The case of an arbitrary $\varphi \in \mathcal{M}$ is derived with the use of interpolation. We perform it, e.g., for assertion (i). The proof of assertion (ii) is analogous.

First, assume that $s \notin E_r$. Let the positive number $\varepsilon = \delta$ be the same as in Theorem 4.11(ii). The mapping $u \mapsto Ku$, where $u \in C^\infty(\overline{\Omega})$, is extended by continuity to the bounded linear operators

$$K : H^{s\mp\varepsilon,(r)}(\Omega) \to H^{s\mp\varepsilon-\varkappa,(0)}(\Omega).$$

Now we apply the interpolation with the function parameter ψ from Theorem 1.14. By virtue of Theorems 4.11(ii) and 3.10, we obtain assertion (i) in the case of $s \notin E_r$.

Now assume that $s \in E_r$. We choose an arbitrary number $\varepsilon \in (0,1)$. Taking $s \mp \varepsilon \notin E_r$ into account and using the proved result, we obtain the bounded linear operators

$$K : H^{s\mp\varepsilon,\varphi,(r)}(\Omega) \to H^{s\mp\varepsilon-\varkappa,\varphi,(0)}(\Omega).$$

Applying the interpolation with the power parameter $t^{1/2}$ and using (4.67) and Lemma 4.3, we obtain assertion (i) in the case of $s \notin E_r$.

Theorem 4.13 is proved.

It follows from Theorem 4.13(i) for $\varkappa = 0$ that the multiplication by an arbitrary function from the class $C^\infty(\overline{\Omega})$ is a bounded linear operator on each of the spaces $H^{s,\varphi,(r)}(\Omega)$.

4.2.3 Roitberg-type theorems on solvability. The complete collection of isomorphisms

We study the regular elliptic boundary-value problem (4.1) in the modified refined scale, for which the order of modification $r = 2q$. It is useful to compare the results of this subsection and their proof with the results given in Subsection 4.2.1.

According to Theorem 4.13, mapping (4.2) is extended by continuity to the bounded linear operator

$$(L,B) : H^{s,\varphi,(2q)}(\Omega) \to H^{s-2q,\varphi,(0)}(\Omega) \oplus \bigoplus_{j=1}^{q} H^{s-m_j-1/2,\varphi}(\Gamma)$$

$$=: \mathcal{H}_{s,\varphi,(0)}(\Omega,\Gamma). \qquad (4.85)$$

Therefore, for an arbitrary element $u \in H^{s,\varphi,(2q)}(\Omega)$, the right-hand sides $f \in H^{s-2q,\varphi,(0)}(\Omega)$ and $g_j \in H^{s-m_j-1/2,\varphi}(\Gamma)$ of the boundary-value problem (4.1) are defined by means of closure. By virtue of Theorem 4.11(i), the equality $(L,B)u=(f,g_1,\dots,g_q)$ is equivalent to the statement that the vector $(u_0,u_1,\dots,u_{2q}):=T_{2q}u$ is a Roitberg generalized solution to this boundary-value problem. We identify the element u with the vector $(u_0,u_1,\dots,u_{2q})$ and also call this vector a Roitberg generalized solution to the boundary-value problem (4.1).

We study properties of operator (4.85).

Theorem 4.14. *For arbitrary parameters $s\in\mathbb{R}$ and $\varphi\in\mathcal{M}$, the bounded operator* (4.85) *is Fredholm. Its kernel is equal to N and the range consists of all vectors $(f,g_1,\dots,g_q)\in\mathcal{H}_{s,\varphi,(0)}(\Omega,\Gamma)$ that satisfy condition* (4.4). *The index of operators* (4.85) *is equal to* $\dim N-\dim N^{+}$ *and does not depend on s and φ.*

Proof. In the Sobolev case, where $\varphi\equiv 1$, this theorem was established by Ya. A. Roitberg in [202], [203]; see also his monograph [209, Theorems 4.1.1 and 5.3.1]. Based on this result and applying interpolation with a function parameter, we will prove the theorem in the general case of arbitrary $\varphi\in\mathcal{M}$.

First, assume that $s\notin E_{2q}$. Let a positive number $\varepsilon=\delta$ be the same as in Theorem 4.11(ii). Mapping 4.2 is extended by continuity to the bounded and Fredholm operators

$$(L,B):\ H^{s\mp\varepsilon,(2q)}(\Omega)\to\mathcal{H}_{s\mp\varepsilon,(0)}(\Omega,\Gamma). \tag{4.86}$$

They have the common kernel N, the same index $\varkappa:=\dim N-\dim N^{+}$, and the range

$$(L,B)(H^{s\mp\varepsilon}(\Omega))=\{(f,g_1,\dots,g_q)\in\mathcal{H}_{s\mp\varepsilon,(0)}(\Omega,\Gamma):\ (4.4)\text{ is true}\}. \tag{4.87}$$

We apply the interpolation with the parameter ψ to the spaces in which operators (4.86) act. By virtue of Theorem 1.7, we obtain the bounded and Fredholm operator

$$\begin{aligned}(L,B):[H^{s-\varepsilon,(2q)}(\Omega),H^{s+\varepsilon,(2q)}(\Omega)]_{\psi}\\ \to[\mathcal{H}_{s-\varepsilon,(0)}(\Omega,\Gamma),\mathcal{H}_{s+\varepsilon,(0)}(\Omega,\Gamma)]_{\psi},\end{aligned} \tag{4.88}$$

which is extension of mapping (4.2) by continuity. Here, by virtue of the interpolation theorems 1.5, 2.2, and 3.10, we have

$$[\mathcal{H}_{s-\varepsilon,(0)}(\Omega,\Gamma),\mathcal{H}_{s+\varepsilon,(0)}(\Omega,\Gamma)]_{\psi}=\mathcal{H}_{s,\varphi,(0)}(\Omega,\Gamma) \tag{4.89}$$

with equivalence of norms. It follows from this and Theorem 4.11(ii) that mapping (4.2) extends by continuity to the bounded operator (4.85), which is equal to (4.88). According to Theorem 1.7, the Fredholm property of operators (4.86) implies the Fredholm property of operator (4.85), which inherits their kernel N and the index $\varkappa$. In addition, the range of operator (4.85) is equal to

$$\mathcal{H}_{s,\varphi,(0)}(\Omega,\Gamma)\cap(L,B)(H^{s-\varepsilon,(2q)}(\Omega)).$$

By virtue of (4.87), this implies that the range is the same as in the statement of the theorem.

Now assume that $s \in E_{2q}$. We arbitrarily choose a number $\varepsilon \in (0,1)$. According to the proved result and in view of $s \mp \varepsilon \notin E_{2q}$, mapping (4.2) is extended by continuity to the bounded and Fredholm operators

$$(L,B) : H^{s\mp\varepsilon,\varphi,(2q)}(\Omega) \to \mathcal{H}_{s\mp\varepsilon,\varphi,(0)}(\Omega,\Gamma).$$

They have the common kernel N and index $\varkappa$. Applying the interpolation with the power parameter $t^{1/2}$ and using Theorem 1.7, equality (4.67), and Lemma 4.3, we obtain the bounded and Fredholm operator (4.85). It extends mapping (4.2) by continuity and has the same kernel N and index $\varkappa$. In addition, the range of this operator is equal to

$$\mathcal{H}_{s,\varphi,(0)}(\Omega,\Gamma) \cap (L,B)(H^{s-\varepsilon,\varphi,(2q)}(\Omega)).$$

It follows directly from this and the results proved above that the range is the same as in the statement of the theorem.

Theorem 4.14 is proved.

Theorem 4.14 is a general theorem on the solvability of the elliptic boundary-value problem (4.1) because the space $H^{s,\varphi,(2q)}(\Omega)$ used as a domain of the operator (L,B) is independent of the elliptic expression L.

If $N = N^{+} = \{0\}$ (the defect of the boundary-value problem is absent), then operator (4.85) is an isomorphism of $H^{s,\varphi,(2q)}(\Omega)$ onto $\mathcal{H}_{s,\varphi,(0)}(\Omega,\Gamma)$. This follows from Theorem 4.14 and the Banach theorem on inverse operator. In the general case, it is convenient to define an isomorphism with the help of the following projectors.

Lemma 4.4. *For arbitrary parameters $s \in \mathbb{R}$ and $\varphi \in \mathcal{M}$, the following decompositions of the spaces $H^{s,\varphi,(2q)}(\Omega)$ and $\mathcal{H}_{s,\varphi,(0)}(\Omega,\Gamma)$ in direct sums of (closed) subspaces are true:*

$$H^{s,\varphi,(2q)}(\Omega) = N \dotplus \big\{u \in H^{s,\varphi,(2q)}(\Omega) : (u_0,w)_\Omega = 0 \ \textit{for all}\ w \in N\big\}, \quad (4.90)$$

$$\mathcal{H}_{s,\varphi,(0)}(\Omega,\Gamma) = \{(v,0,\dots,0) : v \in N^{+}\} \dotplus (L,B)(H^{s,\varphi,(2q)}(\Omega)). \quad (4.91)$$

Here, u_0 is the initial component of the vector $(u_0,u_1,\dots,u_{2q}) := T_{2q}u$. Let P_{2q} denote the oblique projector of the space $H^{s,\varphi,(2q)}(\Omega)$ onto the second summand in (4.90), and let Q_0^{+} denote the oblique projector of the space $\mathcal{H}_{s,\varphi,(0)}(\Omega,\Gamma)$ onto the second summand in (4.91), both parallel to the first summand. These projectors are independent of s and φ.

Proof. First, we prove equality (4.90). It follows from the definition of the space $H^{s,\varphi,(2q)}(\Omega)$ that the mapping $u \mapsto u_0$ is the bounded operator

$$T_0 : H^{s,\varphi,(2q)}(\Omega) \to H^{s,\varphi,(0)}(\Omega).$$

Therefore, the second summand in (4.90) is closed in $H^{s,\varphi,(2q)}(\Omega)$ and has the trivial intersection with N. Therefore, the following decomposition is true:

$$H^{s,\varphi,(0)}(\Omega) = N \dotplus \{u_0 \in H^{s,\varphi,(0)}(\Omega) : \, (u_0, w)_\Omega = 0 \text{ for all } w \in N\}. \quad (4.92)$$

Indeed, the space dual to the subspace N of $H^{-s,1/\varphi,(0)}(\Omega)$ is isomorphic to the factor space of the space $H^{s,\varphi,(0)}(\Omega)$ by the second summand in (4.92). Hence, the codimension of this summand is equal to $\dim N$, which yields (4.92).

Let Π denote the oblique projector of the space $H^{s,\varphi,(0)}(\Omega)$ onto the first summand in (4.92) parallel to the second summand. For a arbitrary element $u \in H^{s,\varphi,(2q)}(\Omega)$, we write $u = u' + u''$, where $u' := \Pi u_0 \in N$ and

$$u'' := u - \Pi u_0 \in H^{s,\varphi,(2q)}(\Omega)$$

satisfies the condition

$$(u_0'', w)_\Omega = (u_0 - \Pi u_0, w)_\Omega = 0 \quad \text{for any} \quad w \in N.$$

Equality (4.90) is proved.

Equality (4.91) follows from the fact that, by virtue of Theorem 4.14, the linear manifolds on the right-hand side of this equality are closed, have the trivial intersection, and the finite dimension of the first of them coincides with the codimension of the second. Finally, inclusions $N, N^+ \subset C^\infty(\overline{\Omega})$ implies that the projectors P_{2q} and Q_0^+ are independent of the parameters s and φ.

Lemma 4.4 is proved.

Theorem 4.15. *For arbitrary parameters $s \in \mathbb{R}$ and $\varphi \in \mathcal{M}$, the restriction of mapping* (4.85) *to the subspace $P_{2q}(H^{s,\varphi}(\Omega))$ is the isomorphism*

$$(L, B) : \, P_{2q}(H^{s,\varphi}(\Omega)) \leftrightarrow Q_0^+(\mathcal{H}_{s,\varphi,(0)}(\Omega, \Gamma)). \quad (4.93)$$

Proof. According to Theorem 4.14, N is the kernel and $Q_0^+(\mathcal{H}_{s,\varphi,(0)}(\Omega, \Gamma))$ is the range of operator (4.85). Hence, the bounded operator (4.93) is a bijection. Therefore, it is an isomorphism by the Banach theorem on inverse operator.

Theorem 4.15 is proved.

For each fixed $\varphi \in \mathcal{M}$, the collection of isomorphisms (4.93) is complete because s runs through the whole real axis. This property differs this collection from the one-sided collection of isomorphisms in Theorem 4.2.

The following *a priori* estimate for the solution to the elliptic boundary-value problem (4.1) ensues from Theorem 4.15.

Theorem 4.16. *Let $s \in \mathbb{R}$ and $\varphi \in \mathcal{M}$, and let the number $\varepsilon > 0$. Suppose that the element $u \in H^{s,\varphi,(2q)}(\Omega)$ is a Roitberg generalized solution to the boundary-value problem* (4.1), *where $(f, g_1, \ldots, g_q) \in \mathcal{H}_{s,\varphi,(0)}(\Omega, \Gamma)$.*

Then we have the estimate

$$\|u\|_{H^{s,\varphi,(2q)}(\Omega)} \leq c\left(\|(f, g_1, \ldots, g_q)\|_{\mathcal{H}_{s,\varphi,(0)}(\Omega,\Gamma)} + \|u\|_{H^{s-\varepsilon,\varphi,(2q)}(\Omega)} \right), \tag{4.94}$$

where the number $c = c(s, \varphi, \varepsilon) > 0$ is independent of u and $(f, g_1, \ldots, g_q)$.

Proof. We use decomposition (4.90) and write the element $u \in H^{s,\varphi,(2q)}(\Omega)$ in the form $u = u' + u''$, where

$$u' := (1 - P_{2q})u \in N \quad \text{and} \quad u'' := P_{2q}u \in P_{2q}(H^{s,\varphi}(\Omega)).$$

By virtue of Theorem 4.15,

$$\|u''\|_{H^{s,\varphi,(2q)}(\Omega)} \leq c_2\|(L, B)u''\|_{\mathcal{H}_{s,\varphi,(0)}(\Omega,\Gamma)}$$

$$= c_2\|(L, B)u\|_{\mathcal{H}_{s,\varphi,(0)}(\Omega,\Gamma)} = c_2\|(f, g_1, \ldots, g_q)\|_{\mathcal{H}_{s,\varphi,(0)}(\Omega,\Gamma)}.$$

Here, c_2 is the norm of the inverse to (4.93). In addition, since the space N is finite-dimensional and $1 - P_{2q}$ is a projector of the space $H^{s-\varepsilon,\varphi,(2q)}(\Omega)$ onto N (Lemma 4.4), we have

$$\|u'\|_{H^{s,\varphi,(2q)}(\Omega)} \leq c_0\|u'\|_{H^{s-\varepsilon,\varphi,(2q)}(\Omega)} \leq c_1\|u\|_{H^{s-\varepsilon,\varphi,(2q)}(\Omega)}.$$

Here, the positive numbers c_0 and c_1 are independent of u', u, and $(f, g_1, \ldots, g_q)$. Summing these inequalities, we obtain estimate (4.94).

Theorem 4.16 is proved.

If $N = \{0\}$, i.e., the boundary-value problem (4.1) has at most one solution, the norm $\|u\|_{H^{s-\varepsilon,\varphi,(2q)}(\Omega)}$ is absent on the right-hand side of (4.94).

In the Sobolev case of $\varphi \equiv 1$, Theorems 4.14–4.16 were proved by Ya. A. Roitberg in [202, 203]; see also monographs [209, Secs. 3.3 and 5.3] and [21, Chap. 3, Sec. 6, Subsec. 8] and review [11, Sec. 7.9 b]. For an arbitrary $\varphi \in \mathcal{M}$, these theorems refines Roitberg's results for the scale (4.68).

By virtue of Theorems 4.12(iii) and 3.9(i), the following equalities of spaces are true up to equivalence of norms in them:

$$H^{s,\varphi,(2q)}(\Omega) = H^{s,\varphi}(\Omega) \quad \text{and} \quad H^{s-2q,\varphi,(0)}(\Omega) = H^{s-2q,\varphi}(\Omega) \tag{4.95}$$
$$\text{for each} \quad s > 2q - 1/2.$$

Therefore, Theorems 4.14–4.16 include Theorems 4.1–4.3 on solvability of the elliptic boundary-value problem (4.1) in the scale of positive Hörmander spaces. Moreover, the following result is true.

Theorem 4.17. *Theorems 4.1–4.3 are true for arbitrary parameters $s > 2q - 1/2$ and $\varphi \in \mathcal{M}$.*

This theorem follows from Theorems 4.14–4.16, relations (4.95), and the equality of the projectors $P_{2q} = P$ on $H^{s,\varphi,(2q)}(\Omega)$.

4.2.4 Smoothness of generalized solutions up to the boundary

We study the smoothness of Roitberg generalized solutions to the boundary-value problem (4.1), considered in the modified refined Sobolev scale. It is useful to compare the results of this section with results obtained in Subsection 4.1.2.

For integer $r \geq 0$, let $H^{-\infty,(r)}(\Omega)$ denote the union of all the spaces $H^{s,\varphi,(r)}(\Omega)$ with $s \in \mathbb{R}$ and $\varphi \in \mathcal{M}$. The topology of inductive limit is introduced in the space $H^{-\infty,(r)}(\Omega)$.

The bounded operators (4.85) considered for all parameters $s \in \mathbb{R}$ and $\varphi \in \mathcal{M}$ generate the continuous linear operator

$$(L,B) : H^{-\infty,(2q)}(\Omega) \to H^{-\infty,(0)}(\Omega) \times (\mathcal{D}'(\Gamma))^q$$
$$=: \mathcal{H}_{-\infty,(0)}(\Omega,\Gamma). \tag{4.96}$$

By virtue of Theorem 4.14, the kernel of this operator is equal to N and the range consists of all vectors $(f, g_1, \dots, g_q) \in \mathcal{H}_{-\infty,(0)}(\Omega,\Gamma)$ that satisfy condition (4.4).

Theorem 4.18. *Suppose that $u \in H^{-\infty,(2q)}(\Omega)$ is a generalized solution to the boundary-value problem (4.1) in which*

$$f \in H^{s-2q,\varphi,(0)}(\Omega) \quad \text{and} \quad g_j \in H^{s-m_j-1/2,\varphi}(\Gamma), \ \ j = 1,\dots,q,$$

for some parameters $s \in \mathbb{R}$ and $\varphi \in \mathcal{M}$. Then $u \in H^{s,\varphi,(2q)}(\Omega)$.

Proof. By virtue of the above-mentioned properties of operator (4.96), the vector $F := (f, g_1, \dots, g_q) = (L,B)u$ satisfies (4.4). By condition, $F \in \mathcal{H}_{s,\varphi}(\Omega,\Gamma)$; hence, by Theorem 4.14, the inclusion $F \in (L,B)(H^{s,\varphi}(\Omega))$ is true. Therefore, parallel with condition $(L,B)u = F$, the equality $(L,B)v = F$ is true for a certain element $v \in H^{s,\varphi}(\Omega)$. This yields, $(L,B)(u-v) = 0$, which implies the inclusion

$$w := u - v \in N \subset H^{s,\varphi}(\Omega).$$

Thus, $u = v + w \in H^{s,\varphi}(\Omega)$.

Theorem 4.18 is proved.

Theorem 4.18 is a statement about the global smoothness of generalized solutions (i.e., on the whole closed domain $\overline{\Omega}$). In the Sobolev case of $\varphi \equiv 1$, this theorem was proved by Ya. A. Roitberg in [202, 203] (see also his monograph [209, Theorem 7.1.1]).

Now consider the case of local smoothness. Let U be an open set in $\mathbb{R}^n$ that has a nonempty intersection with domain Ω. As in Section 4.1.2, we set $\Omega_0 := U \cap \Omega$ and $\Gamma_0 := U \cap \Gamma$ (the case $\Gamma_0 = \varnothing$ is possible). We introduce the following local analog of the space $H^{\sigma,\varphi,(r)}(\Omega)$, where $\sigma \in \mathbb{R}$, $\varphi \in \mathcal{M}$, and the integer $r \geq 0$. We set

$$H^{\sigma,\varphi,(r)}_{\mathrm{loc}}(\Omega_0,\Gamma_0) := \{u \in H^{-\infty,(r)}(\Omega) : \chi u \in H^{\sigma,\varphi,(r)}(\Omega)$$

for all $\chi \in C^\infty(\overline{\Omega})$ such that $\operatorname{supp}\chi \subset \Omega_0 \cup \Gamma_0\}$. By virtue of Theorem 4.13(i), the multiplication by the function $\chi \in C^\infty(\overline{\Omega})$ is a bounded operator on the space $H^{\sigma,\varphi,(r)}(\Omega)$. Therefore, for each element $u \in H^{-\infty,(r)}(\Omega)$, the product $\chi u \in H^{-\infty,(r)}(\Omega)$ is well defined. In addition,

$$H^{\sigma,\varphi,(r)}(\Omega) \subset H^{\sigma,\varphi,(r)}_{\mathrm{loc}}(\Omega_0,\Gamma_0).$$

We also need the local space $H^{\sigma,\varphi}_{\mathrm{loc}}(\Gamma_0)$ introduced in Sec. 2.2.3.

Theorem 4.19. *Suppose that $u \in H^{-\infty,(2q)}(\Omega)$ is a generalized solution to the boundary-value problem (4.1) in which*

$$f \in H^{s-2q,\varphi,(0)}_{\mathrm{loc}}(\Omega_0,\Gamma_0) \quad \textit{and}$$
$$g_j \in H^{s-m_j-1/2,\varphi}_{\mathrm{loc}}(\Gamma_0), \quad j = 1,\dots,q, \tag{4.97}$$

for some parameters $s \in \mathbb{R}$ and $\varphi \in \mathcal{M}$. Then $u \in H^{s,\varphi,(2q)}_{\mathrm{loc}}(\Omega_0,\Gamma_0)$.

Proof. First we prove that, by virtue of the condition of this theorem, the following implication holds for each $r \geq 1$:

$$u \in H^{s-r,\varphi,(2q)}_{\mathrm{loc}}(\Omega_0,\Gamma_0) \;\Rightarrow\; u \in H^{s-r+1,\varphi,(2q)}_{\mathrm{loc}}(\Omega_0,\Gamma_0). \tag{4.98}$$

We choose $r \geq 1$ arbitrarily and assume that the premise of implication (4.98) is true. Let functions $\chi,\eta \in C^\infty(\overline{\Omega})$ be such that $\operatorname{supp}\chi$, $\operatorname{supp}\eta \subset \Omega_0 \cup \Gamma_0$ and $\eta = 1$ in a neighborhood of $\operatorname{supp}\chi$. Interchanging the operator of multiplication by χ with the differential operators L and B_j, $j = 1,\dots,q$, we obtain equality (4.16), where the differential expressions L' and B'_j are the same as those used in the proof of Theorem 4.5. Recall that $\operatorname{ord} L' \leq 2q-1$ and every $\operatorname{ord} B'_j \leq m_j - 1$. Thus,

$$(L,B)(\chi u) = \chi F + (L',B')(\eta u),$$

where $\chi F \in \mathcal{H}_{s,\varphi,(0)}(\Omega,\Gamma)$ by condition (4.97), and

$$(L',B')(\eta u) \in \mathcal{H}_{s-r+1,\varphi,(0)}(\Omega,\Gamma)$$

by Theorem 4.13 and the premise of implication (4.98). Since $(L,B)(\chi u) \in \mathcal{H}_{s-r+1,\varphi,(0)}(\Omega,\Gamma)$, we conclude that $\chi u \in H^{s-r+1,\varphi,(2q)}(\Omega)$ by virtue of Theorem 4.18. Thus, $u \in H^{s-r+1,\varphi,(2q)}_{\mathrm{loc}}(\Omega_0,\Gamma_0)$; implication (4.98) is proved.

Now we may complete the proof of the theorem in the following way. Since $u \in H^{-\infty,(2q)}(\Omega)$, there exists an integer $k \geq 1$ such that $u \in H^{s-k,\varphi,(2q)}(\Omega)$. Therefore, the premise of implication (4.98) is true for $r = k$. Using this implication in succession for values $r = k$, $r = k-1, \ldots,$ and $r = 1$, we conclude that

$$u \in H^{s-k,\varphi,(2q)}(\Omega) \Rightarrow u \in H^{s-k+1,\varphi,(2q)}_{\mathrm{loc}}(\Omega_0, \Gamma_0)$$

$$\Rightarrow \ldots \Rightarrow u \in H^{s-1,\varphi,(2q)}_{\mathrm{loc}}(\Omega_0, \Gamma_0)$$

$$\Rightarrow u \in H^{s,\varphi,(2q)}_{\mathrm{loc}}(\Omega_0, \Gamma_0).$$

Thus, $u \in H^{s,\varphi,(2q)}_{\mathrm{loc}}(\Omega_0, \Gamma_0)$.

Theorem 4.19 is proved.

Theorems 4.18 and 4.19 refines the Roitberg theorems [202, 203] on increase in smoothness of generalized solutions to the elliptic boundary-value problem considered on the modified Sobolev scale (see also Ya. A. Roitberg's monograph [209, Sections 7.1 and 7.2]). It follows from equality (4.95) that Theorems 4.18 and 4.19 contain Theorems 4.4 and 4.5. In Theorem 4.19, we note the $\Gamma_0 = \varnothing$ case, which leads to the statement about increase in local smoothness of the generalized solutions in neighborhoods of inner points of the domain Ω.

As an application of Theorems 4.18 and 4.19, we establish a sufficient condition for a Roitberg generalized solution $u \in H^{-\infty,(2q)}(\Omega)$ of the elliptic boundary-value problem (4.1) to be classical, i.e., to satisfy the condition

$$u \in H^{\sigma+2q}(\Omega) \cap C^{2q}(\Omega) \cap C^{m}(\overline{\Omega}), \tag{4.99}$$

where $\sigma > -1/2$ and $m := \max\{m_1, \ldots, m_q\}$. Let us explain why this condition appears. By virtue of Theorems 4.12(iii) and 3.5, it follows from the inclusion

$$u \in H^{\sigma+2q,(2q)}(\Omega) = H^{\sigma+2q}(\Omega)$$

that the element u is a solution to problem (4.1) in the sense of theory of distributions defined in the domain Ω. Not it is reasonable to consider the inclusion $u \in C^{2q}(\Omega) \cap C^{m}(\overline{\Omega})$. It implies that the functions Lu and $B_j u$ are calculated in (4.1) with the use of classical derivatives, i.e., the solution u is classical.

Theorem 4.20. *Suppose that $u \in H^{-\infty,(2q)}(\Omega)$ is a Roitberg generalized solution to the boundary-value problem* (4.1) *in which*

$$f \in H^{n/2,\varphi,(0)}_{\mathrm{loc}}(\Omega, \varnothing) \cap H^{m-2q+n/2,\varphi,(0)}(\Omega) \cap H^{\sigma,(0)}(\Omega), \tag{4.100}$$

$$g_j \in H^{m-m_j+(n-1)/2,\varphi}(\Gamma) \cap H^{\sigma+2q-m_j-1/2}(\Gamma), \quad j = 1, \ldots, q, \tag{4.101}$$

for some number $\sigma > -1/2$ *and a certain function parameter* $\varphi \in \mathcal{M}$ *that satisfies condition* (1.37). *Then the solution* u *is classical, i.e.,* u *satisfies inclusion* (4.99).

Proof. Applying Theorems 4.18 and 4.19 and using conditions (4.100) and (4.101), we obtain the inclusion

$$u \in H^{2q+n/2,\varphi,(2q)}_{\mathrm{loc}}(\Omega, \varnothing) \cap H^{m+n/2,\varphi,(2q)}(\Omega) \cap H^{\sigma+2q,(2q)}(\Omega).$$

Hence, by Theorems 4.12(iii) and 3.4, we have

$$u \in H^{m+n/2,\varphi,(2q)}(\Omega) \cap H^{\sigma+2q,(2q)}(\Omega) = H^{m+n/2,\varphi}(\Omega) \cap H^{\sigma+2q}(\Omega)$$
$$\subseteq C^{m}(\overline{\Omega}) \cap H^{\sigma+2q}(\Omega).$$

(The last equality becomes clear if we separately consider the cases $m + n/2 \geq \sigma + 2q$ and $m + n/2 < \sigma + 2q$.) In addition,

$$\chi u \in H^{2q+n/2,\varphi,(2q)}(\Omega) = H^{2q+n/2,\varphi}(\Omega) \subset C^{2q}(\overline{\Omega})$$

for any function $\chi \in C_0^{\infty}(\Omega)$, which yields the inclusion $u \in C^{2q}(\Omega)$. Thus, condition (4.99) is satisfied, i.e., u is a classical solution.

Theorem 4.20 is proved.

4.2.5 Interpolation in the modified refined scale

We study two problems concerning interpolation in the modified refined scale. First, we prove that the right-hand side of equality (4.67) is independent of the choice of the parameter ε. Second, we establish that the interpolation formula (4.73) is true under essentially weaker conditions on the parameters involved.

Theorem 4.21. *Let* $r \in \mathbb{N}$, $s \in E_r$, *and* $\varphi \in \mathcal{M}$. *The space*

$$H^{s,\varphi,(r)}(\Omega, \varepsilon) := \left[H^{s-\varepsilon,\varphi,(r)}(\Omega), H^{s+\varepsilon,\varphi,(r)}(\Omega) \right]_{t^{1/2}},$$

is independent of the parameter $\varepsilon \in (0, 1)$ *up to equivalence of norms.*

Proof. First, assume that $r = 2q$ is an even number. Consider the regular elliptic boundary-value problem in the domain Ω:

$$Lu \equiv (1 - \Delta)^q\, u = f \quad \text{in} \quad \Omega, \tag{4.102}$$

$$B_j u \equiv D_\nu^{j-1}\, u = g_j \quad \text{on} \quad \Gamma, \quad j = 1, \ldots, q. \tag{4.103}$$

As usual, Δ is the Laplace operator. This problem is formally self-adjoint and $N = N^+ = \{0\}$ for it. According to Theorem 4.14, we have the isomorphism

$$(L, B) : H^{s,\varphi,(2q)}(\Omega, \varepsilon) \leftrightarrow H^{s-2q,\varphi,(0)}(\Omega) \oplus \bigoplus_{j=1}^{q} H^{s-j+1/2,\varphi}(\Gamma)$$

for $0 < \varepsilon < 1$. This immediately yields the conclusion of theorem for even $r = 2q$.

Further, assume that the number r is odd. By virtue of Theorem 4.11 (i), for any number $\sigma \in (-\infty, r + 1/2) \setminus E_r$, we have the isometric isomorphisms

$$T_r : H^{\sigma,\varphi,(r)}(\Omega) \leftrightarrow K_{\sigma,\varphi,(r)}(\Omega, \Gamma),$$

$$T_{r+1} : H^{\sigma,\varphi,(r+1)}(\Omega) \leftrightarrow K_{\sigma,\varphi,(r+1)}(\Omega, \Gamma) = K_{\sigma,\varphi,(r)}(\Omega, \Gamma) \oplus H^{\sigma-r-1/2,\varphi}(\Gamma).$$

Therefore, the composition of the mappings

$$u \mapsto T_{r+1}\, u =: (u_0, u_1, \ldots, u_r, u_{r+1}) \mapsto (T_r^{-1}(u_0, u_1, \ldots, u_r), u_{r+1}),$$

where $u \in H^{\sigma,\varphi,(r+1)}(\Omega)$, defines the isometric isomorphism

$$T : H^{\sigma,\varphi,(r+1)}(\Omega) \leftrightarrow H^{\sigma,\varphi,(r)}(\Omega) \oplus H^{\sigma-r-1/2,\varphi}(\Gamma). \tag{4.104}$$

Here, we take $\sigma = s \mp \varepsilon$, where $0 < \varepsilon < 1$, and apply the interpolation with the power parameter $t^{1/2}$. In view of Lemma 4.3, we obtain the isomorphism

$$T : H^{s,\varphi,(r+1)}(\Omega, \varepsilon) \leftrightarrow H^{s,\varphi,(r)}(\Omega, \varepsilon) \oplus H^{s-r-1/2,\varphi}(\Gamma) =: X(\varepsilon).$$

Therefore,

$$\|u\|_{H^{s,\varphi,(r)}(\Omega,\varepsilon)} = \|(u, 0)\|_{X(\varepsilon)} \asymp \|T^{-1}(u, 0)\|_{H^{s,\varphi,(r+1)}(\Omega,\varepsilon)}.$$

It follows from this and the theorem proved for even $r + 1$ that the norms in the spaces $H^{s,\varphi,(r)}(\Omega, \varepsilon)$, where $0 < \varepsilon < 1$, are equivalent. Therefore, these spaces are equal because the set $C^\infty(\overline{\Omega})$ is dense in each of them according to Theorem 4.12(ii).

Theorem 4.21 is proved.

Theorem 4.22. *Let $r \in \mathbb{N}$, $s \in \mathbb{R}$, $\varphi \in \mathcal{M}$ and positive numbers ε and δ be arbitrarily chosen. If the number r is odd, then we additionally suppose that at least one of the inequalities $s - \varepsilon > r - 1/2$ and $s + \delta < r + 1/2$ is satisfied. Then the interpolation formula* (4.73) *is true, namely,*

$$[H^{s-\varepsilon,(r)}(\Omega), H^{s+\delta,(r)}(\Omega)]_\psi = H^{s,\varphi,(r)}(\Omega) \tag{4.105}$$

with equivalence of norms. Here, ψ is the interpolation parameter in Theorem 1.14.

Proof. First, assume that $r = 2q$ is an even number. By virtue of Theorem 4.14, for the elliptic boundary-value problem (4.102), (4.103), we have the isomorphisms

$$(L,B):\, H^{s,\varphi,(2q)}(\Omega) \leftrightarrow H^{s-2q,\varphi,(0)}(\Omega) \oplus \bigoplus_{j=1}^{q} H^{s-j+1/2,\varphi}(\Gamma), \tag{4.106}$$

$$(L,B):\, H^{\sigma,(2q)}(\Omega) \leftrightarrow H^{\sigma-2q,(0)}(\Omega) \oplus \bigoplus_{j=1}^{q} H^{\sigma-j+1/2}(\Gamma), \quad \sigma \in \mathbb{R}. \tag{4.107}$$

We apply the interpolation with the parameter ψ to the spaces in which isomorphisms (4.107), with $\sigma \in \{s-\varepsilon,\, s+\delta\}$, act. Taking into account that ψ is an interpolation parameter and using Theorems 3.10 and 2.2, we obtain one more isomorphism

$$\begin{aligned}(L,B):\, &[H^{s-\varepsilon,(2q)}(\Omega), H^{s+\delta,(2q)}(\Omega)]_{\psi}\\ &\leftrightarrow H^{s-2q,\varphi,(0)}(\Omega) \oplus \bigoplus_{j=1}^{q} H^{s-j+1/2,\varphi}(\Gamma).\end{aligned} \tag{4.108}$$

Now, using isomorphisms (4.106) and (4.108), we obtain the equality of spaces (4.105) up to equivalence of norms.

Further, assume that the number r is odd. We separately consider the case of $s-\varepsilon > r-1/2$ and the case of $s+\delta < r+1/2$.

If $s-\varepsilon > r-1/2$, then, by virtue of Theorems 4.12(iii) and 3.2, the following equalities of spaces are true up to equivalence of norms:

$$\begin{aligned}[H^{s-\varepsilon,(r)}(\Omega), H^{s+\delta,(r)}(\Omega)]_{\psi} &= [H^{s-\varepsilon}(\Omega), H^{s+\delta}(\Omega)]_{\psi}\\ &= H^{s,\varphi}(\Omega) = H^{s,\varphi,(r)}(\Omega).\end{aligned}$$

Relation (4.105) is proved in the considered case.

Now consider the case where $s+\delta < r+1/2$. Note that isomorphism (4.104) is true for any $\sigma < r+1/2$. Indeed, if $\sigma \notin E_r$, then this isomorphism was established in the proof of Theorem 4.21. Hence, we deduce it for every $\sigma \in E_r$ if we apply interpolation and use equalities (4.67) and (4.80). In particular, we have the isomorphisms

$$T:\, H^{s,\varphi,(r+1)}(\Omega) \leftrightarrow H^{s,\varphi,(r)}(\Omega) \oplus H^{s-r-1/2,\varphi}(\Gamma), \tag{4.109}$$

$$T:\, H^{\sigma,(r+1)}(\Omega) \leftrightarrow H^{\sigma,(r)}(\Omega) \oplus H^{\sigma-r-1/2}(\Gamma), \quad \sigma \in \{s-\varepsilon,\, s+\delta\}. \tag{4.110}$$

We apply the interpolation with the parameter ψ to (4.110). By virtue of the result proved above (since the number $r+1$ is even) and Theorem 2.2, we obtain one more isomorphism

$$T: H^{s,\varphi,(r+1)}(\Omega) \leftrightarrow [H^{s-\varepsilon,(r)}(\Omega), H^{s+\delta,(r)}(\Omega)]_{\psi} \oplus H^{s-r-1/2,\varphi}(\Gamma). \qquad (4.111)$$

Now, using homeomorphisms (4.109) and (4.111), we can write the following:

$$u \in [H^{s-\varepsilon,(r)}(\Omega), H^{s+\delta,(r)}(\Omega)]_{\psi} \Leftrightarrow T^{-1}(u,0) \in H^{s,\varphi,(r+1)}(\Omega)$$

$$\Leftrightarrow u \in H^{s,\varphi,(r)}(\Omega).$$

Therefore, the equality of spaces (4.105) is true in the case of $s+\delta<r+1/2$. The norms in these spaces are equivalent:

$$\|u\|_{[H^{s-\varepsilon,(r)}(\Omega),H^{s+\delta,(r)}(\Omega)]_{\psi}} \asymp \|T^{-1}(u,0)\|_{H^{s,\varphi,(r+1)}(\Omega)} \asymp \|u\|_{H^{s,\varphi,(r)}(\Omega)},$$

where $u \in H^{s,\varphi,(r)}(\Omega)$.

Theorem 4.22 is proved.

4.3 Some properties of the modified refined scale

In this section, we formulate and prove two important properties of the modified refined Sobolev scale, whose order of modification is an arbitrary even number $2q$, with $q \in \mathbb{N}$. They will be used in Section 4.5.

4.3.1 Statement of results

The first property gives us an equivalent alternative definition of the space $H^{s,\varphi,(2q)}(\Omega)$. It turns out that the norm in this space is equivalent to a certain norm involving the properly elliptic expression L and not using any boundary values of functions. Unlike (4.66), it is not necessary to eliminate the case of $s \in E_{2q}$.

Theorem 4.23. *Let $s \in \mathbb{R}$ and $\varphi \in \mathcal{M}$. The following assertions are true:*

(i) *On the set of all functions $u \in C^{\infty}(\overline{\Omega})$, the norm in the space $H^{s,\varphi,(2q)}(\Omega)$ is equivalent to the graph norm*

$$\left(\|u\|^2_{H^{s,\varphi,(0)}(\Omega)} + \|Lu\|^2_{H^{s-2q,\varphi,(0)}(\Omega)} \right)^{1/2}. \qquad (4.112)$$

Therefore, the space $H^{s,\varphi,(2q)}(\Omega)$ coincides with the completion of $C^{\infty}(\overline{\Omega})$ with respect to the norm (4.112).

(ii) *The mapping*

$$I_L : u \mapsto (u, Lu), \quad u \in C^\infty(\overline{\Omega}), \tag{4.113}$$

extends uniquely (by continuity) to the isomorphism

$$I_L : H^{s,\varphi,(2q)}(\Omega) \leftrightarrow K_{s,\varphi,L}(\Omega). \tag{4.114}$$

Here,

$$K_{s,\varphi,L}(\Omega) := \big\{(u_0, f) : u_0 \in H^{s,\varphi,(0)}(\Omega),\ f \in H^{s-2q,\varphi,(0)}(\Omega),$$

$$(u_0, L^+ w)_\Omega = (f, w)_\Omega \textit{ for all } w \in C^\infty_{\nu,2q}(\overline{\Omega})\big\} \tag{4.115}$$

is a (closed) subspace of

$$H^{s,\varphi,(0)}(\Omega) \oplus H^{s-2q,\varphi,(0)}(\Omega).$$

Recall (see Section 3.5.1) that

$$C^\infty_{\nu,2q}(\overline{\Omega}) := \big\{u \in C^\infty(\overline{\Omega}) : D_\nu^{j-1} u = 0 \text{ on } \Gamma,\ j = 1, \dots, 2q\big\}.$$

If $s > -1/2$, then, for the distributions u_0 and f in Theorem 4.23(ii), condition (4.115) is equivalent to the condition

$$(u_0, L^+ w)_\Omega = (f, w)_\Omega \quad \text{for all} \quad w \in C_0^\infty(\Omega). \tag{4.116}$$

The latter means that $Lu_0 = f$ in the domain Ω. (The equivalence will be proved in Lemma 4.5.) If $s < -1/2$, then this is not true and we need the second property.

Theorem 4.24. *Let $s < -1/2$, $s + 1/2 \notin \mathbb{Z}$, and $\varphi \in \mathcal{M}$. Then, for arbitrary distributions*

$$u_0 \in H^{s,\varphi,(0)}(\Omega), \quad f \in H^{s-2q,\varphi,(0)}(\Omega) \tag{4.117}$$

that satisfy condition (4.116)*, there exists a unique pair $(u_0^*, f) \in K_{s,\varphi,L}(\Omega)$ such that*

$$(u_0, w)_\Omega = (u_0^*, w)_\Omega \quad \textit{for all} \quad w \in C_0^\infty(\Omega). \tag{4.118}$$

Moreover,

$$\|u_0^*\|_{H^{s,\varphi,(0)}(\Omega)} \le c\,\big(\,\|u_0\|^2_{H^{s,\varphi,(0)}(\Omega)} + \|f\|^2_{H^{s-2q,\varphi,(0)}(\Omega)}\,\big)^{1/2}, \tag{4.119}$$

where the number $c = c(s, \varphi) > 0$ is independent of u_0 and f.

In the Sobolev case of $\varphi \equiv 1$, Theorems 4.23 and 4.24 were proved by Ya. A. Roitberg in [207] (see also his monograph [209, Theorems 6.1.1 and 6.2.1]). Note that Ya. A. Roitberg made use of somewhat other statements, which are equivalent to ours. We describe their differences.

In the statement of Theorem 4.23, Ya. A. Roitberg [209, Theorem 6.1.1], instead of (4.115), uses the condition (for $\varphi \equiv 1$)

$$(u_0, L^+ w)_\Omega = (f, w)_\Omega \quad \text{for all} \quad w \in H_0^{2q,1/\varphi}(\Omega) \cap H^{2q-s,1/\varphi,(0)}(\Omega). \tag{4.120}$$

This is a tantamount change, as we will show in Lemma 4.5.

Further, using Theorem 4.23 and setting $u^* := I_L^{-1}(u_0^*, f)$, we can reformulate Theorem 4.24 in the following equivalent form: For arbitrary distributions (4.117) that satisfy condition (4.118), there exists a unique element $u^* \in H^{s,\varphi,(2q)}(\Omega)$ such that

$$u_0^* = u_0 \ \text{ in } \Omega \quad \text{and} \quad Lu^* = f \ \text{ in } H^{s-2q,\varphi,(0)}(\Omega).$$

Moreover,

$$\|u^*\|_{H^{s,\varphi,(2q)}(\Omega)} \leq c\left(\|u_0\|^2_{H^{s,\varphi,(0)}(\Omega)} + \|f\|^2_{H^{s-2q,\varphi,(0)}(\Omega)}\right)^{1/2}.$$

Here, u_0^* is the initial component of the vector $T_{2q}u^*$. This equivalent statement of Theorem 4.24 is used by Ya. A. Roitberg in [209, Theorem 6.2.1] for $\varphi \equiv 1$.

4.3.2 Proof of results

We derive Theorems 4.23 and 4.24 from the Sobolev case of $\varphi \equiv 1$ with the use of interpolation. First, we prove Lemma 4.5 mentioned above.

Lemma 4.5. *Let $s \in \mathbb{R}$ and $\varphi \in \mathcal{M}$. Then, for arbitrarily given distributions* (4.117), *conditions* (4.115) *and* (4.120) *are equivalent. If $s > -1/2$, then conditions* (4.115) *and* (4.116) *are equivalent. They are also equivalent for $s = -1/2$ in the case of $\varphi \equiv 1$.*

Proof. First, we show that (4.115) $\Leftrightarrow$ (4.120). By virtue of Theorem 3.20(i), we have (4.120) $\Rightarrow$ (4.115). We prove the converse. Assume that condition (4.115) is satisfied. We separately consider the case of $s \geq 0$ and the case of $s < 0$.

The case of $s \geq 0$. By virtue of Theorem 3.20(i), we have

$$H_0^{2q,1/\varphi}(\Omega) \cap H^{2q-s,1/\varphi,(0)}(\Omega) = H_0^{2q,1/\varphi}(\Omega) = H_{\nu,2q}^{2q,1/\varphi}(\Omega). \tag{4.121}$$

We approximate an arbitrary distribution w belonging to space (4.121) by a certain sequence of functions $w_j \in C^\infty_{\nu,2q}(\overline{\Omega})$ with respect to the norm in the space $H^{2q,1/\varphi}(\Omega)$. According to condition (4.115), we have

$$(u_0, L^+ w_j)_\Omega = (f, w_j)_\Omega \quad \text{for all} \quad j \in \mathbb{N}. \tag{4.122}$$

Passing to the limit as $j \to \infty$, we obtain condition (4.120). This follows from inclusions (4.117) and the following limits:

$$\lim_{j\to\infty} L^+ w_j = L^+ w \quad \text{in} \quad H^{0,1/\varphi}(\Omega) \hookrightarrow H^{-s,1/\varphi,(0)}(\Omega),$$

$$\lim_{j\to\infty} w_j = w \quad \text{in} \quad H^{2q,1/\varphi}(\Omega) \hookrightarrow H^{2q-s,1/\varphi,(0)}(\Omega).$$

The case of $s < 0$. By virtue of Theorem 3.20(i) and relation (3.119), we have

$$H_0^{2q,1/\varphi}(\Omega) \cap H^{2q-s,1/\varphi,(0)}(\Omega) = H_{\nu,2q}^{2q-s,1/\varphi}(\Omega). \tag{4.123}$$

We approximate an arbitrary distribution w belonging to space (4.123) by a certain sequence of functions $w_j \in C^\infty_{\nu,2q}(\,\overline{\Omega}\,)$ with respect to the norm in the space $H^{2q-s,1/\varphi}(\Omega)$. According to condition (4.115), equalities (4.122) hold. Passing in them to the limit as $j \to \infty$, we again obtain (4.120). This follows from (4.117) and the limits

$$\lim_{j\to\infty} L^+ w_j = L^+ w \quad \text{in} \quad H^{-s,1/\varphi}(\Omega) = H^{-s,1/\varphi,(0)}(\Omega),$$

$$\lim_{j\to\infty} w_j = w \quad \text{in} \quad H^{2q-s,1/\varphi}(\Omega) = H^{2q-s,1/\varphi,(0)}(\Omega).$$

We have proved that (4.115) $\Leftrightarrow$ (4.120) for any $s \in \mathbb{R}$.

Now, assuming that $s > -1/2$, we prove the equivalence of conditions (4.115) and (4.116). Implication (4.115) $\Rightarrow$ (4.116) is obvious. Let us prove the converse. Assume that condition (4.116) is satisfied. By virtue of the result proved above, it suffices to show that (4.116) $\Rightarrow$ (4.120). Using Theorem 3.20(i) and the inequality $s > -1/2$, we obtain the equality

$$H_0^{2q,1/\varphi}(\Omega) \cap H^{2q-s,1/\varphi,(0)}(\Omega) = H_0^{\lambda,1/\varphi}(\Omega), \tag{4.124}$$

where $\lambda := \max\{2q,\, 2q - s\}$. Therefore, each distribution w belonging to space (4.124) can be approximated by a certain sequence of functions $w_j \in C_0^\infty(\Omega)$ with respect to the norm in the space $H^{\lambda,1/\varphi}(\Omega)$. According to condition (4.116), equality (4.122) is true. Hence, passing to the limit as $j \to \infty$, we obtain condition (4.120). This follows from inclusions (4.117) and the limits

$$\lim_{j\to\infty} L^+ w_j = L^+ w \quad \text{in } H^{\lambda-2q,1/\varphi}(\Omega) \hookrightarrow H^{-s,1/\varphi,(0)}(\Omega),$$

$$\lim_{j\to\infty} w_j = w \quad \text{in } H^{\lambda,1/\varphi}(\Omega) \hookrightarrow H^{2q-s,1/\varphi,(0)}(\Omega).$$

Thus, we have proved that (4.115) $\Leftrightarrow$ (4.116) for $s > -1/2$.

Finally, if $s = -1/2$ and $\varphi \equiv 1$, then equality (4.124) remains true by virtue of [258, Theorem 4.7.1 (a)]. Repeating the reasonings given in the previous paragraph, we obtain equivalence (4.115) $\Leftrightarrow$ (4.116) in this case as well.

Lemma 4.5 is proved.

To prove Theorem 4.23, we need the following lemma on projectors.

Lemma 4.6. *For each number $\sigma \in \mathbb{R}$, there exists a projector Π_σ of the space $H^{\sigma,(0)}(\Omega) \times H^{\sigma-2q,(0)}(\Omega)$ onto the subspace $K_{\sigma,L}(\Omega)$ such that Π_σ is an extension of the map Π_λ for $\sigma < \lambda$.*

Proof. Let $\sigma \in \mathbb{R}$. First, we establish a useful equality. Passing to the limit and using Theorem 4.13, we conclude that the Green formula (3.6) remains true for arbitrary distributions $u \in H^{\sigma,(2q)}(\Omega)$ and $v \in H^{2q-\sigma,(2q)}(\Omega)$, namely

$$(Lu, v_0)_\Omega + \sum_{j=1}^{q} (B_j u,\, C_j^+ v)_\Gamma = (u_0, L^+ v)_\Omega + \sum_{j=1}^{q} (C_j u,\, B_j^+ v)_\Gamma. \tag{4.125}$$

Here, the components of the sesquilinear forms $(\cdot,\cdot)_\Omega$ and $(\cdot,\cdot)_\Gamma$ belong to the following spaces, which are mutually dual with respect to these forms:

$$Lu \in H^{\sigma-2q,(0)}(\Omega), \quad (v_0, v_1, \ldots, v_{2q}) := T_{2q} v, \quad v_0 \in H^{2q-\sigma,(0)}(\Omega),$$

$$(u_0, u_1, \ldots, u_{2q}) := T_{2q} u, \quad u_0 \in H^{\sigma,(0)}(\Omega), \quad L^+ v \in H^{-\sigma,(0)}(\Omega),$$

$$B_j u \in H^{\sigma-m_j-1/2}(\Gamma), \quad C_j^+ v \in H^{-\sigma+m_j+1/2}(\Gamma),$$

$$C_j u \in H^{\sigma-2q+m_j^++1/2}(\Gamma), \quad B_j^+ v \in H^{2q-\sigma-m_j^+-1/2}(\Gamma).$$

Recall that $\operatorname{ord} L = \operatorname{ord} L^+ = 2q$ and, by virtue of (3.7),

$$\operatorname{ord} B_j = m_j \le 2q-1, \quad \operatorname{ord} C_j^+ = 2q-1-m_j,$$

$$\operatorname{ord} B_j^+ =: m_j^+ \le 2q-1, \quad \operatorname{ord} C_j = 2q-1-m_j^+.$$

Further, since $\{B_1^+, \ldots, B_q^+, C_1^+, \ldots, C_q^+\}$ is the Dirichlet system of order $2q$ (see [121, Part 2, Theorem 2.1]), the following statement [209, Lemma 6.1.2] is true for it. The bounded operator

$$\bigl(B_1^+, \ldots, B_q^+, C_1^+, \ldots, C_q^+\bigr) : H^{2q-\sigma,(2q)}(\Omega)$$

$$\to \bigoplus_{j=1}^{q} H^{2q-\sigma-m_j^+-1/2}(\Gamma) \oplus \bigoplus_{j=1}^{q} H^{-\sigma+m_j+1/2}(\Gamma)$$

has the linear bounded right inverse operator

$$\Phi_\sigma : \bigoplus_{j=1}^{q} H^{2q-\sigma-m_j^+-1/2}(\Gamma) \oplus \bigoplus_{j=1}^{q} H^{-\sigma+m_j+1/2}(\Gamma) \to H^{2q-\sigma,(2q)}(\Omega) \quad (4.126)$$

such that Φ_σ is a restriction of the operator Φ_λ for $\sigma < \lambda$.

For an arbitrary vector

$$h := (0, \dots, 0, h_1, \dots, h_q) \in \{0\}^q \oplus \bigoplus_{j=1}^{q} H^{-\sigma+m_j+1/2}(\Gamma), \quad (4.127)$$

we set $v := \Phi_\sigma h \in H^{2q-\sigma,(2q)}(\Omega)$ in the Green formula (4.125). Since

$$B_j^+ v = 0, \quad C_j^+ v = h_j \quad \text{for each} \quad j \in \{1, \dots, q\},$$

we obtain the useful equality

$$(u_0, L^+\Phi_\sigma h)_\Omega - (Lu, (\Phi_\sigma h)_0)_\Omega = \sum_{j=1}^{q} (B_j u, h_j)_\Gamma. \quad (4.128)$$

Here, the element $u \in H^{\sigma,(2q)}(\Omega)$ and the vector h of the form (4.127) are arbitrary. As before,

$$u_0 \in H^{\sigma,(0)}(\Omega) \quad \text{and} \quad (\Phi_\sigma h)_0 = v_0 \in H^{2q-\sigma,(0)}(\Omega)$$

are the initial components of the vectors $T_{2q}u$ and $T_{2q}\Phi_\sigma h = T_{2q}v$, respectively.

Now we construct the projector Π_σ. To this end, we use the following five mappings. Let an arbitrary vector

$$(u_0, f) \in H^{\sigma,(0)}(\Omega) \oplus H^{\sigma-2q,(0)}(\Omega) \quad (4.129)$$

be chosen. Let $\|\cdot\|_\sigma$ denote the norm in space (4.129).

Mappings 1 and 2. Consider the following decomposition of the space

$$H^{\sigma,(0)}(\Omega)$$

in the direct sum of (closed) subspaces (see the proof of Lemma 4.4):

$$H^{\sigma,(0)}(\Omega) = N \dotplus \left\{ u_0' \in H^{\sigma,(0)}(\Omega) : (u_0', w)_\Omega = 0 \text{ for all } w \in N \right\}. \quad (4.130)$$

It exists because N is a finite-dimensional subspace of $H^{\sigma,(0)}(\Omega)$. Let Ψ_σ denote the projector of the space $H^{\sigma(0)}(\Omega)$ onto the subspace N parallel to the second term of sum (4.130). Since $N \subset C^\infty(\overline{\Omega})$, Ψ_σ is the extension of the projector Ψ_λ for $\sigma < \lambda$. We define linear mappings 1 and 2 by the relations

$$(u_0, f) \mapsto (\Psi_\sigma u_0, 0) \in K_{\sigma,L}(\Omega), \tag{4.131}$$

$$(u_0, f) \mapsto (u_0 - \Psi_\sigma u_0, f) =: (u_0', f). \tag{4.132}$$

The inclusion in (4.131) follows from the inclusions $\Psi_\sigma u_0 \in N \subset C^\infty(\overline{\Omega})$ and the definition of the space $K_{\sigma,L}(\Omega)$. Namely,

$$\bigl((\Psi_\sigma u_0, L^+ w)_\Omega = (L\Psi_\sigma u_0, w)_\Omega = (0, w)_\Omega \text{ for all } w \in C^\infty_{\nu,2q}(\overline{\Omega})\bigr)$$

$$\Rightarrow (\Psi_\sigma u_0, 0) \in K_{\sigma,L}(\Omega).$$

Here, the first equality is obtained by the Green formula (4.61), integrals over Γ being absent because of $w \in C^\infty_{\nu,2q}(\overline{\Omega})$.

Mapping 3. Consider relation (4.128). For the pair (u_0', f), we construct the functional

$$l_\sigma(h) := (u_0', L^+\Phi_\sigma h)_\Omega - (f, (\Phi_\sigma h)_0)_\Omega, \quad h \text{ being vector (4.127).} \tag{4.133}$$

This functional is bounded by virtue of the following chain of inequalities:

$$\begin{aligned}
|l_\sigma(h)| &\le |(u_0', L^+\Phi_\sigma h)_\Omega| + |(f, (\Phi_\sigma h)_0)_\Omega| \\
&\le \|u_0'\|_{H^{\sigma,(0)}(\Omega)} \, \|L^+\Phi_\sigma h\|_{H^{-\sigma,(0)}(\Omega)} \\
&\quad + \|f\|_{H^{\sigma-2q,(0)}(\Omega)} \, \|(\Phi_\sigma h)_0\|_{H^{-\sigma+2q,(0)}(\Omega)} \\
&\le \bigl(c_1 \|u_0'\|_{H^{\sigma,(0)}(\Omega)} + c_2 \|f\|_{H^{\sigma-2q,(0)}(\Omega)}\bigr) \, \|\Phi_\sigma h\|_{H^{-\sigma+2q,(2q)}(\Omega)} \\
&\le (c_1 + c_2)\, c_3 \, \|(u_0', f)\|_\sigma \left(\sum_{j=1}^{q} \|h_j\|^2_{H^{-\sigma+m_j+1/2}(\Gamma)} \right)^{1/2}.
\end{aligned}$$

Here, c_1, c_2, and c_3 are, respectively, the norms of the operators

$$L^+ : H^{-\sigma+2q,(2q)}(\Omega) \to H^{-\sigma,(0)}(\Omega),$$

$$T_{2q} : H^{-\sigma+2q,(2q)}(\Omega) \to H^{-\sigma+2q,(0)}(\Omega) \oplus \bigoplus_{j=1}^{2q} H^{-\sigma+2q-j+1/2}(\Gamma),$$

and operator (4.126). Thus, l_σ is an antilinear bounded functional on the space $\bigoplus_{j=1}^{q} H^{-\sigma+m_j+1/2}(\Gamma)$ and, furthermore, its norm satisfies the inequality

$\|l_\sigma\| \le c_4 \, \|(u_0', f)\|_\sigma$, where $c_4 := (c_1 + c_2)\, c_3$. By virtue of Theorem 2.3(v),

$$\left(\exists !\ g = (g_1, \ldots, g_q) \in \bigoplus_{j=1}^{q} H^{\sigma - m_j - 1/2}(\Gamma)\right) : \ l_\sigma(h) = \sum_{j=1}^{q} (g_j, h_j)_\Gamma. \qquad (4.134)$$

Moreover,

$$\left(\sum_{j=1}^{q} \|g_j\|^2_{H^{\sigma - m_j - 1/2}(\Gamma)}\right)^{1/2} \asymp \|l_\sigma\| \le c_4 \, \|(u_0', f)\|_\sigma. \qquad (4.135)$$

We define linear mapping 3 by the relation

$$R_\sigma : \ (u_0', f) \mapsto l_\sigma \mapsto g. \qquad (4.136)$$

We show that R_σ is an extension of the mapping R_λ for $\sigma < \lambda$. If

$$(u_0', f) \in H^{\lambda,(0)}(\Omega) \oplus H^{\lambda - 2q,(0)}(\Omega),$$

then, parallel with l_σ, the functional l_λ is defined. Moreover, l_σ is a restriction of the operator l_λ because Φ_σ is a restriction of the operator Φ_λ. In particular, $l_\sigma(h) = l_\lambda(h)$ for any vector $(h_1, \ldots, h_q) \in (C^\infty(\Gamma))^q$. By virtue of (4.134) and (4.136), this yields $R_\sigma(u_0', f) = R_\lambda(u_0', f)$, i.e., R_σ is an extension of the mapping R_λ.

We define mapping 4 with the use of isomorphism (4.93) (Theorem 4.15) as follows:

$$(f, g) \mapsto (L, B)^{-1} Q^+(f, g) =: \omega \in P(H^{\sigma,(2q)}(\Omega)). \qquad (4.137)$$

Recall that $f \in H^{\sigma - 2q,(0)}(\Omega)$ and that the vector g satisfies condition (4.134). The linear mapping (4.137) does not depend on σ and is bounded; namely,

$$\|\omega\|_{H^{\sigma,(2q)}(\Omega)} \le c_5 \, \|Q^+(f, g)\|_{\mathcal{H}_{\sigma,(0)}(\Omega,\Gamma)} \le c_5 \, c_6 \, \|(f, g)\|_{\mathcal{H}_{\sigma,(0)}(\Omega,\Gamma)}. \qquad (4.138)$$

Here, c_5 is the norm of the inverse of (4.93), and c_6 is the norm of the projector Q^+ acting on the space $\mathcal{H}_{\sigma,(0)}(\Omega, \Gamma)$.

We construct mapping 5 based on the $\varphi \equiv 1$ case of Theorem 4.23 proved by Ya. A. Roitberg [209, Theorem 6.1.1]:

$$\omega \mapsto I_L \, \omega \in K_{\sigma,L}(\Omega), \quad \omega \in P(H^{\sigma,(2q)}(\Omega)). \qquad (4.139)$$

This mapping is independent of σ and satisfies the two-sided estimate

$$\|I_L \, \omega\|_\sigma \asymp \|\omega\|_{H^{\sigma,(2q)}(\Omega)}. \qquad (4.140)$$

Now, using mappings 1–5, we define the operator Π_σ on vectors (4.129) as follows:

$$\Pi_\sigma : \ (u_0, f) \mapsto (\Psi_\sigma u_0, 0) + I_L \, \omega. \qquad (4.141)$$

Here, we recall that

$$\omega = (L,B)^{-1}Q^{+}(f,g), \quad g = R_\sigma(u_0', f), \quad \text{and} \quad u_0' = u_0 - \Psi_\sigma u_0. \tag{4.142}$$

The operator Π_σ is linear because mappings 1–5 are linear. It is bounded on space (4.129) by virtue of estimates (4.135), (4.138), and (4.140) and the boundedness of the projector Ψ_σ on the space $H^{\sigma,(0)}(\Omega)$. In addition, inclusions (4.131) and (4.139) yield the property $\Pi_\sigma(u_0, f) \in K_{\sigma,L}(\Omega)$. Thus, we have the bounded linear operator

$$\Pi_\sigma : H^{\sigma,(0)}(\Omega) \oplus H^{\sigma-2q,(0)}(\Omega) \to K_{\sigma,L}(\Omega).$$

Since, with the decrease in the parameter σ, mappings 1–5 extend, Π_σ is an extension of the operator Π_λ for $\sigma < \lambda$.

It remains to show that Π_σ is a projector onto the subspace $K_{\sigma,L}(\Omega)$, i.e., $\Pi_\sigma(u_0, f) = (u_0, f)$ for any vector $(u_0, f) \in K_{\sigma,L}(\Omega)$. We arbitrarily choose such a vector. By virtue of (4.131) and (4.132), we have the inclusion $(u_0', f) \in K_{\sigma,L}(\Omega)$. As above, $u_0' = u_0 - \Psi_\sigma u_0$. Therefore, according to the $\varphi \equiv 1$ case of Theorem 4.23,

$$\bigl(\exists!\ u' \in H^{\sigma,(2q)}(\Omega)\bigr) : I_L\, u' = (u_0', f). \tag{4.143}$$

This equality means the following:

$$u_0' \text{ is the initial component of } T_{2q}u' = (u_0', u_1', \dots, u_{2q}'), \tag{4.144}$$

$$Lu' = f \quad \text{in} \quad H^{\sigma-2q,(0)}(\Omega). \tag{4.145}$$

We show that $u' = \omega$, where the element $\omega \in P(H^{\sigma,(2q)}(\Omega))$ is defined according to (4.142). Recall that, by virtue of (4.133) and (4.134), the following inequality is true for an arbitrary vector h of the form (4.127):

$$l_\sigma(h) := (u_0', L^{+}\Phi_\sigma h)_\Omega - (f, (\Phi_\sigma h)_0)_\Omega = \sum_{j=1}^{q} (g_j, h_j)_\Gamma.$$

On the other hand, setting $u := u'$ in formula (4.128) and substituting relations (4.144) and (4.145) into this formula we obtain one more equality

$$(u_0', L^{+}\Phi_\sigma h)_\Omega - (f, (\Phi_\sigma h)_0)_\Omega = \sum_{j=1}^{q} (B_j u', h_j)_\Gamma.$$

It follows from these equalities that $B_j u' = g_j$ for every $j \in \{1, \dots, q\}$. Together with (4.145), this means the equality $(L,B)\,u' = (f,g)$. Therefore, by virtue of Theorem 4.15, the equality $Q^{+}(f,g) = (f,g)$ is true. In addition, by virtue of Theorem 4.15, it follows from relations (4.144) and (4.132) and the definition of the projector Ψ_σ that $Pu' = u'$. Using isomorphism (4.93), we can write the equality $u' = (L, B)^{-1} Q^{+}(f, g)$. Together with (4.142), it yields the required equality $u' = \omega$.

Now, by virtue of (4.143), we have $I_L\,\omega = I_L\,u' = (u_0', f)$. It follows from this result and relations (4.141) and (4.142) that

$$\Pi_\sigma(u_0, f) = (\Psi_\sigma u_0, 0) + I_L\,\omega = (\Psi_\sigma u_0, 0) + (u_0', f) = (u_0, f)$$

for an arbitrary vector $(u_0, f) \in K_{\sigma,L}(\Omega)$. Thus, the required projector Π_σ is constructed.

Lemma 4.6 is proved.

Based on this lemma, we prove Theorem 4.23.

Proof of Theorem 4.23. Let $s \in \mathbb{R}$, $\varphi \in \mathcal{M}$, and $\varepsilon > 0$. In the Sobolev case of $\varphi \equiv 1$, this theorem is proved by Ya. A. Roitberg [209, Theorem 6.1.1]. Thus, mapping (4.113) extends by continuity to the isomorphisms

$$I_L : H^{s\mp\varepsilon,(2q)}(\Omega) \leftrightarrow K_{s\mp\varepsilon,L}(\Omega).$$

Let us use the interpolation with the function parameter ψ from Theorem 1.14, where $\varepsilon = \delta$. We obtain one more isomorphism

$$I_L : [H^{s-\varepsilon,(2q)}(\Omega), H^{s+\varepsilon,(2q)}(\Omega)]_\psi \leftrightarrow [K_{s-\varepsilon,L}(\Omega), K_{s+\varepsilon,L}(\Omega)]_\psi. \tag{4.146}$$

We describe spaces in which it acts. By Theorem 4.22,

$$\bigl[H^{s-\varepsilon,(2q)}(\Omega), H^{s+\varepsilon,(2q)}(\Omega)\bigr]_\psi = H^{s,\varphi,(2q)}(\Omega). \tag{4.147}$$

Further, by virtue of Lemma 4.6 and Theorem 1.6, we can interpolate a pair of subspaces $K_{s\mp\varepsilon,L}(\Omega)$ as follows:

$$\begin{aligned}
&\bigl[K_{s-\varepsilon,L}(\Omega), K_{s+\varepsilon,L}(\Omega)\bigr]_\psi \\
&\quad = \bigl[H^{s-\varepsilon,(0)}(\Omega) \oplus H^{s-\varepsilon-2q,(0)}(\Omega), H^{s+\varepsilon,(0)}(\Omega) \oplus H^{s+\varepsilon-2q,(0)}(\Omega)\bigr]_\psi \\
&\qquad \cap K_{s-\varepsilon,L}(\Omega) \\
&\quad = H^{s,\varphi,(0)}(\Omega) \oplus H^{s-2q,\varphi,(0)}(\Omega) \cap K_{s-\varepsilon,L}(\Omega) = K_{s,\varphi,L}(\Omega).
\end{aligned}$$

Here, we also used Theorems 1.5 and 3.10 and the definition of the sets $K_{s-\varepsilon,L}(\Omega)$ and $K_{s,\varphi,L}(\Omega)$. By virtue of Theorem 1.6, $K_{s,\varphi,L}(\Omega)$ is a subspace of the space $H^{s,\varphi,(0)}(\Omega) \oplus H^{s-2q,\varphi,(0)}(\Omega)$, and, moreover, equalities (4.147) and

$$\bigl[K_{s-\varepsilon,L}(\Omega), K_{s+\varepsilon,L}(\Omega)\bigr]_\psi = K_{s,\varphi,L}(\Omega) \tag{4.148}$$

are true up to equivalence of norms. Substituting these equalities into (4.146), we obtain isomorphism (4.114). Assertion (ii) of Theorem 4.23 is proved and, hence, assertion (i) is also proved.

Theorem 4.23 is proved.

To prove Theorem 4.24, we need one more lemma on projectors. Let $\sigma \in \mathbb{R}$ and $\varphi \in \mathcal{M}$. We set

$$K^{0}_{\sigma,\varphi,L}(\Omega) := \big\{(u_0, f) \in H^{\sigma,\varphi,(0)}(\Omega) \oplus H^{\sigma-2q,\varphi,(0)}(\Omega) : \ (4.116) \text{ is true}\big\}.$$

By virtue of Theorem 3.9(iii), $K^{0}_{\sigma,\varphi,L}(\Omega)$ is a subspace of the space $H^{\sigma,\varphi,(0)}(\Omega) \oplus H^{\sigma-2q,\varphi,(0)}(\Omega)$. Moreover, $K_{\sigma,\varphi,L}(\Omega) \subseteq K^{0}_{\sigma,\varphi,L}(\Omega)$.

Lemma 4.7. *Let numbers $r \in \mathbb{N}$ and $\sigma \in \mathbb{R}$ satisfy the condition*

$$-r - 1/2 \leq \sigma < -r + 1/2. \tag{4.149}$$

Then there exist a projector $\Pi^{(r)}_{\sigma}$ of the space $H^{\sigma,(0)}(\Omega) \oplus H^{\sigma-2q,(0)}(\Omega)$ onto the subspace $K^{0}_{\sigma,L}(\Omega)$ such that $\Pi^{(r)}_{\sigma}$ is an extension of the mapping $\Pi^{(r)}_{\lambda}$ if $\sigma < \lambda < -r + 1/2$.

Proof. First, we establish a useful equality. Let $(u_0, f) \in K^{0}_{\sigma,L}(\Omega)$. In the Sobolev case of $\varphi \equiv 1$, Theorem 4.24 is proved by Ya. A. Roitberg [209, Theorem 6.2.1] for any parameter $s < -1/2$. Therefore, there exists a unique pair $(u_0^*, f) \in K_{\sigma,L}(\Omega)$ such that $u_0 = u_0^*$ in the domain Ω. Since $u_0, u_0^* \in H^{\sigma,(0)}(\Omega) = H^{\sigma}_{\overline{\Omega}}(\mathbb{R}^n)$ and $\operatorname{supp}(u_0 - u_0^*) \subseteq \Gamma$, inequality (4.149) yields the following representation for the distribution $u_0 - u_0^*$ (see, e.g., [209, Lemma 6.2.2]). There exists a unique vector

$$\omega^* = (\omega_1^*, \ldots, \omega_r^*) \in \bigoplus_{j=1}^{r} H^{\sigma+j-1/2}(\Gamma) \tag{4.150}$$

such that

$$(u_0 - u_0^*, w)_\Omega = \sum_{j=1}^{r} (\omega_j^*, D_\nu^{j-1} w)_\Gamma \tag{4.151}$$

for any $w \in H^{-\sigma,(0)}(\Omega) = H^{-\sigma}(\Omega)$.

In the Green formula (4.125), we set

$$u := u^* := I_L^{-1}(u_0^*, f) \in H^{\sigma,(2q)}(\Omega)$$

and use the property

$$Lu = Lu^* = f \in H^{\sigma-2q,(0)}(\Omega) = H^{\sigma-2q}_{\overline{\Omega}}(\mathbb{R}^n).$$

For any $v \in H^{2q-\sigma,(2q)}(\Omega) = H^{2q-\sigma}(\Omega)$ (see Theorem 4.12(iii)), we obtain the equality

$$(f,v)_\Omega + \sum_{j=1}^{q} (B_j u^*,\, C_j^+ v)_\Gamma = (u_0^*, L^+ v)_\Omega + \sum_{j=1}^{q} (C_j u^*,\, B_j^+ v)_\Gamma. \tag{4.152}$$

Here, u_0^* and $v = v_0$ are the initial components of the vectors $T_{2q}u^*$ and $T_{2q}v$, respectively.

Using relations (4.152) and (4.151) with $w := L^+ v \in H^{-\sigma}(\Omega)$, we obtain the equality

$$\begin{aligned}(f,v)_\Omega &+ \sum_{j=1}^{q} (B_j u^*,\, C_j^+ v)_\Gamma \\ &= (u_0, L^+ v)_\Omega - \sum_{j=1}^{r} (\omega_j^*, D_\nu^{j-1} L^+ v)_\Gamma + \sum_{j=1}^{q} (C_j u^*,\, B_j^+ v)_\Gamma. \end{aligned}\tag{4.153}$$

The boundary operators B_j^+, C_j^+, with $j = 1,\dots,q$, and $(D_\nu^{j-1} L^+ \cdot) \restriction \Gamma$, with $j = 1,\dots,r$, form a Dirichlet system of order $2q + r$. According to [209, Lemma 6.1.2], this system has the following property. The bounded linear operator

$$\begin{aligned}&(B_1^+,\dots,B_q^+, C_1^+,\dots,C_q^+, (L^+ \cdot)\restriction\Gamma,\dots,(D_\nu^{r-1} L^+ \cdot)\restriction\Gamma) : H^{2q-\sigma,(2q+r)}(\Omega) \\ &\to \bigoplus_{j=1}^{q} H^{2q-\sigma-m_j^+-1/2}(\Gamma) \oplus \bigoplus_{j=1}^{q} H^{-\sigma+m_j+1/2}(\Gamma) \oplus \bigoplus_{j=1}^{r} H^{-\sigma-j+1/2}(\Gamma)\end{aligned}$$

has the bounded right inverse operator

$$\begin{aligned}\Phi_\sigma^{(r)} : &\bigoplus_{j=1}^{q} H^{2q-\sigma-m_j^+-1/2}(\Gamma) \oplus \bigoplus_{j=1}^{q} H^{-\sigma+m_j+1/2}(\Gamma) \oplus \bigoplus_{j=1}^{r} H^{-\sigma-j+1/2}(\Gamma) \\ &\to H^{2q-\sigma,(2q+r)}(\Omega) = H^{2q-\sigma}(\Omega)\end{aligned}\tag{4.154}$$

such that $\Phi_\sigma^{(r)}$ is a restriction of the operator $\Phi_\lambda^{(r)}$ for $\sigma < \lambda < -r + 1/2$. Here, $\operatorname{ord} C_j^+ = 2q - 1 - m_j$ and $\operatorname{ord} B_j^+ =: m_j^+$. Moreover, the equality in (4.154) follows from Theorem 4.12(iii) in view of the condition $\sigma < -r + 1/2$.

For an arbitrary vector

$$h := (0,\dots,0,h_1,\dots,h_r) \in \{0\}^{2q} \oplus \bigoplus_{j=1}^{r} H^{-\sigma-j+1/2}(\Gamma), \tag{4.155}$$

we set $v := \Phi_\sigma^{(r)} h \in H^{2q-\sigma}(\Omega)$ in relation (4.153). Since

$$B_j^+ v = 0, \quad C_j^+ v = 0 \quad \text{on} \quad \Gamma \quad \text{for every} \quad j \in \{1, \dots, q\},$$

$$D_\nu^{j-1} L^+ v = h_j \quad \text{on} \quad \Gamma \quad \text{for every} \quad j \in \{1, \dots, r\},$$

we obtain the equality

$$(u_0, L^+ \Phi_\sigma^{(r)} h)_\Omega - (f, \Phi_\sigma^{(r)} h)_\Omega = \sum_{j=1}^{r} (\omega_j^*, h_j)_\Gamma. \tag{4.156}$$

Recall that, here, the pair $(u_0, f) \in K^0_{\sigma,L}(\Omega)$ and the vector h of the form (4.155) are arbitrary, while the vector $(\omega_1^*, \dots, \omega_r^*)$ satisfies conditions (4.150) and (4.151) and is uniquely defined by them on the basis of the pair (u_0, f).

Now we proceed to the construction of the projector $\Pi_\sigma^{(r)}$. We arbitrarily chose a vector (u_0, f) that satisfies inclusion (4.129). As before, let $\|\cdot\|_\sigma$ denote the norm in space (4.129). Consider the functional

$$l_\sigma^{(r)}(h) := (u_0, L^+ \Phi_\sigma^{(r)} h)_\Omega - (f, \Phi_\sigma^{(r)} h)_\Omega, \quad h \text{ being vector (4.155).} \tag{4.157}$$

The functional is bounded by virtue of the chain of inequalities

$$\begin{aligned}
|l_\sigma^{(r)}(h)| &\le |(u_0, L^+ \Phi_\sigma^{(r)} h)_\Omega| + |(f, \Phi_\sigma^{(r)} h)_\Omega| \\
&\le \|u_0\|_{H^{\sigma,(0)}(\Omega)} \, \|L^+ \Phi_\sigma^{(r)} h\|_{H^{-\sigma,(0)}(\Omega)} \\
&\quad + \|f\|_{H^{\sigma-2q,(0)}(\Omega)} \, \|\Phi_\sigma^{(r)} h\|_{H^{-\sigma+2q,(0)}(\Omega)} \\
&\le \big(c_1 \|u_0\|_{H^{\sigma,(0)}(\Omega)} + \|f\|_{H^{\sigma-2q,(0)}(\Omega)}\big) \, \|\Phi_\sigma^{(r)} h\|_{H^{-\sigma+2q}(\Omega)} \\
&\le (c_1 + 1)\, c_2 \, \|(u_0, f)\|_\sigma \left(\sum_{j=1}^{r} \|h_j\|^2_{H^{-\sigma-j+1/2}(\Gamma)} \right)^{1/2}.
\end{aligned}$$

Here, c_1 is the norm of the operator

$$L^+ : H^{-\sigma+2q}(\Omega) \to H^{-\sigma}(\Omega) = H^{-\sigma,(0)}(\Omega),$$

and c_2 is the norm of operator (4.154). Thus, $l_\sigma^{(r)}$ is a bounded antilinear functional on the space $\bigoplus_{j=1}^{r} H^{-\sigma-j+1/2}(\Gamma)$ and, furthermore, its norm satisfies the inequality $\|l_\sigma^{(r)}\| \le c_3 \|(u_0, f)\|_\sigma$, where $c_3 := (c_1 + 1)\, c_2$. By virtue of Theorem 2.3(v),

$$\left(\exists!\ (\omega_1, \dots, \omega_r) \in \bigoplus_{j=1}^{r} H^{\sigma+j-1/2}(\Gamma) \right) : \ l_\sigma^{(r)}(h) = \sum_{j=1}^{r} (\omega_j, h_j)_\Gamma. \tag{4.158}$$

Moreover,

$$\left(\sum_{j=1}^{r}\|\omega_j\|^2_{H^{\sigma+j-1/2}(\Gamma)}\right)^{1/2}\asymp\|l^{(r)}_\sigma\|\le c_3\,\|(u_0,f)\|_\sigma. \tag{4.159}$$

Based on the vector $(\omega_1,\ldots,\omega_r)$, we form the distribution $\omega'\in H^{\sigma,(0)}(\Omega)$ by the formula

$$(\omega',w)_\Omega:=\sum_{j=1}^{r}(\omega_j,D_\nu^{j-1}w)_\Gamma, \tag{4.160}$$

where $w\in H^{-\sigma,(0)}(\Omega)=H^{-\sigma}(\Omega)$ is an arbitrary distribution. This definition is reasonable by virtue of Theorems 2.3(v), 3.5, and 3.8(iii). Indeed, since $-\sigma>r-1/2$, by virtue of condition (4.149), we have the bounded operators

$$D_\nu^{j-1}:H^{-\sigma}(\Omega)\to H^{-\sigma-j+1/2}(\Gamma)\quad\text{for}\quad j=1,\ldots,r.$$

Therefore, $(\omega',\cdot)_\Omega$ is a bounded antilinear functional on the space $H^{-\sigma}(\Omega)$, which yields the inclusion

$$\omega'\in H^{\sigma,(0)}(\Omega)=H^{\sigma}_{\overline{\Omega}}(\mathbb{R}^n).$$

Moreover, the inclusion $\operatorname{supp}\omega'\subseteq\Gamma$ and the two-sided estimate

$$\|\omega'\|_{H^{\sigma,(0)}(\Omega)}\asymp\left(\sum_{j=1}^{r}\|\omega_j\|^2_{H^{\sigma+j-1/2}(\Gamma)}\right)^{1/2} \tag{4.161}$$

are true (see, e.g., [209, Lemma 6.2.2]).

On vectors (4.129), we define the linear mapping $\Upsilon^{(r)}_\sigma$ by the formula

$$\Upsilon^{(r)}_\sigma:(u_0,f)\mapsto l^{(r)}_\sigma\mapsto(\omega_1,\ldots,\omega_r)\mapsto\omega'. \tag{4.162}$$

By virtue of estimates (4.159) and (4.161), we have the bounded operator

$$\Upsilon^{(r)}_\sigma:H^{\sigma,(0)}(\Omega)\oplus H^{\sigma-2q,(0)}(\Omega)\to H^{\sigma,(0)}(\Omega). \tag{4.163}$$

Let us show that $\Upsilon^{(r)}_\sigma$ is an extension of the mapping $\Upsilon^{(r)}_\lambda$ for $\sigma<\lambda<-r+1/2$. If

$$(u_0,f)\in H^{\lambda,(0)}(\Omega)\oplus H^{\lambda-2q,(0)}(\Omega),$$

then, parallel with $l^{(r)}_\sigma$, the functional $l^{(r)}_\lambda$ is defined. Moreover, $l^{(r)}_\sigma$ is a restriction of the functional $l^{(r)}_\lambda$ because $\Phi^{(r)}_\sigma$ is a restriction of the operator $\Phi^{(r)}_\lambda$. In particular, $l^{(r)}_\sigma(h)=l^{(r)}_\lambda(h)$ for any vector $(h_1,\ldots,h_q)\in(C^\infty(\Gamma))^q$. By virtue of (4.158) and (4.160), this yields $\Upsilon^{(r)}_\sigma(u_0,f)=\Upsilon^{(r)}_\lambda(u_0,f)$, i.e., $\Upsilon^{(r)}_\sigma$ is an extension of the mapping $\Upsilon^{(r)}_\lambda$.

We define the linear mapping $\Pi_\sigma^{(r)}$ with the use of the operator $\Upsilon_\sigma^{(r)}$ and the projector Π_σ in Lemma 4.6 as follows:

$$\Pi_\sigma^{(r)}(u_0, f) := (\omega', 0) + \Pi_\sigma(u_0 - \omega', f), \quad \omega' := \Upsilon_\sigma^{(r)}(u_0, f). \tag{4.164}$$

Here, (u_0, f) is an arbitrary vector (4.129). As was mentioned above, the inclusions $\omega' \in H^{\sigma,(0)}(\Omega)$ and $\operatorname{supp}\omega' \subseteq \Gamma$ are true. Therefore, $(\omega', 0) \in K^0_{\sigma,L}(\Omega)$. This fact, together with the boundedness of operator (4.163), Lemma 4.6, and the inclusion $K_{\sigma,L}(\Omega) \subseteq K^0_{\sigma,L}(\Omega)$, yields the boundedness of the operator

$$\Pi_\sigma^{(r)} : H^{\sigma,(0)}(\Omega) \oplus H^{\sigma-2q,(0)}(\Omega) \to K^0_{\sigma,L}(\Omega).$$

Moreover, since both the operators $\Upsilon_\sigma^{(r)}$ and Π_σ extend provided the parameter σ decreases, we conclude that $\Pi_\sigma^{(r)}$ is an extension of the operator $\Pi_\lambda^{(r)}$ for $\sigma < \lambda < -r + 1/2$.

It remains to show that $\Pi_\sigma^{(r)}$ is a projector onto the subspace $K^0_{\sigma,L}(\Omega)$, i.e.,

$$\Pi_\sigma^{(r)}(u_0, f) = (u_0, f)$$

for any vector $(u_0, f) \in K^0_{\sigma,L}(\Omega)$. We arbitrarily choose such a vector. According to Theorem 4.24 in the case of $\varphi \equiv 1$, we state that, for (u_0, f), there exists a unique pair $(u_0^*, f) \in K_{\sigma,L}(\Omega)$ satisfying condition (4.118). Let us show that $u_0 - u_0^* = \omega'$, where $\omega' := \Upsilon_\sigma^{(r)}(u_0, f)$. For an arbitrary distribution $w \in H^{-\sigma}(\Omega)$, we define the vector h by formula (4.155) with

$$h_j := (D_\nu^{j-1} w) \upharpoonright \Gamma \in H^{-\sigma-j+1/2}(\Gamma) \quad \text{for each} \quad j \in \{1, \ldots, r\}. \tag{4.165}$$

As was mentioned above, the traces h_j exist in view of condition (4.149). We also recall that condition (4.118) means the inclusion $\operatorname{supp}(u_0 - u_0^*) \subseteq \Gamma$, which yields equality (4.151) for a certain (unique) vector (4.150). Now, using relations (4.151), (4.156), and (4.165), we obtain the equality

$$(u_0 - u_0^*, w)_\Omega = (u_0, L^+ \Phi_\sigma^{(r)} h)_\Omega - (f, \Phi_\sigma^{(r)} h)_\Omega.$$

On the other hand, by virtue of (4.157), (4.158), (4.160), and (4.165), we get

$$(\omega', w)_\Omega = l_\sigma^{(r)}(h) = (u_0, L^+ \Phi_\sigma^{(r)} h)_\Omega - (f, \Phi_\sigma^{(r)} h)_\Omega.$$

Therefore,

$$(u_0 - u_0^*, w)_\Omega = (\omega', w)_\Omega \quad \text{for any} \quad w \in H^{-\sigma}(\Omega).$$

By virtue of Theorem 3.8(iii), this is equivalent to the equality

$$u_0 - u_0^* = \omega' := \Upsilon_\sigma^{(r)}(u_0, f) \quad \text{in} \quad H^{\sigma,(0)}(\Omega).$$

Using this result and (4.164), we obtain

$$\begin{aligned}\Pi_\sigma^{(r)}(u_0, f) &= (\omega', 0) + \Pi_\sigma(u_0 - \omega', f)\\ &= (\omega', 0) + \Pi_\sigma(u_0^*, f)\\ &= (\omega', 0) + (u_0^*, f)\\ &= (u_0, f)\end{aligned}$$

for an arbitrary vector $(u_0, f) \in K^0_{\sigma,L}(\Omega)$. Here, we also used the inclusion $(u_0^*, f) \in K_{\sigma,L}(\Omega)$ and the property of Π_σ to be a projector onto the subspace $K_{\sigma,L}(\Omega)$. Thus, the required projector $\Pi_\sigma^{(r)}$ is constructed.

Lemma 4.7 is proved.

Using this lemma, we prove Theorem 4.24.

Proof of Theorem 4.24. Let $s < -1/2$, $s + 1/2 \notin \mathbb{Z}$, and $\varphi \in \mathcal{M}$. We choose numbers $r \in \mathbb{N}$ and $\varepsilon \in (0, 1/2)$ such that

$$-r - 1/2 < s \mp \varepsilon < -r + 1/2. \tag{4.166}$$

As was mentioned above, Theorem 4.24 was proved by Ya. A. Roitberg [209, Theorem 6.2.1] in the Sobolev case of $\varphi \equiv 1$,. Therefore, we can introduce the linear mapping $G : (u_0, f) \mapsto (u_0^*, f)$, where $(u_0, f) \in K^0_{s-\varepsilon,L}(\Omega)$ and, moreover, $(u_0^*, f) \in K_{s-\varepsilon,L}(\Omega)$ satisfies (4.118). This mapping defines the bounded operators

$$G : K^0_{s\mp\varepsilon,L}(\Omega) \to K_{s\mp\varepsilon,L}(\Omega). \tag{4.167}$$

Theorem 4.24 will be proved for any $\varphi \in \mathcal{M}$ if we establish that G is a bounded operator from $K^0_{s,\varphi,L}(\Omega)$ to $K_{s,\varphi,L}(\Omega)$. We do this using the interpolation with the parameter ψ in Theorem 1.14 for $\varepsilon = \delta$. Applying this theorem to (4.167), we obtain one more bounded operator

$$G : [K^0_{s-\varepsilon,L}(\Omega), K^0_{s+\varepsilon,L}(\Omega)]_\psi \to [K_{s-\varepsilon,L}(\Omega), K_{s+\varepsilon,L}(\Omega)]_\psi. \tag{4.168}$$

Let us describe the spaces between which this operator acts. Based on Lemma 4.7, inequality (4.166), and Theorem 1.6, we can interpolate a pair of the subspaces $K^0_{s\mp\varepsilon,L}(\Omega)$ as follows:

$$\begin{aligned}&[K^0_{s-\varepsilon,L}(\Omega), K^0_{s+\varepsilon,L}(\Omega)]_\psi\\ &\qquad = [H^{s-\varepsilon,(0)}(\Omega) \oplus H^{s-\varepsilon-2q,(0)}(\Omega), H^{s+\varepsilon,(0)}(\Omega) \oplus H^{s+\varepsilon-2q,(0)}(\Omega)]_\psi\\ &\qquad\quad \cap K^0_{s-\varepsilon,L}(\Omega)\end{aligned}$$

$$= H^{s,\varphi,(0)}(\Omega) \oplus H^{s-2q,\varphi,(0)}(\Omega) \cap K^0_{s-\varepsilon,L}(\Omega)$$

$$= K^0_{s,\varphi,L}(\Omega).$$

Here, we also used Theorems 1.5 and 3.10 and the definition of the spaces $K^0_{s-\varepsilon,L}(\Omega)$ and $K^0_{s,\varphi,L}(\Omega)$. The obtained equality of spaces

$$[K^0_{s-\varepsilon,L}(\Omega), K^0_{s+\varepsilon,L}(\Omega)]_\psi = K^0_{s,\varphi,L}(\Omega) \tag{4.169}$$

is true up to equivalence of norms. Substituting equalities (4.169) and (4.148) into relation (4.167), we obtain the bounded operator

$$G : K^0_{s,\varphi,L}(\Omega) \to K_{s,\varphi,L}(\Omega),$$

which is what to be proved.

Theorem 4.24 is proved.

Remark 4.8. Theorems 4.23, 4.24 and Lemmas 4.6, 4.7 are proved under the assumption that problem (4.1) is regularly elliptic. Actually, they are true for an arbitrary differential expression L which is properly elliptic in the domain $\overline{\Omega}$. This follows from the fact that, for L, there exists a regular elliptic boundary-value problem, e.g., the Dirichlet problem,.

4.4 Generalization of the Lions–Magenes theorems

In this section, we establish individual theorems on solvability of the regular elliptic boundary-value problem (4.1) in scales of Sobolev spaces. Unlike the general theorems 4.1 and 4.14, in individual theorems, the domain of the operator (L, B) depends on the coefficients of the elliptic expression L. J.-L. Lions and E. Magenes [119, 120, 121, 126], proved a number of individual theorems for the operator (L, B) acting in Sobolev spaces containing nonregular distributions. We prove a certain general form of the Lions–Magenes theorems and determine a general condition, for the space of the right-hand sides of the elliptic equation, under which the operator (L, B) is bounded and Fredholm in the corresponding pair of Hilbert spaces. We also indicate wide classes of spaces that satisfy this condition. They contain the spaces used by J.-L. Lions and E. Magenes and many other spaces including some Hördmander spaces. Unlike the general theorem 4.14, we consider individual theorems in which the solution and the right-hand side of the elliptic equation are distributions in the domain Ω.

The results of this section simulate statements and methods for the proof of individual theorems on solvability of elliptic boundary-value problems in scales of Hörmander spaces. These theorems will be established in Section 4.5.

4.4.1 Lions–Magenes theorems

First, we indicate an important difference in the definitions of the negative order Sobolev spaces over a Euclidean domain used here and in J.-L. Lions and E. Magenes' papers cited above. Recall that, following [258, 256], we define the Sobolev space of an arbitrary order $s \in \mathbb{R}$ over the domain Ω by the formulas:

$$H^s(\Omega) := \{u := w\restriction\Omega : w \in H^s(\mathbb{R}^n)\}, \tag{4.170}$$

$$\|u\|_{H^s(\Omega)} := \inf\{\|w\|_{H^s(\mathbb{R}^n)} : w \in H^s(\mathbb{R}^n),\ w = u \text{ on } \Omega\} \tag{4.171}$$

(see Definition 3.2 in the case of $\varphi \equiv 1$). J.-L. Lions and E. Magenes use this definition only for $s \geq 0$. For $s < 0$, they consider the dual space $(H_0^{-s}(\Omega))'$ as the Sobolev space of order s over the domain Ω. Recall that $H_0^{-s}(\Omega)$ is a closure of the set $C_0^\infty(\Omega)$ in the space $H^{-s}(\Omega)$, the duality of spaces being considered with respect to the inner product in $L_2(\Omega)$. It is known [258, Theorem 4.8.2] that these definitions give the same space (up to equivalence of norms) if the order $s < 0$ is not half-integer. For half-integer $s < 0$, different spaces are obtained.

In this section, we follow J.-L. Lions and E. Magenes and define the Sobolev spaces over the domain Ω by the formula

$$\mathrm{H}^s(\Omega) := \begin{cases} H^s(\Omega) & \text{for } s \geq 0, \\ (H_0^{-s}(\Omega))' & \text{for } s < 0. \end{cases} \tag{4.172}$$

On the left-hand side of the definition, we use the Roman upright type of letter H rather than italic used in definition (4.170). Here, the dual space $(H_0^{-s}(\Omega))'$ consists of antilinear functionals.

The functionals belonging to the space $\mathrm{H}^s(\Omega)$ with $s < 0$ are uniquely defined by their values on the test functions in $C_0^\infty(\Omega)$. Therefore, it is correctly to identify these functionals with distributions given in the domain Ω. Moreover, for any $s < 0$,

$$\mathrm{H}^s(\Omega) = \{w\restriction\Omega : w \in H^s_{\overline{\Omega}}(\mathbb{R}^n)\}, \tag{4.173}$$

$$\|u\|_{\mathrm{H}^s(\Omega)} = \inf\{\|w\|_{H^s(\mathbb{R}^n)} : w \in H^s_{\overline{\Omega}}(\mathbb{R}^n),\ w = u \text{ in } \Omega\} \tag{4.174}$$

(see [121, Chap. 1, Remark 12.5]). It follows from these relations and formulas (4.170) and (4.171) that $\mathrm{H}^s(\Omega) \hookrightarrow H^s(\Omega)$ continuously and that $C_0^\infty(\Omega)$ is dense in $\mathrm{H}^s(\Omega)$ for each $s < 0$.

As was mentioned above,

$$\mathrm{H}^s(\Omega) = H^s(\Omega) \Leftrightarrow s \in \mathbb{R} \setminus \{-1/2, -3/2, -5/2, \ldots\}. \tag{4.175}$$

If the parameter $s < 0$ is half-integer, then the space $\mathrm{H}^s(\Omega)$ is narrower than the space $H^s(\Omega)$.

Note that, for arbitrary $s \in \mathbb{R}$ and $\varepsilon > 0$, the compact and dense embedding $\mathrm{H}^{s+\varepsilon}(\Omega) \hookrightarrow \mathrm{H}^s(\Omega)$ is true.

Now we proceed to the Lions–Magenes theorems. To this end, we first recall the classical general theorem (Proposition 3.1) on solvability of the boundary-value problem (4.1) in the scale of positive Sobolev spaces.

Theorem A. *The mapping*

$$u \mapsto (Lu, Bu), \quad u \in C^\infty(\overline{\Omega}), \tag{4.176}$$

extends uniquely (by continuity) to the Fredholm bounded operator

$$(L, B) : \mathrm{H}^{\sigma+2q}(\Omega) \to \mathrm{H}^{\sigma}(\Omega) \oplus \bigoplus_{j=1}^{q} H^{\sigma+2q-m_j-1/2}(\Gamma) =: \mathbf{H}_\sigma(\Omega, \Gamma) \tag{4.177}$$

for any real $\sigma \geq 0$. The kernel of this operator coincides with N and the range consists of all vectors $(f, g_1, \ldots, g_q) \in \mathbf{H}_\sigma(\Omega, \Gamma)$ that satisfy condition (4.4). *The index of operator* (4.177) *is equal to* $\dim N - \dim N^+$ *and does not depend on σ.*

Here and below, unlike the previous theorems on solvability of the boundary-value problem (4.1), it is convenient to represent the index s, which defines the smoothness of the domain of (L, B), in the form of the sum $s = \sigma + 2q$. Note that Theorem A has been already extended over the refined Sobolev scale in Section 4.1.1 (Theorem 4.1).

As has been mentioned in Section 4.2.1, Theorem A is not true if σ runs through the negative semiaxis. J.-L. Lions and E. Magenes suggested to replace $\mathrm{H}^{\sigma+2q}(\Omega)$, as the domain of (L, B), with the narrower space

$$D^{\sigma+2q}_{L,X}(\Omega) := \{u \in \mathrm{H}^{\sigma+2q}(\Omega) : Lu \in X^\sigma(\Omega)\}, \tag{4.178}$$

where $X^\sigma(\Omega)$ is a certain Hilbert space continuously embedded in $\mathrm{H}^\sigma(\Omega)$. Here and below, the image Lu of $u \in \mathcal{D}'(\Omega)$ is understood in the sense of the theory of distributions. In space (4.178), we introduce the graph inner product

$$(u_1, u_2)_{D^{\sigma+2q}_{L,X}(\Omega)} := (u_1, u_2)_{\mathrm{H}^{\sigma+2q}(\Omega)} + (Lu_1, Lu_2)_{X^\sigma(\Omega)} \tag{4.179}$$

and the corresponding norm.

The space $D^{\sigma+2q}_{L,X}(\Omega)$ with inner product (4.179) is complete. Indeed, if a sequence (u_k) is fundamental in $D^{\sigma+2q}_{L,X}(\Omega)$, then, since the sets $\mathrm{H}^{\sigma+2q}(\Omega)$ and $X^\sigma(\Omega)$ are complete, there exist limits $u := \lim u_k$ in $\mathrm{H}^{\sigma+2q}(\Omega) \hookrightarrow \mathcal{D}'(\Omega)$ and $f := \lim Lu_k$ in $X^\sigma(\Omega) \hookrightarrow \mathcal{D}'(\Omega)$ (embeddings are continuous). Since the differential operator L is continuous on $\mathcal{D}'(\Omega)$, we have $Lu = \lim Lu_k$ in $\mathcal{D}'(\Omega)$ by virtue of the first limit. Using the second limit, we obtain the equality $Lu = f \in X^\sigma(\Omega)$. Therefore, $u \in D^{\sigma+2q}_{L,X}(\Omega)$ and $\lim u_k = u$ in the space $D^{\sigma+2q}_{L,X}(\Omega)$, i.e., this space is complete.

J.-L. Lions and E. Magenes [119, 120, 121, 126] give some important examples of spaces $X^\sigma(\Omega)$ such that the mapping (4.176) extends by continuity to the Fredholm bounded operator

$$(L,B): D^{\sigma+2q}_{L,X}(\Omega) \to X^\sigma(\Omega) \oplus \bigoplus_{j=1}^{q} H^{\sigma+2q-m_j-1/2}(\Gamma)$$
$$=: \mathbf{X}_\sigma(\Omega,\Gamma) \tag{4.180}$$

if $\sigma < 0$. Unlike Theorem A, the domain of operator (4.180) depends, together with topology, on coefficients of the elliptic expression L. Therefore, theorems on properties of operator (4.180) are individual theorems on solvability of the boundary-value problem (4.1).

We formulate two individual theorems established by J.-L. Lions and E. Magenes.

Theorem LM$_1$ [119, 120]. *Let $\sigma < 0$ and let $X^\sigma(\Omega) := L_2(\Omega)$. Then mapping* (4.176) *extends uniquely (by continuity) to the Fredholm bounded operator* (4.180). *Its kernel coincides with N and the range consists of all vectors $(f, g_1, \dots, g_q) \in \mathbf{X}_\sigma(\Omega,\Gamma)$ that satisfy condition* (4.4). *The index of operator* (4.180) *is equal to* $\dim N - \dim N^+$ *and does not depend on σ.*

Here, we should select the case of $\sigma = -2q$, which is important in the spectral theory of elliptic operators [67, 68, 69, 137, 138]. In this case, the space

$$D^0_{L,L_2}(\Omega) = \{u \in L_2(\Omega) : Lu \in L_2(\Omega)\} \tag{4.181}$$

is the domain of the maximum operator corresponding to the differential expression L acting in the space $L_2(\Omega)$. Note that even if all coefficients of the expression L are constant, then space (4.181) essentially depends on each of them. This follows from L. Hörmander's result [80, Theorem 3.1] cited below.

Let L and M be two linear differential expressions with constant coefficients. Then if $D^0_{L,L_2}(\Omega) \subseteq D^0_{M,L_2}(\Omega)$, then either $M = \alpha L + \beta$ for certain $\alpha, \beta \in \mathbb{C}$ or L and M are polynomials in the operator of differentiation along a certain vector e and, furthermore, $\operatorname{ord} M \le \operatorname{ord} L$. Note that the second possibility is eliminated for elliptic operators.

To formulate the second Lions–Magenes theorem, we consider the weight space

$$\varrho \mathrm{H}^\sigma(\Omega) := \{f = \varrho v : v \in \mathrm{H}^\sigma(\Omega)\}, \tag{4.182}$$

where $\sigma < 0$, while the function $\varrho \in C^\infty(\Omega)$ is positive. We endow this space with the inner product

$$(f_1, f_2)_{\varrho \mathrm{H}^\sigma(\Omega)} := (\varrho^{-1} f_1, \varrho^{-1} f_2)_{\mathrm{H}^\sigma(\Omega)} \tag{4.183}$$

and the corresponding norm. The space $\varrho\mathrm{H}^\sigma(\Omega)$ is complete and embedded continuously in $\mathcal{D}'(\Omega)$. This follows from the fact that the operator of multiplication by the function ϱ is continuous on $\mathcal{D}'(\Omega)$ and sets an isomorphism of the complete space $\mathrm{H}^\sigma(\Omega)$ onto $\varrho\mathrm{H}^\sigma(\Omega)$.

We consider weight functions of the form $\varrho := \varrho_1^{-\sigma}$, where

$$\begin{gathered} \varrho_1 \in C^\infty(\overline{\Omega}), \quad \varrho_1 > 0 \text{ in } \Omega, \\ \varrho_1(x) = \operatorname{dist}(x, \Gamma) \text{ in a neighborhood of } \Gamma. \end{gathered} \tag{4.184}$$

Theorem LM$_2$ [121, Chap. 2, Sec. 2.3]. *Let $\sigma < 0$ and let*

$$X^\sigma(\Omega) := \begin{cases} \varrho_1^{-\sigma}\mathrm{H}^\sigma(\Omega) & \text{if } \sigma + 1/2 \notin \mathbb{Z}, \\ [\varrho_1^{-\sigma+1/2}\mathrm{H}^{\sigma-1/2}(\Omega), \varrho_1^{-\sigma-1/2}\mathrm{H}^{\sigma+1/2}(\Omega)]_{t^{1/2}} & \\ \qquad \text{if } \sigma + 1/2 \in \mathbb{Z}. \end{cases} \tag{4.185}$$

Then the conclusion of Theorem LM$_1$ *remains true.*

Remark 4.9. J.-L. Lions and E. Magenes [121, Chap. 2, Sec. 6.3] use the following Hilbert space $\Xi^\sigma(\Omega)$ as $X^\sigma(\Omega)$. For integer $\sigma \geq 0$, the space $\Xi^\sigma(\Omega)$ is defined by (3.53). For fractional $\sigma > 0$, it is defined by interpolation with power parameter

$$\Xi^\sigma(\Omega) := \left[\Xi^{[\sigma]}(\Omega), \Xi^{[\sigma]+1}(\Omega)\right]_{t^{\{\sigma\}}}.$$

Finally, for negative $\sigma < 0$, it is defined by passing to the dual space (with respect to the inner product in $L_2(\Omega)$), namely, $\Xi^\sigma(\Omega) := (\Xi^{-\sigma}(\Omega))'$. For an arbitrary $\sigma < 0$, the space $\Xi^\sigma(\Omega)$ coincides (up to equivalence of norms) with the right-hand side of relation (4.185). This follows from J.-L. Lions and E. Magenes' result [121, Chap. 2, Corollary 7.4].

4.4.2 Key individual theorem

Here, we prove the key individual theorem on solvability of the boundary-value problem (4.1). According to this theorem, operator (4.180) is well-defined, bounded, and Fredholm for every $\sigma < 0$ if the Hilbert space $X^\sigma(\Omega) \hookrightarrow \mathcal{D}'(\Omega)$ satisfies condition I$_\sigma$ given below. This theorem is a key for proving other individual theorems.

Condition I_σ. The set $X^\infty(\Omega) := X^\sigma(\Omega) \cap C^\infty(\overline{\Omega})$ is dense in $X^\sigma(\Omega)$, and there exists a number $c > 0$ such that

$$\|\mathcal{O}f\|_{H^\sigma(\mathbb{R}^n)} \le c\,\|f\|_{X^\sigma(\Omega)} \quad \text{for any} \quad f \in X^\infty(\Omega). \tag{4.186}$$

Here, $\mathcal{O}f(x) = f(x)$ for every $x \in \overline{\Omega}$, and $\mathcal{O}f(x) = 0$ for every $x \in \mathbb{R}^n \setminus \overline{\Omega}$.

Note that the smaller σ is the weaker condition I_σ will be, for the same space $X^\sigma(\Omega)$.

Remark 4.10. Ya. A. Roitberg [207, Sec. 2.4] considers a condition for the space $X^\sigma(\Omega)$ which is somewhat stronger than our condition I_σ. In addition, he assumes that $C^\infty(\overline{\Omega}) \subset X^\sigma(\Omega)$. Under this condition, Ya. A. Roitberg proves the boundedness of operator (4.180) for all $\sigma < 0$ (see [207, Sec. 2.4] and [209, Remark 6.2.2]). Note that this condition does not include the important case $X^\sigma(\Omega) = \{0\}$ and several weight spaces $X^\sigma(\Omega) = \varrho \mathrm{H}^\sigma(\Omega)$ considered below.

Let us formulate the key individual theorem.

Theorem 4.25. *Let $\sigma < 0$ and let $X^\sigma(\Omega)$ be an arbitrary Hilbert space that is continuously embedded in $\mathcal{D}'(\Omega)$ and satisfies condition I_σ. Then the following assertions are true:*

(i) *The set*
$$D^\infty_{L,X}(\Omega) := \{u \in C^\infty(\overline{\Omega}) : Lu \in X^\sigma(\Omega)\}$$
is dense in the space $D^{\sigma+2q}_{L,X}(\Omega)$.

(ii) *The mapping $u \to (Lu, Bu)$, where $u \in D^\infty_{L,X}(\Omega)$, extends uniquely (by continuity) to the bounded linear operator* (4.180).

(iii) *Operator* (4.180) *is Fredholm. Its kernel coincides with N, and the range consists of all vectors $(f, g_1, \dots, g_q) \in \mathbf{X}_\sigma(\Omega, \Gamma)$ that satisfy condition* (4.4).

(iv) *If the set $\mathcal{O}(X^\infty(\Omega))$ is dense in the space $H^\sigma_{\overline{\Omega}}(\mathbb{R}^n)$, then the index of operator* (4.180) *is equal to* $\dim N - \dim N^+$.

Proof. The proof is based on Theorems 4.14, 4.18, 4.23, and 4.24 for the Sobolev case of $\varphi \equiv 1$, we taking $s = \sigma + 2q$. Recall that, in this case, they are proved by Ya. A. Roitberg [209], Theorem 4.24 being established for any $s < -1/2$.

It follows from the condition of the theorem that the mapping $f \mapsto \mathcal{O}f$, where $f \in X^\infty(\Omega)$, is extended by continuity to the bounded linear operator

$$\mathcal{O} : X^\sigma(\Omega) \to H^\sigma_{\overline{\Omega}}(\mathbb{R}^n) = H^{\sigma,(0)}(\Omega). \tag{4.187}$$

This operator is injective. Indeed, let $\mathcal{O}f = 0$ for a certain distribution $f \in X^\sigma(\Omega)$. We choose a sequence $(f_k) \subset X^\infty(\Omega)$ such that $f_k \to f$ in $X^\sigma(\Omega) \hookrightarrow \mathcal{D}'(\Omega)$. Then $\mathcal{O}f_k \to 0$ in $H^\sigma_{\overline{\Omega}}(\mathbb{R}^n) \hookrightarrow \mathcal{S}'(\mathbb{R}^n)$, which yields

$$(f, v)_\Omega = \lim_{k\to\infty} (f_k, v)_\Omega = \lim_{k\to\infty} (\mathcal{O}f_k, v)_\Omega = 0 \quad \text{for all} \quad v \in C_0^\infty(\Omega).$$

Thus, $f = 0$ as a distribution from the space $X^\sigma(\Omega) \hookrightarrow \mathcal{D}'(\Omega)$, i.e., the operator (4.187) is injective. It defines the continuous embedding $X^\sigma(\Omega) \hookrightarrow H^{\sigma,(0)}(\Omega)$.

According to Theorem 4.13, for every element $u \in H^{\sigma+2q,(2q)}(\Omega)$, we define $Lu \in H^{\sigma,(0)}(\Omega)$ passing to the limit. We set

$$D^{\sigma+2q,(2q)}_{L,X}(\Omega) := \big\{u \in H^{\sigma+2q,(2q)}(\Omega) : Lu \in X^\sigma(\Omega)\big\}.$$

In the space $D^{\sigma+2q,(2q)}_{L,X}(\Omega)$, we introduce the graph inner product

$$(u_1, u_2)_{D^{\sigma+2q,(2q)}_{L,X}(\Omega)} := (u_1, u_2)_{H^{\sigma+2q,(2q)}(\Omega)} + (Lu_1, Lu_2)_{X^\sigma(\Omega)}.$$

The space $D^{\sigma+2q,(2q)}_{L,X}(\Omega)$ is complete with respect to this inner product. Indeed, let a sequence (u_k) be fundamental in $D^{\sigma+2q,(2q)}_{L,X}(\Omega)$. Since both the spaces $H^{\sigma+2q,(2q)}(\Omega)$ and $X^\sigma(\Omega)$ are complete, there exist limits $u := \lim u_k$ in $H^{\sigma+2q,(2q)}(\Omega)$ and $f := \lim Lu_k$ in $X^\sigma(\Omega)$. The first limit yields the convergence $\lim Lu_k = Lu$ in $H^{\sigma,(0)}(\Omega)$. Hence, using the second limit and continuous embedding (4.187), we conclude that $Lu = f \in X^\sigma(\Omega)$. Therefore, $u \in D^{\sigma+2q,(2q)}_{L,X}(\Omega)$ and $\lim u_k = u$ in $D^{\sigma+2q,(2q)}_{L,X}(\Omega)$, i.e., the space $D^{\sigma+2q,(2q)}_{L,X}(\Omega)$ is complete.

Based on Theorem 4.14, we conclude that the restriction of the operator

$$(L, B) : H^{\sigma+2q,(2q)}(\Omega) \to H^{\sigma,(0)}(\Omega) \oplus \bigoplus_{j=1}^{q} H^{\sigma+2q-m_j-1/2}(\Gamma)$$

$$=: \mathbf{H}_{\sigma,(0)}(\Omega, \Gamma) \tag{4.188}$$

onto the space $D^{\sigma+2q,(2q)}_{L,X}(\Omega)$ is the bounded operator

$$(L, B) : D^{\sigma+2q,(2q)}_{L,X}(\Omega) \to \mathbf{X}_\sigma(\Omega, \Gamma). \tag{4.189}$$

The kernel of operator (4.189) is equal to N, and the range consists of all vectors $(f, g_1, \dots, g_q) \in \mathbf{X}_\sigma(\Omega, \Gamma)$ that satisfy condition (4.4). Therefore, operator (4.189) is Fredholm, its kernel being of a dimension $\beta \le \dim N^+$.

In addition, if the set $\mathcal{O}(X^\infty(\Omega)$ is dense in the space $H^\sigma_{\overline{\Omega}}(\mathbb{R}^n)$, then $\beta = \dim N^+$. Indeed, let Λ denote operator (4.188) and let Λ_0 denote the narrower operator (4.189); consider the adjoint operators Λ^* and Λ_0^*. Since the continuous embedding $\mathbf{X}_\sigma(\Omega,\Gamma) \hookrightarrow \mathbf{H}_{\sigma,(0)}(\Omega,\Gamma)$ is dense, we have $\ker \Lambda_0^* \supseteq \ker \Lambda^*$. Therefore,

$$\beta = \dim \operatorname{coker} \Lambda_0 = \dim \ker \Lambda_0^* \geq \dim \ker \Lambda^* = \dim \operatorname{coker} \Lambda = \dim N^+.$$

Hence, $\beta = \dim N^+$, and the index of operator (4.189) is equal to $\dim N - \dim N^+$ in the considered case.

Let us show that the set $D^{\infty}_{L,X}(\Omega)$ is dense in the space $D^{\sigma+2q,(2q)}_{L,X}(\Omega)$. Using the density of $X^\infty(\Omega) \times (C^\infty(\Gamma))^q$ in $\mathbf{X}_\sigma(\Omega,\Gamma)$ and applying the Gohberg–Krein lemma [65, Lemma 2.1], we can write

$$\mathbf{X}_\sigma(\Omega,\Gamma) = (L,B)\big(D^{\sigma+2q,(2q)}_{L,X}(\Omega)\big) \dotplus \mathcal{Q}(\Omega,\Gamma), \tag{4.190}$$

where $\mathcal{Q}(\Omega,\Gamma)$ is a certain finite-dimensional space that satisfies the condition

$$\mathcal{Q}(\Omega,\Gamma) \subset X^\infty(\Omega) \times (C^\infty(\Gamma))^q. \tag{4.191}$$

Let Π denote the projection operator of the space $\mathcal{X}_\sigma(\Omega,\Gamma)$ onto the first summand in (4.190) parallel to the second.

Let $u \in D^{\sigma+2q,(2q)}_{L,X}(\Omega)$. We approximate $F := (L,B)u$ by a certain sequence $(F_k) \subset X^\infty(\Omega) \times (C^\infty(\Gamma))^q$ with respect to the norm in the space $\mathbf{X}_\sigma(\Omega,\Gamma)$. Then

$$\lim_{k\to\infty} \Pi F_k = \Pi F = F = (L,B)u \quad \text{in} \quad \mathbf{X}_\sigma(\Omega,\Gamma), \tag{4.192}$$

where, by virtue of (4.191),

$$(\Pi F_k) \subset X^\infty(\Omega) \times (C^\infty(\Gamma))^q \tag{4.193}$$

The Fredholm operator (4.189) naturally generates the isomorphism

$$\Lambda_0 := (L,B) : D^{\sigma+2q,(2q)}_{L,X}(\Omega)/N \leftrightarrow \Pi(\mathbf{X}_\sigma(\Omega,\Gamma)).$$

By virtue of (4.192), we have

$$\lim_{k\to\infty} \Lambda_0^{-1} \Pi F_k = \{u + w : w \in N\} \quad \text{in} \quad D^{\sigma+2q,(2q)}_{L,X}(\Omega)/N.$$

Therefore, there exists a sequence of representatives $u_k \in D^{\sigma+2q,(2q)}_{L,X}(\Omega)$ of the cosets $\Lambda_0^{-1} \Pi F_k$ such that

$$\lim_{k\to\infty} u_k = u \quad \text{in} \quad D^{\sigma+2q,(2q)}_{L,X}(\Omega). \tag{4.194}$$

Moreover, by virtue of (4.193), we have

$$(L,B)u_k = \Pi F_k \in C^\infty(\overline{\Omega}) \times (C^\infty(\Gamma))^q.$$

Hence, using Theorem 4.18 and the Sobolev embedding theorem, we get

$$u_k \in \bigcap_{s>2q} H^{s,(2q)}(\Omega) = \bigcap_{s>2q} H^s(\Omega) = C^\infty(\overline{\Omega}).$$

Thus, $(u_k) \subset D^\infty_{L,X}(\Omega)$ in (4.194). This proves that $D^\infty_{L,X}(\Omega)$ is dense in $D^{\sigma+2q,(2q)}_{L,X}(\Omega)$.

Further, we separately consider the case of $-2q-1/2 \le \sigma < 0$ and the case of $\sigma < -2q-1/2$.

The first case, namely, $-2q-1/2 \le \sigma < 0$. Then $H^{\sigma+2q,(0)}(\Omega) = \mathrm{H}^{\sigma+2q}(\Omega)$. Indeed, since $H^\lambda_0(\Omega) = H^\lambda(\Omega)$ for $0 \le \lambda \le 1/2$ (see [121, Chap. 1, Theorem 11.1]), we have

$$H^{s,(0)}(\Omega) = (H^{-s}(\Omega))' = (H^{-s}_0(\Omega))' = \mathrm{H}^s(\Omega) \quad \text{for} \quad -1/2 \le s < 0.$$

This yields

$$H^{s,(0)}(\Omega) = \mathrm{H}^s(\Omega) \quad \text{for} \quad s \ge -1/2 \tag{4.195}$$

with equality of norms.

We use Theorem 4.23 and consider the mapping $I_0 : u \mapsto u_0$, where $u \in D^{\sigma+2q,(2q)}_{L,X}(\Omega)$ and $(u_0, f) := I_L u$. The mapping defines the isomorphism

$$I_0 : D^{\sigma+2q,(2q)}_{L,X}(\Omega) \leftrightarrow D^{\sigma+2q}_{L,X}(\Omega). \tag{4.196}$$

Indeed, for arbitrary distributions

$$u_0 \in H^{\sigma+2q,(0)}(\Omega) = \mathrm{H}^{\sigma+2q}(\Omega) \quad \text{and} \quad f \in X^\sigma(\Omega) \hookrightarrow H^{\sigma,(0)}(\Omega),$$

conditions (4.115) and (4.116) are equivalent by virtue of Lemma 4.5. Condition (4.116) means that $Lu_0 = f$ in the sense of the equality of distributions given in the domain Ω. Hence, using Theorem 4.23, we obtain the equality $I_0(D^{\sigma+2q,(2q)}_{L,X}(\Omega)) = D^{\sigma+2q}_{L,X}(\Omega)$. In addition, for $u \in D^{\sigma+2q,(2q)}_{L,X}(\Omega)$ and $(u_0, f) = I_L u$, we have the following equivalence of norms:

$$\begin{aligned} \|u\|^2_{D^{\sigma+2q,(2q)}_{L,X}(\Omega)} &= \|u\|^2_{H^{\sigma+2q,(2q)}(\Omega)} + \|f\|^2_{X^\sigma(\Omega)} \\ &\asymp \|u_0\|^2_{H^{\sigma+2q,(0)}(\Omega)} + \|f\|^2_{H^{\sigma,(0)}(\Omega)} + \|f\|^2_{X^\sigma(\Omega)} \\ &\asymp \|u_0\|^2_{\mathrm{H}^{\sigma+2q}(\Omega)} + \|f\|^2_{X^\sigma(\Omega)} = \|u_0\|^2_{D^{\sigma+2q}_{L,X}(\Omega)}. \end{aligned}$$

Therefore, the mapping I_0 defines isomorphism (4.196).

It follows from properties of operators (4.196) and (4.189) (recall that the latter is denoted by Λ_0) that the operator

$$\Lambda_0 I_0^{-1} : D^{\sigma+2q}_{L,X}(\Omega) \to \mathbf{X}_\sigma(\Omega, \Gamma) \tag{4.197}$$

is bounded and Fredholm. Moreover, its kernel, the range and index are the same as for operator (4.189). Since the operator I_0 establishes a one-to-one mapping of the set $D^{\infty}_{L,X}(\Omega)$ onto itself, this set is dense in the space $D^{\sigma+2q}_{L,X}(\Omega)$ and operator (4.197) is an extension by continuity of the mapping $u \to (Lu, Bu)$, where $u \in D^{\infty}_{L,X}(\Omega)$. Theorem 4.25 is proved in the first case.

The second case, namely, $\sigma < -2q - 1/2$. Then $H^{\sigma+2q,(0)}(\Omega) = H^{\sigma+2q}_{\overline{\Omega}}(\mathbb{R}^n)$. Set $Rw := w \restriction \Omega$ for $w \in \mathcal{D}'(\mathbb{R}^n)$. Let us prove that the mapping $I_0 : u \mapsto Ru_0$, where $u \in D^{\sigma+2q,(2q)}_{L,X}(\Omega)$ and $(u_0, f) := I_L u$, defines isomorphism (4.196) in the considered case (in the first case, $Ru_0 = u_0$). We use Theorem 4.23 and note that (4.115) $\Rightarrow$ (4.116). For an arbitrary $u \in D^{\sigma+2q,(2q)}_{L,X}(\Omega)$, we have $Ru_0 \in \mathrm{H}^{\sigma+2q}(\Omega)$ [see (4.173)], $f = Lu \in X^{\sigma}(\Omega)$ and

$$(Ru_0, L^{+}v)_{\Omega} = (u_0, L^{+}v)_{\Omega} = (f, v)_{\Omega} \quad \text{for all} \quad v \in C^{\infty}_0(\Omega),$$

i.e., $LRu_0 = f$ in the sense of the equality of distributions given in the domain Ω. Therefore, $I_0 u = Ru_0 \in D^{\sigma+2q}_{L,X}(\Omega)$. In addition, by virtue of (4.174) and the definition of the space $H^{\sigma+2q,(2q)}(\Omega)$, we have the estimate

$$\|I_0 u\|^2_{D^{\sigma+2q}_{L,X}(\Omega)} = \|Ru_0\|^2_{\mathrm{H}^{\sigma+2q}(\Omega)} + \|f\|^2_{X^{\sigma}(\Omega)}$$

$$\leq \|u_0\|^2_{H^{\sigma+2q}(\mathbb{R}^n)} + \|f\|^2_{X^{\sigma}(\Omega)} \leq \|u\|^2_{D^{\sigma+2q,(2q)}_{L,X}(\Omega)}.$$

Therefore, the operator $I_0 : D^{\sigma+2q,(2q)}_{L,X}(\Omega) \to D^{\sigma+2q}_{L,X}(\Omega)$ is bounded.

We show that this operator is bijective. Let $\omega \in D^{\sigma+2q}_{L,X}(\Omega)$ and let $f := L\omega \in X^{\sigma}(\Omega)$. By virtue of (4.173), there exists a distribution $u_0 \in H^{\sigma+2q}_{\overline{\Omega}}(\mathbb{R}^n)$ such that $\omega = Ru_0$. In this case, equality (4.116) is true, namely,

$$(u_0, L^{+}v)_{\Omega} = (\omega, L^{+}v)_{\Omega} = (f, v)_{\Omega} \quad \text{for all} \quad v \in C^{\infty}_0(\Omega).$$

According to Theorem 4.24, for the distributions $u_0 \in H^{\sigma+2q,(0)}(\Omega)$ and $f \in X^{\sigma}(\Omega) \hookrightarrow H^{\sigma,(0)}(\Omega)$, there exists a unique pair $(u_0^*, f) \in K_{\sigma+2q,L}(\Omega)$ that satisfies condition (4.118). (In the Sobolev case of $\varphi \equiv 1$, this theorem is true for every $s \in \mathbb{R}$ as was shown by Ya. A. Roitberg [209, Theorem 6.2.1].) Using Theorem 4.23, we can write

$$u^* := I_L^{-1}(u_0^*, f) \in D^{\sigma+2q,(2q)}_{L,X}(\Omega) \quad \text{and} \quad I_0 u^* = Ru_0^* = Ru_0 = \omega.$$

The element u^* is a unique preimage of the distribution ω in the mapping I_0. Indeed, if $I_0 u' = \omega$ for a certain $u' \in D^{\sigma+2q,(2q)}_{L,X}(\Omega)$, then the pair $(u_0', f') := I_L u' \in K_{\sigma+2q,L}(\Omega)$ satisfies the conditions

$$f' = LRu_0' = L\omega = f \quad \text{and}$$

$$(u_0', v)_{\Omega} = (\omega, v)_{\Omega} = (u_0, v)_{\Omega} \quad \text{for all} \quad v \in C^{\infty}_0(\Omega).$$

Therefore, by virtue of Theorem 4.24, the pairs $(u_0', f') = (u_0', f)$ and (u_0^*, f) are equal and, hence, their preimages u' and u^* in the mapping I_L are equal.

Thus, the linear bounded operator (4.196) is bijective in the considered case. Therefore, by the Banach theorem on inverse operator, it is an isomorphism Now, using the Fredholm property of operator (4.189) and reasoning as in the first case, we complete the proof in the second case.

Theorem 4.25 is proved.

Remark 4.11. A statement analogous to Theorem 4.25 is proved in [126, Theorem 6.16] in the case of half-integer $\sigma \leq -2q$ and the Dirichlet boundary-value problem. Moreover, certain other conditions depending on the considered boundary-value problem are imposed on the space $X^\sigma(\Omega)$. Our condition I_σ is independent of this problem.

Remark 4.12. In Theorem 4.25, as in the individual theorems LM_1 and LM_2, the solution and the right-hand side of the elliptic equation $Lu = f$ are distributions given in Ω. Other individual theorems are proved in the papers by Ya. A. Roitberg [204] and Yu. V. Kostarchuk, Ya. A. Roitberg [97]. In these theorems, u and/or f are not distributions in the domain Ω.

It is obvious that the space $X^\sigma(\Omega) := \{0\}$ satisfies condition I_σ. In this case, Theorem 4.25 describes properties of the semihomogeneous boundary-value problem (4.1), where $f = 0$, and is true for any $\sigma \in \mathbb{R}$ in view of Theorem 3.11. Despite the fact that $\mathrm{H}^s(\Omega) \neq H^s(\Omega)$ for half-integer $s = \sigma + 2q < 0$, we have

$$\{u \in \mathrm{H}^s(\Omega) : \, Lu = 0 \text{ in } \Omega\} = \{u \in H^s(\Omega) : \, Lu = 0 \text{ in } \Omega\}. \tag{4.198}$$

In addition, the norms in the spaces $\mathrm{H}^s(\Omega)$ and $H^s(\Omega)$ are equivalent on the distributions u indicated in (4.198).

Further, we consider various applications of Theorem 4.25 caused by a specific choice of the space $X^\sigma(\Omega)$. We separately study the cases where inner product Sobolev spaces or their weight analogs are taken as $X^\sigma(\Omega)$.

4.4.3 Individual theorem for Sobolev spaces

The theorem given below describes all inner product (nonweight) Sobolev spaces that satisfy condition I_σ.

Theorem 4.26. *Let $\sigma < 0$ and $\lambda \in \mathbb{R}$. The space $X^\sigma(\Omega) := \mathrm{H}^\lambda(\Omega)$ satisfies condition I_σ if and only if*

$$\lambda \geq \max\{\sigma, -1/2\}. \tag{4.199}$$

Proof. Note that, since $X^\sigma(\Omega) = \mathrm{H}^\lambda(\Omega)$, the set $X^\infty(\Omega) = C^\infty(\overline{\Omega})$ is dense in the space $X^\sigma(\Omega)$.

Sufficiency. Assume that inequality (4.199) is true. Then, in view of (4.195),

$$\mathrm{H}^\lambda(\Omega) = H^{\lambda,(0)}(\Omega) \hookrightarrow H^{\sigma,(0)}(\Omega)$$

together with topology. Recall that each function $f \in C^\infty(\overline{\Omega})$ is identified with the functional $(f, \cdot)_\Omega$, which, in turn, is identified with the function $\mathcal{O}f$ considered as an element of the space $H^\sigma_{\overline{\Omega}}(\mathbb{R}^n) = H^{\sigma,(0)}(\Omega)$. Therefore,

$$\|\mathcal{O}f\|_{H^\sigma(\mathbb{R}^n)} \le c\,\|f\|_{\mathrm{H}^\lambda(\Omega)} \quad \text{for any} \quad f \in C^\infty(\overline{\Omega}),$$

where the number $c > 0$ is independent of f. Sufficiency is proved.

Necessity. Assume that the space $X^\sigma(\Omega) := \mathrm{H}^\lambda(\Omega)$ satisfies condition I_σ. If $\lambda \ge 0$, then inequality (4.199) is true. For this reason, we restrict ourselves to the condition $\lambda < 0$. Operator (4.187) defines the dense continuous embedding $\mathrm{H}^\lambda(\Omega) \hookrightarrow H^{\sigma,(0)}(\Omega)$. This yields

$$H^{-\sigma}(\Omega) = (H^{\sigma,(0)}(\Omega))' \subseteq (\mathrm{H}^\lambda(\Omega))' = H_0^{-\lambda}(\Omega).$$

Therefore, $-\sigma \ge -\lambda$. In addition, $-\lambda \le 1/2$ because if $-\lambda > 1/2$, then the function $f \equiv 1 \in H^{-\sigma}(\Omega)$ does not belong to the space $H_0^{-\lambda}(\Omega)$ by virtue of Theorem 3.20(i). Thus, σ satisfies inequality (4.199). Necessity is proved.

Theorem 4.26 is proved.

The following individual theorem on solvability of the boundary-value problem (4.1) in Sobolev spaces follows from Theorems 4.25 and 4.26.

Theorem 4.27. *Let $\sigma < 0$ and let $\lambda \ge \max\{\sigma, -1/2\}$. Then mapping* (4.176) *extends uniquely (by continuity) to the Fredholm bounded operator*

$$\begin{aligned}(L, B) : \{u \in \mathrm{H}^{\sigma+2q}(\Omega) : Lu \in \mathrm{H}^\lambda(\Omega)\}&\\ \to \mathrm{H}^\lambda(\Omega) \oplus \bigoplus_{j=1}^{q} H^{\sigma+2q-m_j-1/2}(\Gamma)&\end{aligned} \tag{4.200}$$

whose domain is a Hilbert space with respect to the graph norm

$$\left(\|u\|^2_{\mathrm{H}^{\sigma+2q}(\Omega)} + \|Lu\|^2_{\mathrm{H}^\lambda(\Omega)}\right)^{1/2}.$$

The index of operator (4.200) *is equal to* $\dim\, N - \dim N^+$ *and does not depend on σ and λ.*

Proof. The boundedness and the Fredholm property of operator (4.200) immediately follow from Theorems 4.25 and 4.26, where $X^\sigma(\Omega) := H^\lambda(\Omega)$ and $D^\infty_{L,X}(\Omega) = C^\infty(\overline{\Omega})$. In addition, since the set $\mathcal{O}(C^\infty(\overline{\Omega}))$ identified with $C^\infty(\overline{\Omega})$ is dense in the space $H^\sigma_{\overline{\Omega}}(\mathbb{R}^n) = H^{\sigma,(0)}(\Omega)$, it implies from Theorem 4.25(iv) that the index of operator (4.200) is equal to $\dim N - \dim N^+$ and does not depend on σ and λ.

Theorem 4.27 is proved.

Theorem LM_1 is the special case of Theorem 4.27 where $\sigma = 0$, i.e., $X^\sigma(\Omega) = L_2(\Omega)$.

In Theorem 4.27, certain spaces $X^\sigma(\Omega)$ containing $L_2(\Omega)$ are admitted. The space $X^\sigma(\Omega) = \mathrm{H}^{-1/2}(\Omega)$ with $\sigma \le -1/2$ is the widest among them.

Note that, in the case of $-1/2 \le \sigma = \lambda < 0$, the domain of operator (4.200) is independent of L. Indeed, if $-1/2 < \sigma = \lambda < 0$, then, since the number σ is not half-integer, we have the bounded operator $L : \mathrm{H}^{\sigma+2q}(\Omega) \to \mathrm{H}^\sigma(\Omega)$ (see [121, Chap. 1, Proposition 12.1]). This implies that the domain of operator (4.200) coincides with $\mathrm{H}^{\sigma+2q}(\Omega)$ and is independent of L. In this case, Theorem 4.27 coincides with J.-L. Lions and E. Magenes' theorem [121, Chap. 2, Theorem 7.5], which is proved under assumption that $N = N^+ = \{0\}$.

If $\sigma = \lambda = -1/2$, then the previous arguments are not correct because the space $\mathrm{H}^{-1/2}(\Omega)$ is narrower than $L(\mathrm{H}^{2q-1/2}(\Omega))$. However, it follows from Theorem 4.25(i), equality (4.195), and Theorem 4.23 that the domain of operator (4.200) is the completion of the set of functions $u \in C^\infty(\overline{\Omega})$ with respect to the norm

$$\|u\|^2_{\mathrm{H}^{2q-1/2}(\Omega)} + \|Lu\|^2_{\mathrm{H}^{-1/2}(\Omega)} = \|u\|^2_{H^{2q-1/2,(0)}(\Omega)} + \|Lu\|^2_{H^{-1/2,(0)}(\Omega)}$$

$$\asymp \|u\|_{H^{2q-1/2,(2q)}(\Omega)}.$$

Therefore, it coincides with the space $H^{2q-1/2,(2q)}(\Omega)$, which is independent of L.

In the conclusion of this section, note that the Hörmander spaces $H^{\lambda,\varphi}(\Omega)$, where $\lambda > \max\{\sigma, -1/2\}$ and $\varphi \in \mathcal{M}$, also satisfy condition I_σ for $\sigma < 0$. Individual theorems involving Hörmander spaces will be considered in Section 4.5.

4.4.4 Individual theorem for weight spaces

In Theorem 4.26, always $X^\sigma(\Omega) \subseteq \mathrm{H}^{-1/2}(\Omega)$. The space $X^\sigma(\Omega)$ that contains a wide class of distributions $f \notin \mathrm{H}^{-1/2}(\Omega)$ and satisfies condition I_σ can be obtained using certain weight spaces $\varrho\mathrm{H}^\sigma(\Omega)$.

The function $\varrho : \Omega \to \mathbb{C}$ is called a *multiplier* in the space $\mathrm{H}^\lambda(\Omega)$, where $\lambda \ge 0$, if the operator of multiplication by ϱ continuously maps this space into itself. We denote this operator by M_ϱ. The description of the class of all multipliers in the space $\mathrm{H}^{\lambda}(\Omega) = H^{\lambda}(\Omega)$ for $\lambda \geq 0$ is given in the monograph by V. G. Maz'ya and T. O. Shaposhnikova [131, Sec. 6.3.3].

Let $\sigma < -1/2$. Consider the following condition on the function ϱ.

Condition II_σ. The function ϱ is a multiplier in the space $\mathrm{H}^{-\sigma}(\Omega)$, and

$$D_\nu^j \varrho = 0 \text{ on } \Gamma \text{ for all } j \in \mathbb{Z} \text{ with } 0 \leq j < -\sigma - 1/2. \tag{4.201}$$

Note that if ϱ is a multiplier in the space $\mathrm{H}^{-\sigma}(\Omega)$, then obviously $\varrho \in \mathrm{H}^{-\sigma}(\Omega)$. Therefore, by virtue of Theorem 3.5, there exists a trace

$$(D_\nu^j \varrho)\restriction\Gamma \in H^{-\sigma-j-1/2}(\Gamma)$$

for any integer $j \geq 0$ such that $-\sigma - j - 1/2 > 0$. Therefore, condition II_σ is well-formulated.

Theorem 4.28. *Let $\sigma < -1/2$ and let a function $\varrho \in C^\infty(\Omega)$ be positive. The space $X^\sigma(\Omega) := \varrho\mathrm{H}^\sigma(\Omega)$ satisfies condition I_σ if and only if the function ϱ satisfies condition II_σ.*

To prove this theorem, we need the following lemma.

Lemma 4.8. *Let $\sigma < -1/2$. Multiplication by a function ϱ is a bounded operator*

$$M_\varrho : H^{-\sigma}(\Omega) \to H_0^{-\sigma}(\Omega) \tag{4.202}$$

if and only if condition II_σ is satisfied.

Proof. *Necessity.* If the multiplication by ϱ defines the bounded operator (4.202), then ϱ is a multiplier in the space $H^{-\sigma}(\Omega)$ and, in addition, $\varrho \in H_0^{-\sigma}(\Omega)$. As known, [258, Theorem 4.7.1 (a)],

$$\varrho \in H_0^{-\sigma}(\Omega) \Leftrightarrow (\varrho \in H^{-\sigma}(\Omega) \text{ and (4.201) is true}) \tag{4.203}$$

This yields condition II_σ. Necessity is proved.

Sufficiency. Let the function ϱ satisfy condition II_σ. It is only necessary to prove that $\varrho u \in H_0^{-\sigma}(\Omega)$ for any $u \in H^{-\sigma}(\Omega)$. By virtue of (4.203), condition II_σ yields inclusion $\varrho \in H_0^{-\sigma}(\Omega)$. Let sequences $(u_k) \subset C^\infty(\overline{\Omega})$ and $(\varrho_j) \subset C_0^\infty(\Omega)$ be such that $u_k \to u$ and $\varrho_j \to \varrho$ in $H^{-\sigma}(\Omega)$. Since the functions ϱ and u_k are multipliers in the space $H^{-\sigma}(\Omega)$, we have

$$\lim_{k\to\infty} (\varrho u_k) = \varrho u \quad \text{and} \quad \lim_{j\to\infty} (\varrho_j u_k) = \varrho u_k \quad \text{for all } k \quad \text{in } H^{-\sigma}(\Omega).$$

In view of $\varrho_j u_k \in C_0^\infty(\Omega)$, this implies that $\varrho u \in H_0^{-\sigma}(\Omega)$. Sufficiency is proved.

Lemma 4.8 is proved.

Proof of Theorem 4.28. By condition, $\sigma < -1/2$ and the function $\varrho \in C^\infty(\Omega)$ is positive. Let M_ϱ and $M_{\varrho^{-1}}$ denote the operators of multiplication by ϱ and ϱ^{-1}, respectively. We have the isomorphism $M_\varrho : \mathrm{H}^\sigma(\Omega) \leftrightarrow \varrho\mathrm{H}^\sigma(\Omega)$. Using this result and the density of $C_0^\infty(\Omega)$ in $\mathrm{H}^\sigma(\Omega)$, we conclude that $C_0^\infty(\Omega)$ is dense in $X^\sigma(\Omega) := \varrho\mathrm{H}^\sigma(\Omega)$. Therefore, the wider set $X^\infty(\Omega)$ is dense in $X^\sigma(\Omega)$.

Let us define the inner product space

$$\varrho^{-1}H_0^{-\sigma}(\Omega) := \{f = \varrho^{-1}v : v \in H_0^{-\sigma}(\Omega)\},$$

$$(f_1, f_2)_{\varrho^{-1}H_0^{-\sigma}(\Omega)} := (\varrho f_1, \varrho f_2)_{H^{-\sigma}(\Omega)}.$$

We have the isomorphism

$$M_{\varrho^{-1}} : H_0^{-\sigma}(\Omega) \leftrightarrow \varrho^{-1}H_0^{-\sigma}(\Omega). \tag{4.204}$$

Therefore, the space $\varrho^{-1}H_0^{-\sigma}(\Omega)$ is complete, and the set $C_0^\infty(\Omega)$ is dense in this space.

Note that

$$(\varrho^{-1}H_0^{-\sigma}(\Omega))' = \varrho\mathrm{H}^\sigma(\Omega) \quad \text{with equality of norms.} \tag{4.205}$$

Indeed, passing in (4.204) to the adjoint operator, we obtain the isomorphism

$$M_{\varrho^{-1}} : (\varrho^{-1}H_0^{-\sigma}(\Omega))' \leftrightarrow (H_0^{-\sigma}(\Omega))' = \mathrm{H}^\sigma(\Omega).$$

Using this result and the definition of the space $\varrho\mathrm{H}^\sigma(\Omega)$, we obtain one more isomorphism

$$I = M_\varrho M_{\varrho^{-1}} : (\varrho^{-1}H_0^{-\sigma}(\Omega))' \leftrightarrow \varrho\mathrm{H}^\sigma(\Omega),$$

where I is the identity operator. Thus, we prove relation (4.205) because the used isomorphisms are isometric.

Now we can complete the proof by using the following arguments. By virtue of Lemma 4.8, condition II_σ is equivalent to the boundedness of operator (4.202), which, in view of (4.204), is equivalent to the continuous embedding $H^{-\sigma}(\Omega) \hookrightarrow \varrho^{-1}H_0^{-\sigma}(\Omega)$. This embedding is dense. By virtue of (4.205), it is equivalent to the dense continuous embedding

$$\varrho\mathrm{H}^\sigma(\Omega) = (\varrho^{-1}H_0^{-\sigma}(\Omega))' \hookrightarrow (H^{-\sigma}(\Omega))' = H_{\overline{\Omega}}^\sigma(\mathbb{R}^n).$$

Finally, the continuous embedding $\varrho\mathrm{H}^\sigma(\Omega) \hookrightarrow H_{\overline{\Omega}}^\sigma(\mathbb{R}^n)$ is equivalent to condition I_σ. Note that the last embedding is dense because the set $C_0^\infty(\Omega)$ is dense in the space $H_{\overline{\Omega}}^\sigma(\mathbb{R}^n)$. Thus, the equivalence of conditions II_σ and I_σ is proved for the space $X^\sigma(\Omega) = \varrho\mathrm{H}^\sigma(\Omega)$.

Theorem 4.28 is proved.

Theorems 4.25 and 4.28 lead to the following individual theorem on solvability of the boundary-value problem (4.1) in weight Sobolev spaces.

Theorem 4.29. *Let $\sigma < -1/2$ and let a positive function $\varrho \in C^\infty(\Omega)$ satisfy condition II_σ. Then the mapping $u \to (Lu, Bu)$, where $u \in C^\infty(\overline{\Omega})$ and $Lu \in \varrho\mathrm{H}^\sigma(\Omega)$, extends uniquely (by continuity) to the Fredholm bounded operator*

$$(L,B) : \{u \in \mathrm{H}^{\sigma+2q}(\Omega) : Lu \in \varrho\mathrm{H}^\sigma(\Omega)\}$$

$$\to \varrho\mathrm{H}^\sigma(\Omega) \oplus \bigoplus_{j=1}^{q} H^{\sigma+2q-m_j-1/2}(\Gamma) \tag{4.206}$$

whose domain is a Hilbert space with respect to the graph norm

$$\big(\|u\|^2_{\mathrm{H}^{\sigma+2q}(\Omega)} + \|\varrho^{-1}Lu\|^2_{\mathrm{H}^\sigma(\Omega)}\big)^{1/2}.$$

The index of operator (4.206) *is equal to* $\dim N - \dim N^+$ *and does not depend on σ and ϱ.*

Proof. The boundedness and the Fredholm property of operator (4.206) result immediately from Theorems 4.25 and 4.28, where $X^\sigma(\Omega) := \varrho\mathrm{H}^\sigma(\Omega)$. In addition, since the set $\mathcal{O}(X^\infty(\Omega))$ containing $C_0^\infty(\Omega)$ is dense in the space $H^\sigma_{\overline{\Omega}}(\mathbb{R}^n)$, it follows by Theorem 4.25(iv) that the index of operator (4.206) is equal to $\dim N - \dim N^+$ and does not depend on σ and ϱ.

Theorem 4.29 is proved.

The following theorem gives an important example of a positive function $\varrho \in C^\infty(\Omega)$ that satisfies condition II_σ.

Theorem 4.30. *Let a number $\sigma < -1/2$ and a function ϱ_1 that satisfies condition* (4.184) *be given. Assume that either $\delta \geq -\sigma - 1/2 \in \mathbb{Z}$ or $\delta > -\sigma - 1/2 \notin \mathbb{Z}$. Then the function $\varrho := \varrho_1^\delta$ satisfies condition II_σ.*

This implies that Theorem LM_2 for half-integer $\sigma < -1/2$ is a special case of the individual theorem 4.29. Recall that, in Theorem LM_2, the weight function $\varrho := \varrho_1^{-\sigma}$ is used.

Proof of Theorem 4.30. The function $\varrho = \varrho_1^\delta$ satisfies condition (4.201) because $\varrho_1 = 0$ on Γ and $\delta \geq -\sigma - 1/2$. Therefore, it remains to prove that ϱ_1^δ is a multiplier in the space $\mathrm{H}^{-\sigma}(\Omega) = H^{-\sigma}(\Omega)$. If the positive number δ is integer, then the function ϱ_1^δ belongs to the space $C^\infty(\overline{\Omega})$ and, hence, is a multiplier in $H^{-\sigma}(\Omega)$. Further, assume that $\delta \notin \mathbb{Z}$. Then, by condition, $\delta > -\sigma - 1/2$.

It is easy to verify that the function $\eta_\delta(t) := t^\delta$, $0 \leq t \leq 1$, belongs to the Sobolev space $H^{-\sigma}((0,1))$ (this will be shown in the next paragraph). Then this function has a certain extension over $\mathbb{R}$ belonging to $H^{-\sigma}(\mathbb{R})$. We preserve the notation η_δ for the extension. By R. Strichartz's theorem [250] (see also [131, Sec. 2.2.9]) any function from the space $H^{-\sigma}(\mathbb{R})$ is a multiplier in this space if $-\sigma > 1/2$. Therefore, η_δ is a multiplier in $H^{-\sigma}(\mathbb{R})$. Then the function

$\eta_{\delta,n}(t',t_n) := \eta_\delta(t_n)$ of arguments $t' \in \mathbb{R}^{n-1}$ and $t_n \in \mathbb{R}$ is a multiplier in $H^{-\sigma}(\mathbb{R}^n)$ [131, Sec. 2.4, Proposition 5]. It coincides with ϱ_1^δ in the special local coordinates (x',t_n) near the boundary Γ. Here, x' is the coordinate of a point in the local chart of the surface Γ, and t_n is the distance to Γ. This implies that ϱ_1^δ is a multiplier in each space $H^{-\sigma}(\Omega \cap V_j)$, where $\{V_j : j = 1,\dots,r\}$ is a finite system of balls in $\mathbb{R}^n$, of a sufficiently small radius ε, covering the boundary Γ [131, Sec. 6.4.1, Lemma 3]. We complement this system with the set $V_0 := \{x \in \Omega : \mathrm{dist}(x,\Gamma) > \varepsilon/2\}$ and obtain the finite open covering of the closed domain $\overline{\Omega}$. Let functions $\chi_j \in C_0^\infty(V_j)$, $j = 0,1,\dots,r$, form a partition of unity on $\overline{\Omega}$ subordinated to this covering. Since the multiplication by a function in the class $C_0^\infty(V_j)$ is a bounded operator on the space $H^{-\sigma}(\Omega\cap V_j)$, the function $\chi_j\varrho_1^\delta$ is a multiplier in this space. Therefore, the function $\varrho_1^\delta = \sum_{j=0}^r \chi_j\varrho_1^\delta$ is a multiplier in $H^{-\sigma}(\Omega)$.

It remains to show that $\eta_\delta \in H^{-\sigma}((0,1))$. We use the internal description of the space $H^{-\sigma}((0,1))$. If $-\sigma \in \mathbb{Z}$, then the inclusion $\eta_\delta \in H^{-\sigma}((0,1))$ is equivalent to the pair of inclusions $\eta_\delta \in L_2((0,1))$ and $\eta_\delta^{(-\sigma)} \in L_2((0,1))$. It is obvious that the last two inclusions are true for $\delta > -\sigma - 1/2$. Therefore, $\eta_\delta \in H^{-\sigma}((0,1))$ in the considered case. If $-\sigma \notin \mathbb{Z}$, then the inclusion $\eta_\delta \in H^{-\sigma}((0,1))$ is equivalent to the following: $\eta_\delta \in H^{[-\sigma]}((0,1))$ and

$$\int_0^1\int_0^1 \frac{|D_t^{[-\sigma]}t^\delta - D_\tau^{[-\sigma]}\tau^\delta|^2}{|t-\tau|^{1+2\{-\sigma\}}}\,dt\,d\tau < \infty \tag{4.207}$$

(see, e.g., [1, Theorem 7.48]). As usual, $[-\sigma]$ and $\{-\sigma\}$ are, respectively, the integral and fractional parts of the number $-\sigma$. Since $\delta > [-\sigma] - 1/2$, we conclude that $\eta_\delta \in H^{[-\sigma]}((0,1))$ by the result proved above. In addition, inequality (4.207) is true by virtue of the following elementary lemma, which will be proved at the end of this subsection.

Lemma 4.9. *Let $\alpha,\beta,\gamma \in \mathbb{R}$ and, furthermore, $\alpha \neq 0$ and $\gamma > 0$. Then*

$$I(\alpha,\beta,\gamma) := \int_0^1\int_0^1 \frac{|t^\alpha - \tau^\alpha|^\gamma}{|t-\tau|^\beta}\,dt\,d\tau < \infty \tag{4.208}$$

if and only if all three inequalities

$$\alpha\gamma - \beta > -2, \quad \gamma - \beta > -1, \quad \alpha\gamma > -1 \tag{4.209}$$

are true.

Indeed, the double integral in (4.207) is equal to $c\,I(\alpha,\beta,\gamma)$, where c is a certain positive number and $\alpha = \delta - [-\sigma]$, $\beta = 1 + 2\{-\sigma\}$, and $\gamma = 2$. For the numbers α, β, and γ, inequalities (4.209) are true, namely,

$$\alpha\gamma - \beta = 2(\delta + \sigma) - 1 > -2,$$

$$\gamma - \beta = 1 - 2\{-\sigma\} > -1,$$

$$\alpha\gamma = 2(\delta - [-\sigma]) > -1.$$

In the first and third inequalities, we use the condition $\delta > -\sigma - 1/2$. Thus, $\eta_\delta \in H^{-\sigma}((0,1))$ in the case of noninteger $\sigma - 1/2$ as well.

Theorem 4.30 is proved.

It remains to prove Lemma 4.9.

Proof of Lemma 4.9. Changing variable $\lambda := \tau/t$ in the inner integral, we conclude after obvious transformations that

$$I(\alpha,\beta,\gamma) = 2\int_0^1 dt \int_0^t \frac{|t^\alpha - \tau^\alpha|^\gamma}{|t-\tau|^\beta}\, d\tau = 2\int_0^1 t^{\alpha\gamma-\beta+1} dt \int_0^1 \frac{|1-\lambda^\alpha|^\gamma}{|1-\lambda|^\beta}\, d\lambda.$$

Here, the integral with respect to t is finite if and only if $\alpha\gamma - \beta > -2$, and the integral with respect to λ is finite if and only if $\alpha\gamma > -1$ and $\gamma - \beta > -1$. Therefore, (4.208) $\Leftrightarrow$ (4.209).

Lemma 4.9 is proved.

4.5 Hörmander spaces and individual theorems on solvability

We discuss analogs of individual theorems 4.25, 4.27, and 4.29 for Hörmander spaces that form the refined scale.

4.5.1 Key individual theorem for the refined scale

Let $\sigma < 0$ and let $\varphi \in \mathcal{M}$. Assume that a Hilbert space $X^{\sigma,\varphi}(\Omega)$ embedded continuously in $\mathcal{D}'(\Omega)$ be given. Consider the following analog of condition I_σ:

Condition $\mathbf{I}_{\sigma,\varphi}$. The set $X^\infty(\Omega) = X^{\sigma,\varphi}(\Omega) \cap C^\infty(\overline{\Omega})$ is dense in $X^{\sigma,\varphi}(\Omega)$, and there exists a number $c > 0$ such that

$$\|\mathcal{O}f\|_{H^{\sigma,\varphi}(\mathbb{R}^n)} \le c\,\|f\|_{X^{\sigma,\varphi}(\Omega)} \quad \text{for any} \quad f \in X^\infty(\Omega). \tag{4.210}$$

We set

$$D^{\sigma+2q,\varphi}_{L,X}(\Omega) := \{u \in H^{\sigma+2q,\varphi}(\Omega) : Lu \in X^{\sigma,\varphi}(\Omega)\}.$$

The space $D^{\sigma+2q,\varphi}_{L,X}(\Omega)$ is endowed with the graph inner product

$$(u_1,u_2)_{D^{\sigma+2q,\varphi}_{L,X}(\Omega)} := (u_1,u_2)_{H^{\sigma+2q,\varphi}(\Omega)} + (Lu_1,Lu_2)_{X^{\sigma,\varphi}(\Omega)}.$$

This space is complete, which is proved by analogy with the case of $\varphi\equiv 1$ (see Subsection 4.4.1).

Let us formulate the key individual theorem.

Theorem 4.31. *Let $\varphi\in\mathcal{M}$ and let a number $\sigma<0$ be such that*

$$\sigma+2q\neq 1/2-k \quad \textit{for any} \quad k\in\mathbb{N}. \tag{4.211}$$

Let $X^{\sigma,\varphi}(\Omega)$ be an arbitrary Hilbert space that is continuously embedded in $\mathcal{D}'(\Omega)$ and satisfies condition $\mathrm{I}_{\sigma,\varphi}$. Then the following assertions are true:

(i) *The set*

$$D^{\infty}_{L,X}(\Omega) := \{u\in C^{\infty}(\overline{\Omega}) : Lu\in X^{\sigma,\varphi}(\Omega)\}$$

is dense in the space $D^{\sigma+2q,\varphi}_{L,X}(\Omega)$.

(ii) *The mapping $u\to(Lu,Bu)$, where $u\in D^{\infty}_{L,X}(\Omega)$, extends uniquely (by continuity) to the bounded linear operator*

$$\begin{aligned}(L,B):\ D^{\sigma+2q,\varphi}_{L,X}(\Omega)\to X^{\sigma,\varphi}(\Omega)\oplus\bigoplus_{j=1}^{q} H^{\sigma+2q-m_j-1/2,\varphi}(\Gamma)\\ =: \mathbf{X}_{\sigma,\varphi}(\Omega,\Gamma).\end{aligned} \tag{4.212}$$

(iii) *Operator* (4.212) *is Fredholm. Its kernel coincides with N, and the range consists of all vectors $(f,g_1,\ldots,g_q)\in\mathbf{X}_{\sigma,\varphi}(\Omega,\Gamma)$ that satisfy condition* (4.4).

(iv) *If the set $\mathcal{O}(X^{\infty}(\Omega))$ is dense in the space $H^{\sigma,\varphi}_{\overline{\Omega}}(\mathbb{R}^n)$, then the index of operator* (4.212) *is equal to* $\dim N-\dim N^{+}$.

This theorem can be proved by analogy with Theorem 4.25 if we use Theorems 4.14, 4.18, 4.23, and 4.24 for an arbitrary $\varphi\in\mathcal{M}$ and, moreover, apply Theorem 3.21 instead of relations (4.173) and (4.174). The additional condition (4.211) being absent in Theorem 4.25 is caused by the circumstance that the parameter $s=\sigma+2q$ is not half-integer in Theorems 3.21 and 4.24.

4.5.2 Other individual theorems

We give important applications of Theorem 4.31 caused by a specific choice of the space $X^{\sigma,\varphi}(\Omega)$. Let $\sigma<0$ and $\varphi\in\mathcal{M}$.

The space $X^{\sigma,\varphi}(\Omega) := \{0\}$ satisfies condition $\mathrm{I}_{\sigma,\varphi}$. In this case, Theorem 4.31 describes properties of the semihomogeneous boundary-value problem (4.1), where $f = 0$, and is true for any $s \in \mathbb{R}$, which was established in Theorem 3.11.

The space $X^{\sigma,\varphi}(\Omega) := L_2(\Omega)$ satisfies condition $\mathrm{I}_{\sigma,\varphi}$ as well. As has been mentioned in Subsection 4.4.1, this choice of the space $X^{\sigma,\varphi}(\Omega)$ is important in the spectral theory of elliptic operators

As $X^{\sigma,\varphi}(\Omega)$, we admit the choice of several Hörmander spaces. In view of Theorem 4.17, the case $\sigma < -1/2$ is interesting here. Every Hörmander space $X^{\sigma,\varphi}(\Omega) := H^{\lambda,\eta}(\Omega)$, where $\lambda > -1/2$ and $\eta \in \mathcal{M}$, satisfies condition $\mathrm{I}_{\sigma,\varphi}$ for $\sigma < -1/2$. Indeed, by virtue of Theorem 3.9(i) and (iv), the space $H^{\lambda,\eta}(\Omega) = H^{\lambda,\eta,(0)}(\Omega)$ is continuously embedded in $H^{\sigma,\varphi,(0)}(\Omega) = H^{\sigma,\varphi}_{\overline{\Omega}}(\mathbb{R}^n)$. This yields inequality (4.210) in which $X^{\infty}(\Omega) = C^{\infty}(\overline{\Omega})$. Setting $X^{\sigma,\varphi}(\Omega) := H^{\lambda,\eta}(\Omega)$ in the key theorem 4.31, we obtain the following individual theorem on solvability of the boundary-value problem 4.1 in Hörmander spaces.

Theorem 4.32. *Let a number $\sigma < -1/2$ satisfy condition* (4.211) *and let $\lambda > -1/2$ and $\varphi, \eta \in \mathcal{M}$. Then the mapping $u \mapsto (Lu, Bu)$, where $u \in C^{\infty}(\overline{\Omega})$, extends uniquely (by continuity) to the Fredholm bounded operator*

$$(L, B) : \{u \in H^{\sigma+2q,\varphi}(\Omega) : Lu \in H^{\lambda,\eta}(\Omega)\}$$

$$\to H^{\lambda,\eta}(\Omega) \oplus \bigoplus_{j=1}^{q} H^{\sigma+2q-m_j-1/2,\varphi}(\Gamma) \tag{4.213}$$

whose domain is a Hilbert space with respect to the graph norm

$$\left(\|u\|^2_{H^{\sigma+2q,\varphi}(\Omega)} + \|Lu\|^2_{H^{\lambda,\eta}(\Omega)} \right)^{1/2}.$$

The index of operator (4.213) *is equal to* $\dim N - \dim N^{+}$ *and does not depend on the parameters σ, φ and λ, η.*

Note that, in Theorem 4.32, the solution and the right-hand side of the elliptic equation $Lu = f$ can have different additional smoothness φ and η.

In Theorem 4.32, the space $X^{\sigma,\varphi}(\Omega) := H^{\lambda,\eta}(\Omega)$ lies in $H^{-1/2}(\Omega)$ because $\lambda > -1/2$. A space $X^{\sigma,\varphi}(\Omega)$ that contains a wide class of distributions $f \notin H^{-1/2}(\Omega)$ and satisfies condition $\mathrm{I}_{\sigma,\varphi}$ can be obtained by using the weight Hörmander spaces.

Let $\sigma < -1/2$ and $\varphi \in \mathcal{M}$ and let a function $\varrho \in C^{\infty}(\Omega)$ be positive. We define the inner product space

$$\varrho H^{\sigma,\varphi}(\Omega) := \{f = \varrho v : v \in H^{\sigma,\varphi}(\Omega)\},$$

$$(f_1, f_2)_{\varrho H^{\sigma,\varphi}(\Omega)} := (\varrho^{-1} f_1, \varrho^{-1} f_2)_{H^{\sigma,\varphi}(\Omega)}.$$

The multiplication by ϱ defines the isomorphism

$$M_{\varrho} : H^{\sigma,\varphi}(\Omega) \leftrightarrow \varrho H^{\sigma,\varphi}(\Omega).$$

Therefore, the space $\varrho H^{\sigma,\varphi}(\Omega)$ is complete (Hilbert) and continuously embedded in $\mathcal{D}'(\Omega)$.

If the function ϱ is a multiplier in the space $H^{-\sigma,1/\varphi}(\Omega)$ and if (4.201) is true, then the space $X^{\sigma,\varphi}(\Omega) := \varrho H^{\sigma,\varphi}(\Omega)$ satisfies condition $\mathrm{I}_{\sigma,\varphi}$. This is proved by analogy with Theorem 4.28 with the following difference: assertions (i) and (iii) of Theorem 3.20 are used instead of relations (4.203) and (4.172). Setting $X^{\sigma,\varphi}(\Omega) := \varrho H^{\sigma,\varphi}(\Omega)$ in the key theorem 4.31, we obtain the following individual theorem on solvability of the boundary-value problem (4.1) in spaces connected with weight Hörmander spaces.

Theorem 4.33. *Let a number $\sigma < -1/2$ satisfy condition (4.211), let $\varphi \in \mathcal{M}$, and let a function $\varrho \in C^{\infty}(\Omega)$ be positive. Suppose that ϱ is a multiplier in the space $H^{-\sigma,1/\varphi}(\Omega)$ and satisfy condition (4.201). Then the mapping $u \to (Lu, Bu)$, where $u \in C^{\infty}(\overline{\Omega})$ and $Lu \in \varrho H^{\sigma,\varphi}(\Omega)$, extends uniquely (by continuity) to the Fredholm bounded operator*

$$(L,B) : \{u \in H^{\sigma+2q,\varphi}(\Omega) : Lu \in \varrho H^{\sigma,\varphi}(\Omega)\}$$
$$\to \varrho H^{\sigma,\varphi}(\Omega) \oplus \bigoplus_{j=1}^{q} H^{\sigma+2q-m_j-1/2,\varphi}(\Gamma)$$

whose domain is a Hilbert space with respect to the the graph norm

$$\left(\|u\|^2_{H^{\sigma+2q,\varphi}(\Omega)} + \|\varrho^{-1} Lu\|^2_{H^{\sigma,\varphi}(\Omega)} \right)^{1/2}.$$

The index of this operator is equal to $\dim N - \dim N^{+}$ *and does not depend on σ, φ, and ϱ.*

An important example of the function ϱ that satisfies the condition of Theorem 4.33 is obtained by setting $\varrho := \varrho_1^{\delta}$ for an arbitrarily chosen number $\delta > -\sigma - 1/2$. Recall that ϱ_1 is the function of distance, to the boundary Γ, subordinated to condition (4.184). Indeed, the function $\varrho \in C^{\infty}(\Omega)$ is positive and satisfies condition (4.201). By virtue of Theorem 4.30, this function is a multiplier in the space $H^{-\sigma\mp\varepsilon}(\Omega)$, where the number $\varepsilon > 0$ is such that $\delta > -\sigma \mp \varepsilon - 1/2$. By virtue of the interpolation theorem 3.2, this implies that ϱ is a multiplier in the space $H^{-\sigma,1/\varphi}(\Omega)$. Thus, the function $\varrho = \varrho_1^{\delta}$ satisfies the condition of Theorem 4.33.

4.6 Remarks and Comments

Section 4.1. In the fundamental paper [4], S. Agmon, A. Douglis, and L. Nirenberg proved *a priori* estimates for solutions to the elliptic boundary-value problem in the corresponding pairs of Hölder spaces (of noninteger orders) and positive Sobolev spaces under the assumption that the problem is given in a bounded Euclidean domain with smooth boundary. For positive Sobolev spaces, these estimates are proved independently by F. E. Browder [29], L. N. Slobodetskii [238, 239, 241], M. Schechter [222], and others (see references in the M. S. Agranovich's review [11]). *A priori* estimates are equivalent to the Fredholm property of the operator generated by the problem in the mentioned pairs of spaces. If the boundary-value problem is regular elliptic, then the defect subspace and the range of the operator can be described with the use of differential expressions connected with formally adjoint problem. It is also proved that a priory estimates for solutions in pairs of Sobolev inner product spaces lead to the validity of the Lopatinskii complementing condition for a collection of boundary expressions, i.e., to the ellipticity of the boundary-value problem. These questions are systematically discussed, e.g., in the monographs by S. Agmon [3], Yu. M. Berezansky [21], L. Hörmander [81, 86], J.-L. Lions and E. Magenes [121], O. I. Panich [184], M. Schechter [227], H. Triebel [258], and the review by M. S. Agranovich [11]. The index of the elliptic boundary-value problem was calculated by M. F. Atiyah and R. Bott in [17] using the fundamental M. F. Atiyah and I. M. Singer's [18] formula for the index of the elliptic matrix pseudodifferential operator. For the detail description of the theory of index of elliptic boundary-value problems, see the monograph by S. Rempel and B.-W. Schulze [198].

Theorems on solvability of regular elliptic boundary-value problems are also proved for other scales of positive function spaces. G. Shlenzak proved theorems for a certain class of Hörmander inner product spaces [231]. H. Triebel [258, 256] and J. Franke [58] proved theorems for Lizorkin–Triebel and Nikol'skii–Besov both Banach and non-Banach spaces.

Theorems on solvability of elliptic boundary-value problems find various and important applications. Among them, we mention theorems on increase in smoothness of the solutions of the elliptic equation up to the boundary of the domain, applications to the investigation of the Green function of the elliptic boundary-value problem, problems of optimal control, nonlocal elliptic boundary-value problems, some classes of nonlinear elliptic boundary-value problems, *etc.* See the monographs by Yu. M. Berezansky [21], Yu. M. Berezansky, G. F. Us, and Z. G. Sheftel [23], O. A. Ladyzhenskaya and N. N. Ural'tseva [111], J.-L. Lions [118, 117], J.-L. Lions and E. Magenes [121], Ya. A. Roitberg [209, 210], and I. V. Skrypnik [237], the review by M. S. Agranovich [11] and the reference therein.

S. Agmon and L. Nirenberg [2, 6], M. S. Agranovich and M. I. Vishik [13] selected a subclass of elliptic boundary-value problems, depending on the complex-valued parameter, that have the following important property. For sufficiently large absolute values of the parameter, the operator corresponding to the problem sets isomorphisms in appropriate pairs of Sobolev spaces. Moreover, the norm of the operator admits a two-sided estimate with constants not depending on parameter. Not that the index of this problem is equal to zero for all values of the parameter. Parameter-elliptic boundary-value problems have important applications in the theory of parabolic problems and in the spectral theory of differential operators. Various wider or different classes of elliptic operators and elliptic boundary-value problems with parameter are studied by M. S. Agranovich [8, 9], G. Grubb [69, Chap. 2], R. Denk, R. Mennicken, and L. R. Volevich [38, 39, 42], A. N. Kozhevnikov [99, 100, 101], O. I. Panich [182, 183]; see also M. S. Agranovich's surveys [10, Sec. 4] and [11, Sec. 3.6.4] and the reference therein.

All the theorems in Section 4.1 are proved, except for the last theorem 4.10, in our papers [145, 150]. Therein, we consider the more general case where the elliptic boundary-value problem is given on a smooth compact manifold with boundary. Theorem 4.10 is established in [166].

Section 4.2. J.-L. Lions and E. Magenes [119, 120, 126, 121] and Ya. A. Roitberg [202, 203, 209] investigated the solvability of elliptic boundary-value problems in various two-sided scales formed by positive order Sobolev spaces and their negative order analogs. These authors proposed essentially different methods for the construction of the domain of the operator corresponding to the problem, which leads to different types of theorems on solvability: general and individual theorems. In the general Roitberg theorem, the domain does not depend on the coefficients of the elliptic equation and is general for all boundary-value problems of the same order. In the individual Lions–Magenes theorems, it depends on the coefficients. The terms "general" and "individual" theorems on solvability are proposed in [155, 171].

The general Roitberg theorem involves the two-sided scale of the modified Sobolev spaces introduced by Ya. A. Roitberg in [202, 203] and the notion of a generalized solution in these spaces. Ya. A. Roitberg calls them the spaces "of the Sobolev type." We use the term "the Sobolev scale modified in the sense of Roitberg" for the class of these spaces.

Ya. A. Roitberg proved the general theorem on solvability for regular elliptic boundary-value problems [202, 203] at first and then extended it over nonregular elliptic boundary-value problems [205, 206] and boundary-value problems for general elliptic systems [208]. These results are known in literature as "theorems on complete collection of isomorphisms." Ya. A. Roitberg's monograph [209] is devoted to them. They are also set forth in the monograph by Yu. M. Berezansky [21, Chap. 3, Sec. 6, Subsec. 8], handbook [107, Chap. 3, Sec. 6, Subsec. 5], and survey by M. S. Agranovich [11, Sec. 7.9]. In the most general form, the theorem on complete collection of isomorphism is proved by A. N. Kozhevnikov in [102] for general elliptic pseudodifferential boundary-value problems.

Ya. A. Roitberg, Z. G. Sheftel, and their disciples systematically applied the modified Sobolev scale and theorems on complete collection of isomorphisms to the investigation of various classes of elliptic boundary-value problems: the problems with power singularities on right-hand sides, problems with strong degeneration at the boundary, transmission problems, nonlocal problems, the Sobolev problem, *etc*; see the monographs by Ya. A. Roitberg [209, 210] and the reference therein. Ya. A. Roitberg [204] used the theorem on complete collection of isomorphisms to derive a number of other theorems about the solvability of the regular elliptic boundary-value problem. These results were generalized by Ya. A. Roitberg and Yu. V. Kostarchuk [97] to nonregular elliptic boundary-value problems and by I. Ya. Roitberg and Ya. A. Roitberg [200] to elliptic boundary–value problems for mixed order systems (see also the monograph by Ya. A. Roitberg [210, Sec. 1.3]). In [163, 164], A. A. Murach proved theorems on complete collection of isomorphisms for the two-sided scales of the Lizorkin–Triebel and Nikol'skii–Besov spaces modified in the sense of Roitberg.

The concept of the Sobolev scale modified in the sense of Roitberg and the Roitberg generalized solution is used by V. A. Kozlov, V. G. Maz'ya, and J. Rossmann [104] in the theory of elliptic boundary-value problems in domains with nonsmooth boundary, by N. V. Zhitarashu and S. D. Eidel'man [53] in the theory of parabolic equations, and by Ya. A. Roitberg [210] in the theory of hyperbolic equations.

All theorems in Section 4.2 are proved in [151] except for the last theorem 4.22 that has been published for the first time.

Section 4.3. The main result of this section consists of Theorems 4.6 and 4.7. The former endowed the spaces formed the refined Sobolev scale with an equivalent norm associated with the elliptic expression. The latter states that every generalized solution of the elliptic equation has traces on the boundary of the Euclidean domain. In the Sobolev case, these theorems were proved by Ya. A. Roitberg (see [207] and [209, Theorems 6.1.1 and 6.2.1]). For the refined Sobolev scale, they have been established for the first time in the present monograph. Lemmas 4.6 and 4.7 on projectors in certain Sobolev spaces are of interest in their own right.

Section 4.4. J.-L. Lions and E. Magenes [119, 120, 126, 121] proved various individual theorems on the solvability of the regular elliptic boundary-value problem in scales of Sobolev spaces containing nonregular distributions. In this theorems, certain spaces of restrictions of the elliptic differential operator serves as the domain of the operator corresponding to the boundary-value problem.

These spaces are determined by the given right-hand sides of the elliptic equation. In particular, the domain can coincide with the domain of the maximum operator corresponding to the elliptic differential expression. L. Hörmander [80, Theorem 3.1] proved that the letter domain depends essentially on all coefficients of the elliptic expression even if they are constant. This individual theorem is used in the spectral theory of elliptic operators; see the papers by G. Grubb [67, 68, 69] and V. A. Mikhailets [137, 138, 139].

Our condition I_σ for the space of the right-hand sides of the elliptic equation is general enough. This condition holds for the spaces used by J.-L. Lions and E. Magenes. Ya. A. Roitberg [207, Sec. 2.4] used a condition which is somewhat stronger than ours. Under this condition, Ya. A. Roitberg proved that the operator corresponding to the elliptic boundary-value problem is bounded for each $\sigma < 0$ (see [207, Sec. 2.4] and [209, Remark 6.2.4]).

All results in Subsections 4.4.2–4.4.4, specifically, the individual theorems 4.25, 4.27, and 4.29, are proved in [171]. As a special case, they contain the individual Lions–Magenes theorems formulated in Subsection 4.4.1. The key individual theorem 4.25 is close (as to the choice of a wide class of spaces of the right-hand sides) to the two individual theorems discussed in the surveys by E. Magenes [126, Theorem 6.16] and M. S. Agranovich [11, Sec. 7.9, p. 85]. The former is due to J.-L. Lions and E. Magenes, and the latter to Ya. A. Roitberg.

Other individual theorems are proved by Ya. A. Roitberg in [204, Sec. 5, Theorem 4] and by Yu. V. Kostarchuk and Ya. A. Roitberg in [97, Sec. 5, Theorem 4]. In these theorems, solutions and/or right-hand sides of the elliptic equation are not distributions given in a Euclidean domain, which differs them from the individual theorems in Section 4.4

Section 4.5. The results of this section have been announced in [155, Sec. 6]. They generalize the individual theorems proved in Section 4.4 to wide classes of spaces related to Hörmander spaces.

Chapter 5

Elliptic systems

5.1 Uniformly elliptic systems in the refined Sobolev scale

In the present section, we study uniformly elliptic systems of pseudodifferential equations given in the Euclidean space. We prove an *a priori* estimate for their solutions in the refined Sobolev scale and investigate the interior smoothness of the solutions. It is useful to compare these results with the theorems in Section 1.4, where the case of one equation is investigated. The proofs of the results are performed following the same scheme as in the scalar case. For the sake of completeness, we present their full versions.

5.1.1 Uniformly elliptic systems

In the space $\mathbb{R}^n$, we consider a linear system of pseudodifferential equations

$$\sum_{k=1}^{p} A_{j,k}\, u_k = f_j, \quad j = 1, \ldots, p, \tag{5.1}$$

where $n, p \in \mathbb{N}$ and $A_{j,k} \in \Psi^{\infty}(\mathbb{R}^n)$, $j, k = 1, \ldots, p$, are scalar polyhomogeneous PsDOs in $\mathbb{R}^n$. The solutions of equations (5.1) are considered in the class of distributions in $\mathbb{R}^n$.

As an important example of system (5.1), we can mention a system of linear partial differential equations with coefficients in $C^{\infty}_{\mathrm{b}}(\mathbb{R}^n)$.

We study the following class of systems of pseudodifferential equations [10, Sec. 3.2 b]:

Definition 5.1. System (5.1) is called Douglis–Nirenberg *uniformly elliptic* in $\mathbb{R}^n$ if there exist collections of real numbers $\{l_1, \ldots, l_p\}$ and $\{m_1, \ldots, m_p\}$ such that

(i) $\operatorname{ord} A_{j,k} \leq l_j + m_k$ for all $j, k \in \{1, \ldots, p\}$;

(ii) there exists a number $c > 0$ such that

$$\bigl|\det\bigl(a^{(0)}_{j,k}(x,\xi)\bigr)^{p}_{j,k=1}\bigr| \geq c \quad \text{for all} \quad x, \xi \in \mathbb{R}^n \quad \text{with} \quad |\xi| = 1,$$

where $a^{(0)}_{j,k}(x,\xi)$ is the principal symbol of the PsDO $A_{j,k}$ in the case where $\operatorname{ord} A_{j,k} = l_j + m_k$ or $a^{(0)}_{j,k}(x,\xi) \equiv 0$ in the case where $\operatorname{ord} A_{j,k} < l_j + m_k$.

This definition specifies a fairly broad class of elliptic systems of pseudodifferential equations. For differential equations, this class was introduced by Douglis and Nirenberg in [47] who considered ellipticity in Euclidean domains. This class contains homogeneous elliptic systems, for which all $l_j = 0$ and $m_k = m \in \mathbb{R}$, and Petrovskii's elliptic systems [191], for which all $l_j = 0$ but m_k can be different for different k.

In what follows, we assume that system (5.1) satisfies Definition 5.1.

We rewrite this system in the matrix form as follows:

$$Au = f.$$

Here,

$$A := (A_{j,k})_{j,k=1}^{p}$$

is a matrix PsDO, whereas $u = \operatorname{col}(u_1, \dots, u_p)$ and $f = \operatorname{col}(f_1, \dots, f_p)$ are function columns. As the system itself, the matrix PsDO A is called uniformly elliptic in $\mathbb{R}^n$. For this operator, there exists a parametrix B, i.e., the following statement is true [10, Sec. 3.2 b]:

Proposition 5.1. *There exists a matrix classical PsDO $B = (B_{k,j})_{k,j=1}^{p}$ such that all $B_{k,j} \in \Psi^{-m_k - l_j}(\mathbb{R}^n)$, as well as*

$$BA = I + T_1 \quad \text{and} \quad AB = I + T_2, \tag{5.2}$$

where $T_1 = (T_1^{j,k})_{j,k=1}^{p}$ and $T_2 = (T_2^{k,j})_{k,j=1}^{p}$ are matrix PsDOs, all elements of which belong to the class $\Psi^{-\infty}(\mathbb{R}^n)$, and I is the identity operator on $\mathcal{S}'(\mathbb{R}^n)$.

5.1.2 *A priori* estimate for the solutions of the system

We establish an *a priori* estimate for the solutions of the system $Au = f$ considered in the refined Sobolev scale. In view of the inclusion $A_{j,k} \in \Psi^{l_j + m_k}(\mathbb{R}^n)$ and Lemma 1.6, we get the linear bounded operator

$$A : \bigoplus_{k=1}^{p} H^{s+m_k,\varphi}(\mathbb{R}^n) \to \bigoplus_{j=1}^{p} H^{s-l_j,\varphi}(\mathbb{R}^n) \tag{5.3}$$

for any $s \in \mathbb{R}$ and $\varphi \in \mathcal{M}$.

Theorem 5.1. *Let $s \in \mathbb{R}$, $\sigma > 0$, and $\varphi \in \mathcal{M}$. There exists a number $c = c(s, \sigma, \varphi) > 0$ such that, for any vector-valued functions*

$$u = \operatorname{col}(u_1, \dots, u_p) \in \bigoplus_{k=1}^{p} H^{s+m_k,\varphi}(\mathbb{R}^n), \tag{5.4}$$

$$f = \operatorname{col}(f_1, \dots, f_p) \in \bigoplus_{j=1}^{p} H^{s-l_j,\varphi}(\mathbb{R}^n) \tag{5.5}$$

satisfying the equation $Au = f$ in $\mathbb{R}^n$, the following a priori estimate is true:

$$\Big(\sum_{k=1}^{p}\|u_k\|^2_{H^{s+m_k,\varphi}(\mathbb{R}^n)}\Big)^{1/2} \le c\Big(\sum_{j=1}^{p}\|f_j\|^2_{H^{s-l_j,\varphi}(\mathbb{R}^n)}\Big)^{1/2} + c\Big(\sum_{k=1}^{p}\|u_k\|^2_{H^{s+m_k-\sigma,\varphi}(\mathbb{R}^n)}\Big)^{1/2}. \tag{5.6}$$

Proof. Let $\|\cdot\|'_{s,\varphi}$, $\|\cdot\|''_{s,\varphi}$, and $\|\cdot\|'_{s-\sigma,\varphi}$ denote norms in the spaces

$$\bigoplus_{k=1}^{p} H^{s+m_k,\varphi}(\mathbb{R}^n),\quad \bigoplus_{j=1}^{p} H^{s-l_j,\varphi}(\mathbb{R}^n),\quad\text{and}\quad \bigoplus_{k=1}^{p} H^{s+m_k-\sigma,\varphi}(\mathbb{R}^n),$$

respectively. Let the vector-valued functions (5.4) and (5.5) satisfy the equation $Au = f$ in $\mathbb{R}^n$. Due to the first equality in (5.2), we write $u = Bf - T_1u$. This yields estimate (5.6):

$$\|u\|'_{s,\varphi} = \|Bf - T_1u\|'_{s,\varphi} \le \|Bf\|'_{s,\varphi} + \|T_1u\|'_{s,\varphi} \le c\,\|f\|''_{s,\varphi} + c\,\|u\|'_{s-\sigma,\varphi},$$

where c is the maximum of norms of the operators

$$B : \bigoplus_{j=1}^{p} H^{s-l_j,\varphi}(\mathbb{R}^n) \to \bigoplus_{k=1}^{p} H^{s+m_k,\varphi}(\mathbb{R}^n), \tag{5.7}$$

$$T_1 : \bigoplus_{k=1}^{p} H^{s+m_k-\sigma,\varphi}(\mathbb{R}^n) \to \bigoplus_{k=1}^{p} H^{s+m_k,\varphi}(\mathbb{R}^n). \tag{5.8}$$

These operators are bounded by Lemma 1.6 and Proposition 5.1.

Theorem 5.1 is proved.

Theorem 5.1 refines the *a priori* estimate obtained by Hörmander [83, Sec. 1.0] for the Sobolev scale.

5.1.3 Smoothness of solutions

Assume that the right-hand side of the equation $Au = f$ has a certain interior smoothness in the refined Sobolev scale on an open nonempty set $V \subseteq \mathbb{R}^n$. We study the interior smoothness of the solutions u on this set. First, we consider the case where $V = \mathbb{R}^n$. Recall that $H^{-\infty}(\mathbb{R}^n)$ is the union of all spaces $H^{s,\varphi}(\mathbb{R}^n)$, where $s \in \mathbb{R}$ and $\varphi \in \mathcal{M}$.

Theorem 5.2. *Suppose that $u \in (H^{-\infty}(\mathbb{R}^n))^p$ is a solution of the equation $Au = f$ in $\mathbb{R}^n$ under the condition*

$$f_j \in H^{s-l_j,\varphi}(\mathbb{R}^n) \quad \textit{for every} \quad j \in \{1,\dots,p\}$$

with certain parameters $s \in \mathbb{R}$ *and* $\varphi \in \mathcal{M}$. *Then*

$$u_k \in H^{s+m_k,\varphi}(\mathbb{R}^n) \quad \text{for every} \quad k \in \{1,\dots,p\}.$$

Proof. By Theorem 1.15(i), for the vector-valued function $u \in (H^{-\infty}(\mathbb{R}^n))^p$, there exists a number $\sigma > 0$ such that

$$u \in \bigoplus_{k=1}^{p} H^{s+m_k-\sigma,\varphi}(\mathbb{R}^n). \tag{5.9}$$

Using this result, the condition of the theorem, and relations (5.2), (5.7), and (5.8), we obtain the required property

$$u = BAu - T_1 u = Bf - T_1 u \in \bigoplus_{k=1}^{p} H^{s+m_k,\varphi}(\mathbb{R}^n).$$

Theorem 5.2 is proved.

We now consider the general case where V is an arbitrary open nonempty subset of the space $\mathbb{R}^n$. Recall that the space $H^{\sigma,\varphi}_{\mathrm{int}}(V)$, which consists of distributions with given interior smoothness on V, is defined in Section 1.4.3.

Theorem 5.3. *Suppose that* $u \in (H^{-\infty}(\mathbb{R}^n))^p$ *is a solution of the equation* $Au = f$ *on the set* V *under the condition*

$$f_j \in H^{s-l_j,\varphi}_{\mathrm{int}}(V) \quad \text{for every} \quad j \in \{1,\dots,p\} \tag{5.10}$$

with certain parameters $s \in \mathbb{R}$ *and* $\varphi \in \mathcal{M}$. *Then*

$$u_k \in H^{s+m_k,\varphi}_{\mathrm{int}}(V) \quad \text{for every} \quad k \in \{1,\dots,p\}. \tag{5.11}$$

Proof. First, we prove that condition (5.10) yields the following property of increase in interior smoothness of the solution to $Au = f$:

The following implication is true for any $r \geq 1$:

$$u \in \bigoplus_{k=1}^{p} H^{s-r+m_k,\varphi}_{\mathrm{int}}(V) \Rightarrow u \in \bigoplus_{k=1}^{p} H^{s-r+1+m_k,\varphi}_{\mathrm{int}}(V). \tag{5.12}$$

We arbitrarily choose a function $\chi \in C^{\infty}_{\mathrm{b}}(\mathbb{R}^n)$ such that

$$\operatorname{supp}\chi \subset V \quad \text{and} \quad \operatorname{dist}(\operatorname{supp}\chi, \partial V) > 0. \tag{5.13}$$

For this function, there exists a function $\eta \in C^{\infty}_{\mathrm{b}}(\mathbb{R}^n)$ such that

$$\begin{gathered} \operatorname{supp}\eta \subset V, \quad \operatorname{dist}(\operatorname{supp}\eta, \partial V) > 0, \\ \text{and } \eta = 1 \text{ in a neighborhood of } \operatorname{supp}\chi \end{gathered} \tag{5.14}$$

(this was shown in the proof of Theorem 1.18).

Changing the sequence order for the matrix PsDO A and the operator of multiplication by the function χ, we find

$$\begin{aligned} A\chi u &= A\chi\eta u \\ &= \chi\, A\eta u + A'\eta u \\ &= \chi\, Au + \chi\, A(\eta - 1)u + A'\eta u \\ &= \chi f + \chi\, A(\eta - 1)u + A'\eta u \quad \text{in} \quad \mathbb{R}^n. \end{aligned} \tag{5.15}$$

Here, the matrix PsDO $A' = (\,A'_{j,k}\,)_{j,k=1}^{p}$ is the commutator of the PsDO A and the operator of multiplication by the function χ. In view of the inclusion $A'_{j,k} \in \Psi^{l_j+m_k-1}(\mathbb{R}^n)$ and Lemma 1.6, we get the bounded operator

$$A' : \bigoplus_{k=1}^{p} H^{s-r+m_k,\varphi}(\mathbb{R}^n) \to \bigoplus_{j=1}^{p} H^{s-r+1-l_j,\varphi}(\mathbb{R}^n).$$

Therefore,

$$u \in \bigoplus_{k=1}^{p} H^{s-r+m_k,\varphi}_{\text{int}}(V) \;\Rightarrow\; A'\eta u \in \bigoplus_{j=1}^{p} H^{s-r+1-l_j,\varphi}(\mathbb{R}^n). \tag{5.16}$$

Further, according to condition (5.10) and in view of the inequality $r \geq 1$, we have

$$\chi f \in \bigoplus_{j=1}^{p} H^{s-l_j,\varphi}(\mathbb{R}^n) \hookrightarrow \bigoplus_{j=1}^{p} H^{s-r+1-l_j,\varphi}(\mathbb{R}^n). \tag{5.17}$$

In addition, since supports of functions χ and $\eta - 1$ do not intersect, the PsDO

$$\chi A_{j,k}(\eta - 1) \in \Psi^{-\infty}(\mathbb{R}^n)$$

for all $j, k \in \{1, \dots, p\}$. Since for the vector-valued function $u \in (H^{-\infty}(\mathbb{R}^n))^p$ we have inclusion (5.9) valid for some $\sigma > 0$, by virtue of Lemma 1.6 we obtain the inclusion

$$\chi\, A(\eta - 1)u \in \bigoplus_{j=1}^{p} H^{s-r+1-l_j,\varphi}(\mathbb{R}^n). \tag{5.18}$$

By using relations (5.15)–(5.18) and Theorem 5.2, we conclude that

$$\begin{aligned} u \in \bigoplus_{k=1}^{p} H^{s-r+m_k,\varphi}_{\text{int}}(V) \;&\Rightarrow\; A\chi u \in \bigoplus_{j=1}^{p} H^{s-r+1-l_j,\varphi}(\mathbb{R}^n) \\ &\Rightarrow \chi u \in \bigoplus_{k=1}^{p} H^{s-r+1+m_k,\varphi}(\mathbb{R}^n). \end{aligned}$$

Thus, implication (5.12) is proved since the function $\chi \in C_{\mathrm{b}}^{\infty}(\mathbb{R}^n)$, which satisfies condition (5.13), is chosen arbitrarily.

It is now easy to derive property (5.11) from implication (5.12). We can assume that the number $\sigma > 0$ in relation (5.9) is integer. Hence,

$$u \in \bigoplus_{k=1}^{p} H_{\mathrm{int}}^{s-\sigma+m_k,\varphi}(V).$$

Applying (5.12) successively to $r = \sigma,\ \sigma - 1,\ \ldots,\ 1$, we obtain property (5.11):

$$u \in \bigoplus_{k=1}^{p} H_{\mathrm{int}}^{s-\sigma+m_k,\varphi}(V) \ \Rightarrow\ u \in \bigoplus_{k=1}^{p} H_{\mathrm{int}}^{s-\sigma+1+m_k,\varphi}(V)$$
$$\Rightarrow \ldots \Rightarrow\ u \in \bigoplus_{k=1}^{p} H_{\mathrm{int}}^{s+m_k,\varphi}(V).$$

Theorem 5.3 is proved.

Theorem 5.3 enables us to establish the continuity of generalized derivatives for the solutions of system (5.1). To this end, we also use Theorem 1.15(iii).

Theorem 5.4. *Let integers $k \in \{1,\ldots,p\}$ and $r \geq 0$ and a function $\varphi \in \mathcal{M}$ satisfying condition* (1.37) *be given. Suppose that $u \in (H^{-\infty}(\mathbb{R}^n))^p$ is a solution of the equation $Au = f$ on an open set $V \subseteq \mathbb{R}^n$ under the condition*

$$f_j \in H_{\mathrm{int}}^{r-m_k-l_j+n/2,\varphi}(V) \quad \textit{for every} \quad j \in \{1,\ldots,p\}. \tag{5.19}$$

Then the component u_k of the solution has continuous partial derivatives on the set V up to the order r inclusive, and moreover, these derivatives are bounded on each set $V_0 \subset V$ such that $\mathrm{dist}(V_0, \partial V) > 0$.

In particular, if $V = \mathbb{R}^n$, then $u_k \in C_{\mathrm{b}}^{r}(\mathbb{R}^n)$.

Proof. By virtue of Theorem 5.3, where we set $s := r - m_k + n/2$, the inclusion $u_k \in H_{\mathrm{int}}^{r+n/2,\varphi}(V)$ is true. Let a function $\eta \in C_{\mathrm{b}}^{\infty}(\mathbb{R}^n)$ satisfy the conditions

$$\operatorname{supp}\eta \subset V,\ \ \mathrm{dist}(\operatorname{supp}\eta,\ \partial V) > 0, \ \text{ and } \ \eta = 1 \text{ in a neighborhood of } V_0.$$

For the distribution ηu_k, by virtue of Theorem 1.15(iii) we have

$$\eta u_k \in H^{r+n/2,\varphi}(\mathbb{R}^n) \hookrightarrow C_{\mathrm{b}}^{r}(\mathbb{R}^n).$$

This implies that all partial derivatives of the function u_k, up to the order r inclusive, are continuous and bounded in a certain neighborhood of the set V_0.

Then these derivatives are continuous on the set V as well because we can take $V_0 := \{x_0\}$ for any point $x_0 \in V$.

Theorem 5.4 is proved.

As an application of Theorem 5.4, we give the following sufficient condition for a solution of the system $Au = f$ to be classical in the case where all $A_{j,k}$ are linear differential operators with coefficients from the class $C_{\mathrm{b}}^{\infty}(\mathbb{R}^n)$ and all $l_j = 0$. In other words, we consider the case where $Au = f$ is a Petrovskii uniformly elliptic system of differential equations.

Corollary 5.1. *Suppose that $u \in (H^{-\infty}(\mathbb{R}^n))^p$ is a solution of the equation $Au = f$ in an open set $V \subseteq \mathbb{R}^n$ under the condition that $f_j \in H_{\mathrm{int}}^{n/2,\varphi}(V)$ for each number $j \in \{1, \dots, p\}$ and a parameter $\varphi \in \mathcal{M}$ satisfying* (1.37). *Then the solution u is classical in the set V, i.e., $u_k \in C^{m_k}(V)$ for all $k \in \{1, \dots, p\}$.*

This result immediately follows from Theorem 5.4 for $r = m_k$. Note that for the classical solution u of system (5.1) the left-hand sides of the system are determined by classical derivatives (i.e., derivatives which are not properly generalized in the sense of the theory of distributions) and these derivatives are continuous in the set V.

5.2 Elliptic systems on a closed manifold

In the present section, we study Douglis–Nirenberg systems of pseudodifferential equations defined on an infinitely smooth closed (compact) oriented manifold Γ. We prove that the operator corresponding to these systems is bounded and Fredholm in related pairs of Hörmander spaces from the refined Sobolev scale. In addition, we investigate the class of elliptic systems (parameter-elliptic systems) for which the indicated operator is an isomorphism for large absolute values of the complex parameter. It makes sense to compare the obtained results with theorems presented in Section 2.2 where the case of a single equation is investigated. For the proofs of these results, we use the same scheme that is used for proofs of similar results in the scalar case.

5.2.1 Elliptic Systems

Consider a system of linear pseudodifferential equations

$$\sum_{k=1}^{p} A_{j,k}\, u_k = f_j \quad \text{on} \quad \Gamma, \quad \text{with} \quad j = 1, \dots, p. \tag{5.20}$$

Here, $p \in \mathbb{N}$ and $A_{j,k} \in \Psi^{\infty}(\Gamma)$, $j, k = 1, \dots, p$, are scalar classical PsDOs defined on the manifold Γ. Equations (5.20) are understood in the sense of the theory of distributions.

Definition 5.2. System (5.20) is called Douglis–Nirenberg *elliptic* on Γ if there exist collections of real numbers $\{l_1,\dots,l_p\}$ and $\{m_1,\dots,m_p\}$ such that

(i) $\operatorname{ord} A_{j,k} \le l_j + m_k$ for all $j,k \in \{1,\dots,p\}$;

(ii) $\det\big(a^{(0)}_{j,k}(x,\xi)\big)^{p}_{j,k=1} \ne 0$ for arbitrary point $x \in \Gamma$ and covector $\xi \in T^*_x\Gamma\setminus\{0\}$; here, $a^{(0)}_{j,k}(x,\xi)$ is the principal symbol of the PsDO $A_{j,k}$ in the case where $\operatorname{ord} A_{j,k} = l_j + m_k$ or $a^{(0)}_{j,k}(x,\xi) \equiv 0$ in the case where $\operatorname{ord} A_{j,k} < l_j + m_k$.

Further in Section 5.2, it is assumed that system (5.20) satisfies Definition 5.2. We rewrite this system in the matrix form as follows:

$$Au = f \quad \text{on } \Gamma.$$

Here, $A := (A_{j,k})^p_{j,k=1}$ is a matrix PsDO on Γ and $u = \operatorname{col}(u_1,\dots,u_p)$ and $f = \operatorname{col}(f_1,\dots,f_p)$ are function columns. As the system itself, the corresponding matrix PsDO A is called elliptic on Γ.

By $A^+ = ((A^+_{j,k})^p_{j,k=1})^t$ we denote the matrix PsDO which is formally adjoint to the operator A relative to the C^∞-density dx on Γ. In this case, every PsDO $A^+_{j,k}$ is formally adjoint to $A_{j,k}$. The ellipticity of the system $Au = f$ is equivalent to the (Douglis–Nirenberg) ellipticity of the system $A^+v = g$. We set

$$\mathcal{N} := \{u \in (C^\infty(\Gamma))^p : Au = 0 \text{ on } \Gamma\},$$

$$\mathcal{N}^+ := \{v \in (C^\infty(\Gamma))^p : A^+v = 0 \text{ on } \Gamma\}.$$

Since the systems $Au = f$ and $A^+v = g$ are elliptic, the spaces $\mathcal{N}$ and $\mathcal{N}^+$ are finite-dimensional [10, Sec. 3.2 b].

5.2.2 Operator of the elliptic system on the refined scale

We study properties of the matrix PsDO A on the refined Sobolev scale over the manifold Γ. In view of the inclusion $A_{j,k} \in \Psi^{l_j+m_k}(\Gamma)$ and Lemma 2.5, we get the linear bounded operator

$$A : \bigoplus_{k=1}^{p} H^{s+m_k,\varphi}(\Gamma) \to \bigoplus_{j=1}^{p} H^{s-l_j,\varphi}(\Gamma) \tag{5.21}$$

for any $s \in \mathbb{R}$ and $\varphi \in \mathcal{M}$. Let us study its properties.

Theorem 5.5. *For arbitrary parameters $s \in \mathbb{R}$ and $\varphi \in \mathcal{M}$, the bounded operator* (5.21) *is Fredholm. Its kernel coincides with the space $\mathcal{N}$, and the range consists of all vector-valued functions*

$$f = \operatorname{col}(f_1,\dots,f_p) \in \bigoplus_{j=1}^{p} H^{s-l_j,\varphi}(\Gamma) \tag{5.22}$$

that satisfy the condition

$$\sum_{j=1}^{p}(f_j, w_j)_\Gamma = 0 \quad \textit{for any} \quad w = (w_1, \dots, w_p) \in \mathcal{N}^+. \tag{5.23}$$

The index of operator (5.21) *is equal to* $\dim\mathcal{N} - \dim\mathcal{N}^+$ *and does not depend on* s *and* φ.

Proof. For $\varphi \equiv 1$ (the Sobolev scale), this theorem is known [10, Theorem 3.2.1]. In order to rove this theorem in the general case of $\varphi \in \mathcal{M}$, we use the interpolation with a function parameter. Namely, let $s \in \mathbb{R}$. We have the Fredholm bounded operators

$$A : \bigoplus_{k=1}^{p} H^{s\mp 1+m_k}(\Gamma) \to \bigoplus_{j=1}^{p} H^{s\mp 1-l_j}(\Gamma) \tag{5.24}$$

with the common kernel $\mathcal{N}$ and the same index $\varkappa := \dim\mathcal{N} - \dim\mathcal{N}^+$. Moreover,

$$A\Big(\bigoplus_{k=1}^{p} H^{s\mp 1+m_k}(\Gamma)\Big)$$
$$= \Big\{ f = \operatorname{col}(f_1, \dots, f_p) \in \bigoplus_{j=1}^{p} H^{s\mp 1-l_j}(\Gamma) : \ (5.23) \text{ is true}\Big\}. \tag{5.25}$$

By applying to (5.24) the interpolation with the function parameter ψ from Theorem 2.2, where we set $\varepsilon = \delta = 1$, we obtain the bounded operator

$$A : \Big[\bigoplus_{k=1}^{p} H^{s-1+m_k}(\Gamma), \bigoplus_{k=1}^{p} H^{s+1+m_k}(\Gamma)\Big]_\psi$$
$$\to \Big[\bigoplus_{j=1}^{p} H^{s-1-l_j}(\Gamma), \bigoplus_{j=1}^{p} H^{s+1-l_j}(\Gamma)\Big]_\psi,$$

which coincides with operator (5.21) by virtue of Theorems 1.5 and 2.2. Therefore, in accordance with Theorem 1.7, operator (5.21) is Fredholm with the kernel $\mathcal{N}$ and the index $\varkappa = \dim\mathcal{N} - \dim\mathcal{N}^+$. The range of this operator is equal to

$$\Big(\bigoplus_{j=1}^{p} H^{s-l_j,\varphi}(\Gamma)\Big) \cap A\Big(\bigoplus_{k=1}^{p} H^{s-1+m_k}(\Gamma)\Big).$$

This implies, by virtue of (5.25), that this range coincides with the range indicated in the formulation of the theorem.

Theorem 5.5 is proved.

According to this theorem, $\mathcal{N}^+$ is a defect subspace of operator (5.21). By virtue of Theorem 2.3(v), the operator

$$A^+ : \bigoplus_{j=1}^{p} H^{-s+l_j,1/\varphi}(\Gamma) \to \bigoplus_{k=1}^{p} H^{-s-m_k,1/\varphi}(\Gamma) \tag{5.26}$$

is adjoint to operator (5.21). Since the adjoint system $A^+v = g$ is elliptic, by Theorem 5.5 we conclude that the bounded operator (5.26) is Fredholm and has the kernel $\mathcal{N}^+$ and the defect subspace $\mathcal{N}$.

If the spaces $\mathcal{N}$ and $\mathcal{N}^+$ are trivial, then Theorem 5.5 and the Banach theorem on inverse operator imply that operator (5.21) is an isomorphism. In the general case, it is convenient to define the isomorphism in terms of the following projectors:

The spaces of action of operator (5.21) can be represented in the following form of the direct sum of (closed) subspaces:

$$\bigoplus_{k=1}^{p} H^{s+m_k,\varphi}(\Gamma)$$

$$= \mathcal{N} \dotplus \Big\{ u \in \bigoplus_{k=1}^{p} H^{s+m_k,\varphi}(\Gamma) : \sum_{k=1}^{p} (u_k, v_k)_\Gamma = 0 \text{ for all } v \in \mathcal{N} \Big\},$$

$$\bigoplus_{j=1}^{p} H^{s-l_j,\varphi}(\Gamma)$$

$$= \mathcal{N}^+ \dotplus \Big\{ f \in \bigoplus_{j=1}^{p} H^{s-l_j,\varphi}(\Gamma) : \sum_{j=1}^{p} (f_j, w_j)_\Gamma = 0 \text{ for all } w \in \mathcal{N}^+ \Big\}.$$

Recall here that $u = \mathrm{col}\,(u_1, \ldots, u_p)$, $f = \mathrm{col}\,(f_1, \ldots, f_p)$ and $v = (v_1, \ldots, v_p)$, $w = (w_1, \ldots, w_p)$.

These decompositions in direct sums do exist because their terms have the trivial intersection and, in addition, the finite dimension of the first term is equal to the codimension of the second one. (Indeed, for example, in the first sum, the factor space of the space $\bigoplus_{k=1}^{p} H^{s+m_k,\varphi}(\Gamma)$ by the second term is equal to the space which is dual to the subspace $\mathcal{N}$ of $\bigoplus_{k=1}^{p} H^{-s-m_k,1/\varphi}(\Gamma)$).

Let $\mathcal{P}$ and $\mathcal{P}^+$ denote, respectively, oblique projectors of the spaces

$$\bigoplus_{k=1}^{p} H^{s+m_k,\varphi}(\Gamma) \quad \text{and} \quad \bigoplus_{j=1}^{p} H^{s-l_j,\varphi}(\Gamma),$$

onto the related second terms in the indicated sums taken to be parallel to their first terms. These projectors are independent of s and φ.

Theorem 5.6. *For any $s \in \mathbb{R}$ and $\varphi \in \mathcal{M}$, the restriction of operator* (5.21) *to the subspace $\mathcal{P}\big(\bigoplus_{k=1}^{p} H^{s+m_k,\varphi}(\Gamma)\big)$ is the isomorphism*

$$A : \mathcal{P}\Big(\bigoplus_{k=1}^{p} H^{s+m_k,\varphi}(\Gamma)\Big) \leftrightarrow \mathcal{P}^{+}\Big(\bigoplus_{j=1}^{p} H^{s-l_j,\varphi}(\Gamma)\Big). \tag{5.27}$$

Proof. By Theorem 5.5, $\mathcal{N}$ is the kernel and

$$\mathcal{P}^{+}\Big(\bigoplus_{j=1}^{p} H^{s-l_j,\varphi}(\Gamma)\Big)$$

is the range of operator (5.21). Therefore, operator (5.27) is a bijection. In addition, this operator is bounded. Therefore, it is an isomorphism by virtue of the Banach inverse operator theorem.

Theorem 5.6 is proved.

Theorem 5.6 yields the following *a priori* estimate for the solutions of the equation $Au = f$:

Theorem 5.7. *Let $s \in \mathbb{R}$ and $\varphi \in \mathcal{M}$. Suppose that the vector-valued function*

$$u = \operatorname{col}(u_1, \dots, u_p) \in \bigoplus_{k=1}^{p} H^{s+m_k,\varphi}(\Gamma) \tag{5.28}$$

is a solution of the equation $Au = f$ on Γ under condition (5.22) *for the right-hand side of the equation. Then, for chosen parameters s and φ and an arbitrary number $\sigma > 0$, there exists a number $c > 0$ such that*

$$\begin{aligned}&\Big(\sum_{k=1}^{p} \|u_k\|^2_{H^{s+m_k,\varphi}(\Gamma)}\Big)^{1/2}\\ &\leq c\Big(\sum_{j=1}^{p} \|f_j\|^2_{H^{s-l_j,\varphi}(\Gamma)}\Big)^{1/2} + c\Big(\sum_{k=1}^{p} \|u_k\|^2_{H^{s+m_k-\sigma,\varphi}(\Gamma)}\Big)^{1/2}\end{aligned} \tag{5.29}$$

and c is independent of u and f.

Proof. For the sake of brevity, we denote norms in the spaces

$$\bigoplus_{k=1}^{p} H^{s+m_k,\varphi}(\Gamma), \quad \bigoplus_{j=1}^{p} H^{s-l_j,\varphi}(\Gamma), \quad \text{and} \quad \bigoplus_{k=1}^{p} H^{s-\sigma+m_k,\varphi}(\Gamma),$$

which are used in (5.29), by $\|\cdot\|'_{s,\varphi}$, $\|\cdot\|''_{s,\varphi}$, and $\|\cdot\|'_{s-\sigma,\varphi}$ respectively. Since $\mathcal{N}$ is a finite-dimensional subspace of these spaces, these norms are equivalent on $\mathcal{N}$. In particular, for the vector-valued function $u - \mathcal{P}u \in \mathcal{N}$, we get

$$\|u - \mathcal{P}u\|'_{s,\varphi} \leq c_1 \|u - \mathcal{P}u\|'_{s-\sigma,\varphi}$$

with a constant $c_1 > 0$ which is independent of u. This yields

$$\begin{aligned}\|u\|'_{s,\varphi} &\le \|u - \mathcal{P}u\|'_{s,\varphi} + \|\mathcal{P}u\|'_{s,\varphi}\\ &\le c_1 \|u - \mathcal{P}u\|'_{s-\sigma,\varphi} + \|\mathcal{P}u\|'_{s,\varphi}\\ &\le c_1 c_2 \|u\|'_{s-\sigma,\varphi} + \|\mathcal{P}u\|'_{s,\varphi},\end{aligned}$$

where c_2 is the norm of the projector $1 - \mathcal{P}$ acting in the space

$$\bigoplus_{k=1}^{p} H^{s-\sigma+m_k,\varphi}(\Gamma).$$

Thus,

$$\|u\|'_{s,\varphi} \le \|\mathcal{P}u\|'_{s,\varphi} + c_1 c_2 \|u\|'_{s-\sigma,\varphi}. \tag{5.30}$$

We now use the condition $Au = f$. Since $\mathcal{N}$ is the kernel of operator (5.21) and $u - \mathcal{P}u \in \mathcal{N}$, we have $A\mathcal{P}u = f$. Thus, $\mathcal{P}u$ is the preimage of the vector function f under the action of isomorphism (5.27). Therefore,

$$\|\mathcal{P}u\|'_{s,\varphi} \le c_3 \|f\|''_{s,\varphi},$$

where c_3 is the norm of the operator inverse to (5.27). Using this result and inequality (5.30), we immediately obtain estimate (5.29).

Theorem 5.7 is proved.

Note that if $\mathcal{N} = \{0\}$, i.e., the equation $Au = f$ has at most one solution, then the quantity

$$\sum_{k=1}^{p} \|u_k\|_{H^{s-\sigma+m_k,\varphi}(\Gamma)}$$

on the right-hand side of estimate (5.29) can be omitted. If $\mathcal{N} \neq \{0\}$, then this quantity can be made arbitrarily small for any fixed function u by choosing a sufficiently large number σ.

5.2.3 Local smoothness of solutions

Let Γ_0 be an open nonempty subset of the manifold Γ. We investigate the local smoothness of solutions of the elliptic equation $Au = f$ on Γ_0 in the refined Sobolev scale. First, we consider the case where $\Gamma_0 = \Gamma$.

Theorem 5.8. *Suppose that a vector-valued function $u \in (\mathcal{D}'(\Gamma))^p$ is a solution of the equation $Au = f$ on the manifold Γ under the condition*

$$f_j \in H^{s-l_j,\varphi}(\Gamma) \quad \textit{for every} \quad j \in \{1, \dots, p\} \tag{5.31}$$

with certain parameters $s \in \mathbb{R}$ and $\varphi \in \mathcal{M}$. Then

$$u_k \in H^{s+m_k,\varphi}(\Gamma) \quad \textit{for every} \quad k \in \{1, \dots, p\}. \tag{5.32}$$

Proof. Since the manifold Γ is compact, the space $\mathcal{D}'(\Gamma)$ is the union of the Sobolev spaces $H^{\sigma}(\Gamma)$, where $\sigma \in \mathbb{R}$. Therefore, for the vector-valued function $u \in (\mathcal{D}'(\Gamma))^p$, there exists a number $\sigma < s$ such that

$$u \in \bigoplus_{k=1}^{p} H^{\sigma+m_k}(\Gamma).$$

By virtue of Theorems 5.5 and 2.3(iii), we arrive at the equality

$$\Big(\bigoplus_{j=1}^{p} H^{s-l_j,\varphi}(\Gamma)\Big) \cap A\Big(\bigoplus_{k=1}^{p} H^{\sigma+m_k}(\Gamma)\Big) = A\Big(\bigoplus_{k=1}^{p} H^{s+m_k,\varphi}(\Gamma)\Big).$$

Hence, it follows from condition (5.31) that

$$f = Au \in A\Big(\bigoplus_{k=1}^{p} H^{s+m_k,\varphi}(\Gamma)\Big).$$

Thus, on Γ, in addition to the equality $Au = f$, we get the equality $Av = f$ for a vector-valued function

$$v \in \bigoplus_{k=1}^{p} H^{s+m_k,\varphi}(\Gamma).$$

Therefore, $A(u - v) = 0$ on Γ, and we conclude that $w := u - v \in \mathcal{N}$ by virtue of Theorem 5.5. However,

$$\mathcal{N} \subset (C^{\infty}(\Gamma))^p \subset \bigoplus_{k=1}^{p} H^{s+m_k,\varphi}(\Gamma).$$

Thus,

$$u = v + w \in \bigoplus_{k=1}^{p} H^{s+m_k,\varphi}(\Gamma),$$

i.e., u satisfies property (5.32).

Theorem 5.8 is proved.

We now consider the general case of arbitrary Γ_0. Recall that the space $H^{\sigma,\varphi}_{\mathrm{loc}}(\Gamma_0)$ of distributions, which have a predetermined local smoothness on Γ_0, is defined in Subsection 2.2.3.

Theorem 5.9. *Suppose that $u \in (\mathcal{D}'(\Gamma))^p$ is a solution of the equation $Au = f$ on the set Γ_0 under the condition*

$$f_j \in H^{s-l_j,\varphi}_{\mathrm{loc}}(\Gamma_0) \quad \textit{for every} \quad j \in \{1, \dots, p\}$$

with certain parameters $s \in \mathbb{R}$ and $\varphi \in \mathcal{M}$. Then

$$u_k \in H^{s+m_k,\varphi}_{\mathrm{loc}}(\Gamma_0) \quad \textit{for every} \quad k \in \{1, \dots, p\}.$$

Theorem 5.9 is proved by analogy with Theorem 5.3. However, in this case, we have to apply Theorem 5.8 instead of Theorem 5.2.

Using Theorems 5.9 and 2.8, we immediately obtain the following sufficient condition of continuity of derivatives for the chosen component u_k of the solution to the system $Au = f$:

Corollary 5.2. *Let integers $k \in \{1, \ldots, p\}$ and $r \geq 0$ and a function parameter $\varphi \in \mathcal{M}$ satisfying inequality* (1.37) *be given. Suppose that $u \in (\mathcal{D}'(\Gamma))^p$ is a solution of the equation $Au = f$ on the set Γ_0 under the condition*

$$f_j \in H^{r-m_k-l_j+n/2,\varphi}_{\mathrm{loc}}(\Gamma_0) \quad \textit{for every} \quad j \in \{1, \ldots, p\}.$$

Then $u_k \in C^r(\Gamma_0)$.

5.2.4 Parameter-elliptic systems

Following the survey [10, Sec. 4.3 d], we consider a sufficiently broad class of parameter-elliptic systems on the manifold Γ. Fix arbitrarily numbers $p, q \in \mathbb{N}$, $m > 0$, and $m_1, \ldots, m_p \in \mathbb{R}$. Consider a matrix PsDO $A(\lambda)$ that depends on a complex parameter λ as follows:

$$A(\lambda) := \sum_{r=0}^{q} \lambda^{q-r} A^{(r)}. \tag{5.33}$$

Here, $A^{(r)} := \big(A^{(r)}_{j,k}\big)_{j,k=1}^{p}$ is a square matrix formed by arbitrary scalar polyhomogeneous PsDOs $A^{(r)}_{j,k}$ on Γ of order

$$\operatorname{ord} A^{(r)}_{j,k} \leq mr + m_k - m_j.$$

Assume that $A^{(0)} = -I$, where I is the identity matrix.

Consider a system of linear equations

$$A(\lambda)\, u = f \quad \text{on } \Gamma \tag{5.34}$$

depending on the parameter $\lambda \in \mathbb{C}$. Here, as before, $u = \operatorname{col}(u_1, \ldots, u_p)$ and $f = \operatorname{col}(f_1, \ldots, f_p)$ are function columns whose components are distributions on the manifold Γ.

Let K be a fixed closed angle on the complex plane with the vertex at the origin (we do not exclude the case where K degenerates into a ray).

Definition 5.3. System (5.34) is called *parameter-elliptic* in the angle K if

$$\det \sum_{r=0}^{q} \lambda^{q-r}\, a^{r,0}(x, \xi) \neq 0 \tag{5.35}$$

for any $x \in \Gamma$, $\xi \in T_x^*\Gamma$, and $\lambda \in K$ satisfying the condition $(\xi, \lambda) \neq 0$. Here,

$$a^{r,0}(x,\xi) := \big(a_{j,k}^{r,0}(x,\xi)\big)_{j,k=1}^{p}$$

is a square matrix whose arbitrary element $a_{j,k}^{r,0}(x,\xi)$ is defined as follows: it is either the principal symbol of the PsDO $A_{j,k}^{(r)}$ in the case where $\operatorname{ord} A_{j,k}^{(r)} = mr + m_k - m_j$ or the zero function, otherwise. In the case $r \geq 1$, it is additionally assumed that the function $a_{j,k}^{r,0}(x,\xi)$ is equal to 0 at $\xi = 0$ (because of the principal symbol is not initially defined for $\xi = 0$).

In what follows in the present subsection, we assume that system (5.34) satisfies Definition 5.3.

By virtue of Definition 5.3, system (5.34) is a Douglis–Nirenberg elliptic system for every fixed $\lambda \in \mathbb{C}$. Indeed, due to (5.33), the matrix $A(\lambda)$ is formed by the elements

$$\sum_{r=0}^{q} \lambda^{q-r} A_{j,k}^{(r)}, \quad j,k = 1,\dots,p, \tag{5.36}$$

and these elements are classical scalar PsDOs of orders not greater than $l_j + m'_k$, where $l_j := -m_j$ and $m'_k := mq + m_k$. The principal symbol of the PsDO (5.36) is equal to $a_{j,k}^{q,0}(x,\xi)$ for every fixed λ. According to condition (5.35) with $\lambda := 0$, the following inequality is true:

$$\det\big(a_{j,k}^{q,0}(x,\xi)\big)_{j,k=1}^{p} \neq 0 \quad \text{for all} \quad x \in \Gamma \quad \text{and} \quad \xi \in T_x^*\Gamma \setminus \{0\}.$$

This means that system (5.34) is Douglis–Nirenberg elliptic for every fixed $\lambda \in \mathbb{C}$.

Therefore, Theorem 5.5 is true for the elliptic system (5.34). According to this theorem, the bounded operator

$$A(\lambda) : \bigoplus_{k=1}^{p} H^{s+mq+m_k,\varphi}(\Gamma) \to \bigoplus_{j=1}^{p} H^{s+m_j,\varphi}(\Gamma) \tag{5.37}$$

is Fredholm for arbitrary $\lambda \in \mathbb{C}$, $s \in \mathbb{R}$, and $\varphi \in \mathcal{M}$. Moreover, since system (5.34) is parameter-elliptic in the angle K, this operator has the following additional properties:

Theorem 5.10. *The following assertions are true:*

(i) *There exists a number $\lambda_0 > 0$ such that, for any parameter value $\lambda \in K$ satisfying $|\lambda| \geq \lambda_0$, the matrix PsDO $A(\lambda)$ generates the isomorphism*

$$A(\lambda) : \bigoplus_{k=1}^{p} H^{s+mq+m_k,\varphi}(\Gamma) \leftrightarrow \bigoplus_{j=1}^{p} H^{s+m_j,\varphi}(\Gamma) \tag{5.38}$$

for any $s \in \mathbb{R}$ and $\varphi \in \mathcal{M}$.

(ii) *For any parameters $s \in \mathbb{R}$ and $\varphi \in \mathcal{M}$ chosen arbitrarily, there exists a number $c \geq 1$ such that the two-sided estimate*

$$c^{-1}\sum_{j=1}^{p} \|f_j\|_{H^{s+m_j,\varphi}(\Gamma)}$$

$$\leq \sum_{k=1}^{p} \|u_k\|_{H^{s+mq+m_k,\varphi}(\Gamma)} + |\lambda|^q \sum_{k=1}^{p} \|u_k\|_{H^{s+m_k,\varphi}(\Gamma)}$$

$$\leq c\sum_{j=1}^{p} \|f_j\|_{H^{s+m_j,\varphi}(\Gamma)} \tag{5.39}$$

is valid for any $\lambda \in K$ satisfying the condition $|\lambda| \geq \lambda_0$ and any vector-valued functions

$$u = \mathrm{col}\,(u_1, \ldots, u_p) \in \bigoplus_{k=1}^{p} H^{s+mq+m_k,\varphi}(\Gamma), \tag{5.40}$$

$$f = \mathrm{col}\,(f_1, \ldots, f_p) \in \bigoplus_{j=1}^{p} H^{s+m_j,\varphi}(\Gamma) \tag{5.41}$$

satisfying equation (5.34)*. Here, the number c is independent of λ and vector-valued functions u and f.*

In the case of $\varphi \equiv 1$ (the Sobolev scale), this theorem is known [10, Sec. 4.3 d]. Note that the left inequality in the two-sided estimate (5.39) is true even if we omit the assumption that system (5.34) is parameter-elliptic (cf. [13, Sec. 2, Subsec. 1]). In order to avoid awkward expressions in (5.39), we used equivalent non-Hilbert norms in spaces (5.40) and (5.41).

We prove statements (i) and (ii) of Theorem 5.10 separately. The general case of $\varphi \in \mathcal{M}$ is derived from the Sobolev case of $\varphi \equiv 1$.

Proof of Theorem 5.10. As already indicated, the statement of the theorem is true in the Sobolev case where $\varphi \equiv 1$. Hence, there exists a number $\lambda_0 > 0$ such that the isomorphisms

$$A(\lambda) : \bigoplus_{k=1}^{p} H^{s\mp1+mq+m_k}(\Gamma, |\lambda|^q, mq) \leftrightarrow \bigoplus_{j=1}^{p} H^{s\mp1+m_j}(\Gamma) \tag{5.42}$$

take place for any $\lambda \in K$ with $|\lambda| \geq \lambda_0$ and arbitrary $s \in \mathbb{R}$ and $\varphi \in \mathcal{M}$. Furthermore, the norm of operator (5.42) and the norm of its inverse operator are uniformly bounded in λ. Here, we use the Hilbert space $H^{\sigma,\varphi}(\Gamma, \varrho, \theta)$ with $\sigma \in \mathbb{R}$, $\varrho = |\lambda|^q$, and $\theta = mq$, defined in Subsection 2.2.4. This space is equal to $H^{\sigma,\varphi}(\Gamma)$ up to equivalence of norms.

We choose arbitrary $s \in \mathbb{R}$ and $\varphi \in \mathcal{M}$. Let ψ be the interpolation parameter from Proposition 2.2, where we set $\varepsilon = \delta = 1$. Applying the interpolation with this parameter to (5.42), we obtain the isomorphism

$$A(\lambda) : \left[\bigoplus_{k=1}^{p} H^{s-1+mq+m_k}(\Gamma, |\lambda|^q, mq), \bigoplus_{k=1}^{p} H^{s+1+mq+m_k}(\Gamma, |\lambda|^q, mq)\right]_{\psi}$$
$$\leftrightarrow \left[\bigoplus_{j=1}^{p} H^{s-1+m_j}(\Gamma), \bigoplus_{j=1}^{p} H^{s+1+m_j}(\Gamma)\right]_{\psi}. \tag{5.43}$$

By virtue of Theorem 1.8, the norm of operator (5.43) and the norm of its inverse operator are uniformly bounded in the parameter λ. [In (5.43), the admissible pairs of spaces are normal.] Hence, by using Theorem 1.5 on interpolation of direct sums of spaces, we obtain the isomorphism

$$A(\lambda) : \bigoplus_{k=1}^{p} \left[H^{s-1+mq+m_k}(\Gamma, |\lambda|^q, mq), H^{s+1+mq+m_k}(\Gamma, |\lambda|^q, mq)\right]_{\psi}$$
$$\leftrightarrow \bigoplus_{j=1}^{p} \left[H^{s-1+m_j}(\Gamma), H^{s+1+m_j}(\Gamma)\right]_{\psi}. \tag{5.44}$$

Moreover, the norms of operators (5.43) and (5.44) are equal and the norms of their inverse operators are equal too.

We now use Lemma 2.6 with

$$\sigma := s + mq + m_k, \quad \varrho := |\lambda|^q, \quad \theta := mq, \quad \text{and} \quad \varepsilon = \delta = 1,$$

and apply Theorem 2.2. Due to these results, (5.44) implies the isomorphism

$$A(\lambda) : \bigoplus_{k=1}^{p} H^{s+mq+m_k,\varphi}(\Gamma, |\lambda|^q, mq) \leftrightarrow \bigoplus_{j=1}^{p} H^{s+m_j,\varphi}(\Gamma) \tag{5.45}$$

such that the norm of operator (5.45) and the norm of its inverse operator are uniformly bounded in λ. This yields the required isomorphism (5.38) and the two-sided estimate (5.39).

Theorem 5.10 is proved.

Theorem 5.10(i) yields the following assertion on the index of the operator corresponding to the parameter-elliptic system.

Corollary 5.3. *Suppose that system* (5.34) *is parameter-elliptic on a certain closed ray*

$$K := \{\lambda \in \mathbb{C} : \arg\lambda = \text{const}\}.$$

Then, for any $\lambda \in \mathbb{C}$, *operator* (5.37) *has zero index.*

Proof. For any fixed $\lambda \in \mathbb{C}$, (5.34) is a Douglis–Nirenberg elliptic system. Therefore, by virtue of Theorem 5.5, the index of operator (5.37) is finite and independent of $s \in \mathbb{R}$ and $\varphi \in \mathcal{M}$. Moreover, this index is also independent of the parameter λ. Indeed, by virtue of (5.33), the parameter λ affects only the lowest terms of every element of the matrix PsDO $A(\lambda)$:

$$A(\lambda) - A(0) = \sum_{r=0}^{q-1} \lambda^{q-r} A^{(r)} = \Big(\sum_{r=0}^{q-1} \lambda^{q-r} A^{(r)}_{j,k} \Big)_{j,k=1}^{p},$$

where

$$\operatorname{ord} \sum_{r=0}^{q-1} \lambda^{q-r} A^{(r)}_{j,k} \le m(q-1) + m_k - m_j.$$

Hence, in view of Lemma 2.5, we get the bounded operator

$$A(\lambda) - A(0) : \bigoplus_{k=1}^{p} H^{s+mq+m_k,\varphi}(\Gamma) \to \bigoplus_{j=1}^{p} H^{s+m+m_j,\varphi}(\Gamma).$$

However, by virtue of Theorem 2.3(iii) and the condition $m > 0$, the embedding $H^{s+m+m_j,\varphi}(\Gamma) \hookrightarrow H^{s+m_j,\varphi}(\Gamma)$ is compact. Therefore, the operator

$$A(\lambda) - A(0) : \bigoplus_{k=1}^{p} H^{s+mq+m_k,\varphi}(\Gamma) \to \bigoplus_{j=1}^{p} H^{s+m_j,\varphi}(\Gamma)$$

is compact. This implies (see, e.g., [86, Corollary 19.1.8]) that the operators $A(\lambda)$ and $A(0)$ have the same index, i.e., the index does not depend on the parameter λ. According to Theorem 5.10(i), isomorphism (5.38) holds for sufficiently large absolute values of the parameter $\lambda \in K$. Consequently, the index of the operator $A(\lambda)$ is equal to zero for $\lambda \in K$ satisfying the condition $|\lambda| \gg 1$, and hence, for any $\lambda \in \mathbb{C}$.

Corollary 5.3 is proved.

5.3 Elliptic boundary-value problems for systems of equations

In the present section, we study elliptic boundary-value problems for systems of partial differential equations in the refined scale of spaces. As in the case of a single equation, these problems generate bounded and Fredholm operators in the related pairs of positive Sobolev spaces; see monograph [270, Sec. 9.4], the survey [11, Sec. 6], and the references therein. We extend this result to the refined Sobolev scale applied to Petrovskii elliptic systems.

5.3.1 Statement of the problem

Consider the linear system of $p \geq 2$ partial differential equations

$$\sum_{k=1}^{p} L_{j,k}\, u_k = f_j \quad \text{in} \quad \Omega, \quad j = 1, \dots, p, \tag{5.46}$$

where $L_{j,k} = L_{j,k}(x, D)$, $j, k = 1, \dots, p$, are scalar linear partial differential expressions with complex-valued coefficients from $C^\infty(\overline{\Omega})$. For each number $k = 1, \dots, p$, we set

$$m_k := \max\{\operatorname{ord} L_{j,k} : j = 1, \dots, p\}.$$

Thus, m_k is the maximal differentiation order for the required function u_k. It is assumed that all $m_k \geq 1$.

We associate system (5.46) with the $p \times p$ square matrix

$$\mathbf{L}^{(0)}(x, \xi) := \big(L^{(0)}_{j,k}(x, \xi)\big)_{j,k=1}^{p} \quad \text{of} \quad x \in \overline{\Omega} \quad \text{and} \quad \xi \in \mathbb{C}^n.$$

Here, $L^{(0)}_{j,k}(x, \xi)$ is the principal symbol of the partial differential equation $L_{j,k}$ in the case where $\operatorname{ord} L_{j,k} = m_k$ or $L^{(0)}_{j,k}(x, \xi) \equiv 0$ in the case where $\operatorname{ord} L_{j,k} < m_k$.

Definition 5.4. System (5.46) is called *Petrovskii elliptic* on $\overline{\Omega}$ if

$$\det \mathbf{L}^{(0)}(x, \xi) \neq 0 \quad \text{for all} \quad x \in \overline{\Omega} \quad \text{and} \quad \xi \in \mathbb{R}^n \setminus \{0\}.$$

If system (5.46) is Petrovskii elliptic and $n \geq 3$, then the homogeneous polynomial $\det \mathbf{L}^{(0)}(x, \xi)$ has even degree [11, Sec. 6.1 a]:

$$\sum_{k=1}^{p} m_k = 2q \quad \text{for a certain} \quad q \in \mathbb{N}.$$

We assume that this condition is satisfied for any integer $n \geq 2$.

Consider the solutions of system (5.46) that satisfy the boundary conditions

$$\sum_{k=1}^{p} B_{j,k}\, u_k = g_j \quad \text{on} \quad \Gamma, \quad j = 1, \dots, q. \tag{5.47}$$

Here, $B_{j,k} = B_{j,k}(x, D)$, where $j = 1, \dots, q$ and $k = 1, \dots, p$, are scalar linear boundary differential expressions defined on Γ. The expression $B_{j,k}$ has the order $\operatorname{ord} B_{j,k} \leq m_k - 1$ and infinitely smooth complex-valued coefficients. For each number $j \in \{1, \dots, q\}$, we set

$$r_j := \min\{m_k - \operatorname{ord} B_{j,k} : k = 1, \dots, p\} \geq 1.$$

Here, we define $\operatorname{ord} B_{j,k} := -\infty$ if $B_{j,k} \equiv 0$. Thus, $\operatorname{ord} B_{j,k} \le m_k - r_j$ for all $j \in \{1, \dots, q\}$ and $k \in \{1, \dots, p\}$.

We associate the boundary conditions (5.47) with the $q \times p$ matrix

$$\mathbf{B}^{(0)}(x, \xi) := \big(B^{(0)}_{j,k}(x, \xi)\big)_{\substack{j=1,\dots,q\\ k=1,\dots,p}} \quad \text{of} \quad x \in \Gamma \quad \text{and} \quad \xi \in \mathbb{C}^n.$$

Here, $B^{(0)}_{j,k}(x, \xi)$ is the principal symbol of the differential expression $B_{j,k}$ in the case where $\operatorname{ord} B_{j,k} = m_k - r_j$ or $B^{(0)}_{j,k}(x, \xi) \equiv 0$ in the case where $\operatorname{ord} B_{j,k} < m_k - r_j$.

Definition 5.5. The boundary-value problem (5.46), (5.47) is called Petrovskii elliptic in the domain Ω if the following conditions are satisfied:

(i) System (5.46) is properly elliptic on $\overline{\Omega}$, i.e., for any point $x \in \Gamma$ and any linearly independent vectors $\xi', \xi'' \in \mathbb{R}^n$, the polynomial

$$\det \mathbf{L}^{(0)}(x, \xi' + \tau \xi'')$$

in the variable τ has exactly q roots $\tau^+_j(x; \xi', \xi'')$, $j = 1, \dots, q$, with positive imaginary parts and the same number of roots with negative imaginary parts (taken with regard for multiplicities of the roots).

(ii) The boundary conditions (5.47) satisfy the complementing condition with respect to system (5.46) on Γ, i.e., for any point $x \in \Gamma$ and any vector $\xi \neq 0$ tangent to Γ at the point x, the rows of the matrix

$$\mathbf{B}^{(0)}(x, \xi + \tau\nu(x)) \cdot \mathbf{L}^{(0)}_{\mathrm{c}}(x, \xi + \tau\nu(x)),$$

whose elements are treated as polynomials in τ, are linearly independent modulo the polynomial $\prod_{j=1}^{q}(\tau - \tau^+_j(x; \xi, \nu(x)))$. Here, $\mathbf{L}^{(0)}_{\mathrm{c}}(x, \xi)$ is the transposed matrix of the algebraic complements of elements of the matrix $\mathbf{L}^{(0)}(x, \xi)$.

Remark 5.1. If system (5.46) satisfies condition (i) of Definition 5.5, then it is Petrovskii elliptic on $\overline{\Omega}$. The converse statement is true for $n \ge 3$ [11, Sec. 6.1 a].

We give examples of elliptic boundary-value problems for systems of partial differential equations.

Example 5.1. The elliptic boundary-value problem for the Cauchy–Riemann system

$$\frac{\partial u_1}{\partial x_1} - \frac{\partial u_2}{\partial x_2} = f_1, \quad \frac{\partial u_1}{\partial x_2} + \frac{\partial u_2}{\partial x_1} = f_2 \quad \text{in} \quad \Omega,$$

$$u_1 + u_2 = g \quad \text{on} \quad \Gamma.$$

Here, $n = p = 2$ and $m_1 = m_2 = 1$, and hence, $q = 1$. The Cauchy–Riemann system is an example of a homogeneous elliptic system. These systems satisfy Definition 5.4, with $m_1 = \ldots = m_p$.

Example 5.2. The Petrovskii elliptic boundary-value problem

$$\frac{\partial u_1}{\partial x_1} - \frac{\partial^3 u_2}{\partial x_2^3} = f_1, \quad \frac{\partial u_1}{\partial x_2} + \frac{\partial^3 u_2}{\partial x_1^3} = f_2 \quad \text{in} \quad \Omega,$$

$$u_1 = g_1, \quad u_2 \left(\text{or } \frac{\partial u_2}{\partial \nu}, \text{ or } \frac{\partial^2 u_2}{\partial \nu^2} \right) = g_2 \quad \text{on} \quad \Gamma.$$

Here, $n = p = 2$, $m_1 = 1$, and $m_2 = 3$, and hence, $q = 2$. Note that the analyzed system is not homogeneous elliptic [268, Sec. 1, Subsec. 2 b].

5.3.2 Theorem on solvability

We rewrite the boundary-value problem (5.46), (5.47) in the matrix form

$$\mathbf{L}u = f \text{ in } \boldsymbol{\Omega}, \quad \mathbf{B}u = g \text{ on } \boldsymbol{\Gamma},$$

where $\mathbf{L} := (L_{j,k})_{j,k=1}^{p}$ and $\mathbf{B} := (B_{j,k})_{\substack{j=1,\ldots,q \\ k=1,\ldots,p}}$ are matrix differential expressions and $u := \operatorname{col}(u_1, \ldots, u_p)$, $f := \operatorname{col}(f_1, \ldots, f_p)$, and $g := \operatorname{col}(g_1, \ldots, g_q)$ are function columns.

Theorem 5.11. *Suppose that the boundary-value problem* (5.46), (5.47) *is Petrovskii elliptic in the domain* $\boldsymbol{\Omega}$. *Let* $s > 0$ *and* $\varphi \in \mathcal{M}$. *Then the mapping* $u \mapsto (\mathbf{L}u, \mathbf{B}u)$, *where* $u \in (C^\infty(\overline{\Omega}))^p$, *can be uniquely extended (by continuity) to the bounded Fredholm operator*

$$(\mathbf{L}, \mathbf{B}) : \bigoplus_{k=1}^{p} H^{s+m_k,\varphi}(\Omega) \to (H^{s,\varphi}(\Omega))^p \oplus \bigoplus_{j=1}^{q} H^{s+r_j-1/2,\varphi}(\Gamma)$$

$$=: \mathbb{H}_{s,\varphi}(\Omega, \Gamma). \tag{5.48}$$

The kernel $\mathbf{N}$ *of operator* (5.48) *belongs to* $(C^\infty(\overline{\Omega}))^p$ *and does not depend on* s *and* φ. *The range of this operator consists of all vector-valued functions*

$$(f_1, \ldots, f_p; g_1, \ldots, g_q) \in \mathbb{H}_{s,\varphi}(\Omega, \Gamma)$$

such that

$$\sum_{j=1}^{p} (f_j, w_j)_\Omega + \sum_{j=1}^{q} (g_j, h_j)_\Gamma = 0$$

for any vector-valued function

$$(w_1, \dots, w_p; h_1, \dots, h_q) \in \mathbf{W}.$$

Here, $\mathbf{W}$ *is a finite-dimensional subspace of* $(C^\infty(\overline{\Omega}))^p \times (C^\infty(\Gamma))^q$*, independent of* s *and* φ*. The index of operator* (5.48) *is equal to* $\dim \mathbf{N} - \dim \mathbf{W}$ *and, hence, does not depend on* s *and* φ*.*

In the Sobolev case of $\varphi \equiv 1$, this theorem is a special case of the well-known theorem on solvability of elliptic boundary-value problems for general systems of mixed order; see, e.g., the monograph [270, Sec. 9.4] and the survey [11, Sec. 6.3]. The general case $\varphi \in \mathcal{M}$ is deduced from the Sobolev case with the help of interpolation by analogy with the proof of Theorem 4.1.

5.4 Remarks and comments

Section 5.1. Broad classes of elliptic systems of partial differential equations were introduced and studied by I. G. Petrovskii [191] (see also [192, p. 328]) and A. Douglis and L. Nirenberg [47]. For these systems, the theorems on smoothness of solutions are proved for the Hölder spaces (with noninteger indices) and Sobolev spaces. Moreover, *a priori* estimates of solutions are obtained. See also the monograph by L. Hörmander [81, Sec. 10.6], where the case of Sobolev spaces is considered. L. Hörmander also established *a priori* estimates for the solutions of elliptic systems of pseudodifferential equations [83, Sec. 1.0].

Elliptic systems are encountered in continuum mechanics (e.g., the matrix Lamé equation), in hydrodynamics (the linearized Navier–Stokes system) [11, Sec. 6.2], and in acoustics (the system of Biot equations) [95]. The last two cases are examples of Douglis–Nirenberg elliptic systems. Note that, as a result of reduction of an arbitrary elliptic partial differential equation to the system of first-order equations, we arrive at a Douglis–Nirenberg elliptic system.

It is worth noting that there are classes of Douglis–Nirenberg uniformly elliptic systems generating Fredholm operators on the Sobolev scale over $\mathbb{R}^n$ [196].

All theorems in this section were proved in [169]. They generalize the results presented in Section 1.4 to the case of Douglis–Nirenberg uniformly elliptic systems of pseudodifferential equations defined in the Euclidean space. The case of Petrovskii uniformly elliptic systems was separately considered in [172, 173]. For the extended Sobolev scale, various classes of uniformly elliptic systems were investigated in [173, 272].

Section 5.2. The theory of elliptic systems on closed smooth manifolds is presented, e.g., in the monograph by L. Hörmander [86, Chap. 19] and in the survey by M. S. Agranovich [10, Sec. 3.2]. The *a priori* estimates for the solutions of these systems are equivalent to the assertion that the bounded operator corresponding to the system is a Fredholm operator in appropriate pairs of Sobolev spaces. The index of the operator was found by M. F. Atiyah and I. M. Singer in [18].

For the parameter-elliptic systems, the index is equal to zero and, moreover, the corresponding operator realizes isomorphisms in appropriate pairs of Sobolev spaces provided that the absolute values of the parameter are sufficiently large; see, e.g., the survey by M. S. Agranovich [10, Sec. 4.3]. Various classes of elliptic systems with parameter were studied by A. N. Kozhevnikov [99, 100, 101, 103], M. S. Agranovich [8, 9], R. Denk, R. Mennicken, and L. R. Volevich [38], R. Denk and L. R. Volevich [40, 41], and R. Denk and M. Fairman [36, 37].

There exists a close connection between the elliptic boundary-value problems and matrix elliptic pseudodifferential operators, namely, every elliptic boundary-value problem is equivalent to a certain elliptic system of pseudodifferential equations given on the boundary of the domain. This enables one to use the theory of elliptic systems in proving theorems on solvability of elliptic boundary-value problems; see, e.g., the monographs by J. T. Wloka, B. Rowley, and B. Lawruk [270, Chap. IV], Yu. V. Egorov [52, Chap. III, Sec. 3], and L. Hörmander [86, Chap. 20].

Systems whose properties are close to the properties of elliptic systems were studied by B. R. Vainberg and V. V. Grushin [263] and R. S. Saks [214, 215, 216, 219, 220, 221].

All theorems presented in this section were proved in [170]. They generalize the results presented in Section 2.2 to the case of Douglis–Nirenberg elliptic systems of pseudodifferential equations. The case of Petrovskii elliptic systems was separately studied in [167, 154]. In the extended Sobolev scale, Douglis–Nirenberg elliptic systems were investigated by T. N. Zinchenko in [271].

Section 5.3. The elliptic boundary-value problems for various systems of differential equations were investigated by S. Agmon, A. Douglis, and L. Nirenberg [5], M. S. Agranovich and A. S. Dynin [12], L. R. Volevich [267, 268], L. N. Slobodetskii [238, 241], V. A. Solonnikov [244, 245, 246, 247], and L. Hörmander [81, Sec. 10.6], [86, Sec. 19.5]. See also the monograph by J. T. Wloka, B. Rowley, and B. Lawruk [270] devoted to elliptic boundary-value problems for systems and a survey by M. S. Agranovich [11, Sec. 6] and the references therein. It is well known that the operator corresponding to the problem is bounded and Fredholm in appropriate pairs of positive Sobolev spaces.

In the two-sided modified Sobolev scale, these problems were studied by Ya. A. Roitberg and Z. G. Sheftel [211, 212], I. A. Kovalenko [98], Ya. A. Roitberg [208], and I. Ya. Roitberg and Ya. A. Roitberg [200]. They established the theorems on complete collection of isomorphisms generated by the operator corresponding to the problem. These results are presented in the monographs by Ya. A. Roitberg [209, Chap. 10] and [210, Sec. 1.3] in connection with the Douglis–Nirenberg elliptic systems.

The boundary-value problems for systems whose properties are close to the properties of elliptic systems were studied by B. R. Vainberg and V. V. Grushin [264] and R. S. Saks [213, 217, 218]. These problems are investigated in pairs of positive Sobolev spaces.

The main result of this section (Theorem 5.11) was established in [168]. This result extends Theorem 4.7 (on solvability of scalar elliptic boundary-value problems) to the case of Petrovskii elliptic systems.

Bibliography

[1] Adams RA. Sobolev spaces. New York, Academic Press, 1975.

[2] Agmon S. On the eigenfunctions and on the eigenvalues of general elliptic boundary value problems. Commun Pure Appl Math 1962, 15, No. 2, 119–47.

[3] Agmon S. Lectures on elliptic boundary-value problems. Princenton, van Nostrand Reinhold, 1965.

[4] Agmon S, Douglis A, Nirenberg L. Estimates near the boundary for solutions of elliptic partial differential equations satisfying general boundary conditions. I. Commun Pure Appl Math 1959, 12, No. 4, 623–727.

[5] Agmon S, Douglis A, Nirenberg L. Estimates near the boundary for solutions of elliptic partial differential equations satisfying general boundary conditions. II. Commun Pure Appl Math 1964, 17, No. 1, 35–92.

[6] Agmon S, Nirenberg L. Properties of solutions of ordinary differential equations in Banach space. Commun Pure Appl Math 1963, 16, No. 2, 121–239.

[7] Agranovich MS. Elliptic singular integro-differential operators. Russian Math Surveys 1964, 20, No. 5, 1–121.

[8] Agranovich MS. Non-self-adjoint problems with a parameter that are elliptic in the sence of Agmon-Douglis-Nirenberg. Funct Anal Appl 1990, 24, No 1, 50–3.

[9] Agranovich MS. Moduli of eigenvalues of nonselfadjoint problems with a parameter that are elliptic in the Agmon-Douglis-Nirenberg sense. Funct Anal Appl 1992, 26, No 2, 116–9.

[10] Agranovich MS. Elliptic operators on closed manifolds. in: Encyclopedia of mathematical sciences. Vol. 63, Partial differential equations, VI. Elliptic and parabolic operators. Berlin, Springer, 1994, 1–130.

[11] Agranovich MS. Elliptic boundary problems. in: Encyclopedia of mathematical sciences. Vol. 79, Partial differential equations IX. Berlin, Springer, 1997, 1–144.

[12] Agranovich MS, Dynin AS. General boundary-value problems for elliptic systems in higher-dimensional regions. Soviet Math Dokl 1962, 3, 1323–7.

[13] Agranovich MS, Vishik MI. Elliptic problems with parameter and parabolic problems of general form. Russian Math Surveys 1964, 19, No. 3, 53–157.

[14] Alexits G. Convergence problems of orthogonal series. Budapest, Akadémiai Kiadó, 1961.

[15] Anop AV, Murach AA. Elliptic boundary-value problems in a multiply connected domain on the extended Sobolev scale. in: Collection of Works of the Institute of Mathematics of the Ukrainian National Academy of Sciences 2013, 10, No. 2, 37–59 [in Ukrainian].

[16] Aronszajn N, Milgram AN. Differential equations on Riemannian manifolds. Rend Circ Mat Palermo 1952, 2, 1–61.

[17] Atiyah MF, Bott R. The index theorem for manifolds with boundary. in: Proceedings of Symposium on Differential Analysis. Bombay Coll. London, Oxford University Press, 1964, 175–86.

[18] Atiyah MF, Singer IM. The index of elliptic operators on compact manifolds. Bull Amer Math Soc 1963, 69, No. 3, 422–33.

[19] Avakumović VG. O jednom O-inverznom stavu. Rad Jugoslovenske Akad Znatn Umjetn 1936, 254, 167–86.

[20] Bennet K, Sharpley R. Interpolation of operators. Boston, Academic Press, 1988.

[21] Berezansky YuM. Expansions in eigenfunctions of selfadjoint operators. Providence, Am Math Soc, 1968.

[22] Berezansky YuM, Krein SG, Roitberg YaA. A theorem on homeomorphisms and local increase in smoothness up to the boundary for solutions of elliptic equations. Soviet Math Dokl 1963, 4, 152–5.

[23] Berezansky YuM, Sheftel ZG, Us GF. Functional analysis. Vol. II. Basel, Birkhäser, 1996.

[24] Bergh J, Löfström J. Interpolation spaces. An introduction. Berlin, Springer, 1976.

[25] Besov OV, Il'in VP, Nikol'skii SM. Integral representations of functions and imbedding theorems, Vol I, II. Washington, Winston & Sons, 1978, 1979.

[26] Bingham NH, Goldie CM, Teugels JL. Regular variation. Cambridge, Cambridge University Press, 1989.

[27] Boyd DW. The Hilbert transform on rearrangement-invariant spaces. Canadian J Math 1967, 19, 599–616.

[28] Browder FE. On the regularity properties of solutions of elliptic differential equations. Commun Pure Appl Math 1956, 9, No. 3, 351–61.

[29] Browder FE. Estimates and existence theorems for elliptic boundary-value problems. Proc Nat Acad Sci 1959, 45, No. 3, 365–72.

[30] Brudnyi YuA, Krugljak NYa. Interpolation functors and interpolation spaces. Amsterdam, North-Holland, 1991.

[31] Burenkov V. I. Extension theorems for Sobolev spaces. Oper Theory Adv Appl 1999, 109, 187–200.

[32] Caetano AM, Leopold H-G. Local growth envelops of Triebel–Lizorkin spaces of generalized smoothness. J Fourier Anal Appl 2006, 12, No 4, 427–45.

[33] Calderon AP. Intermediate spaces and interpolation, the complex method. Studia Math 1964, 24, 113–90.

[34] Carro MJ, Cerdà J. On complex interpolation with an analytic functional. Math Scand 1990, 66, No. 2, 264–74.

[35] Cobos F, Fernandez DL. Hardy–Sobolev spaces and Besov spaces with a function parameter. Lecture Notes in Mathematics 1988, 1302, 158–70.

[36] Denk R, Fairman M. Estimates for solutions of a parameter-elliptic multi-order system of differential equations. Integral Equations Operator Theory 2010, 66, No. 3, 327–65.

[37] Denk R, Fairman M. Necessity of parameter-ellipticity for multi-order systems of differential equations. Oper Theory Adv Appl 2012, 221, 255–68.

[38] Denk R, Mennicken R, Volevich LR. The Newton polygon and elliptic problems with parameter. Math Nachr, 1998, 192, No. 1, 125–57.

[39] Denk R, Mennicken R, Volevich LR. On elliptic operator pencils with general boundary conditions. Integral Equations Operator Theory 2001, 39, No. 1, 15–40.

[40] Denk R, Volevich LR. A priori estimates for a singularly perturbed mixed-order boundary value problem. Russ J Math Phys 2000, 7, No. 3, 288–318.

[41] Denk R, Volevich LR. Elliptic boundary value problems with large parameter for mixed order systems. in: Partial differential equations (AMS Transl, Ser. 2, Vol. 206). Providence, Amer Math Soc, 2002, 29–64.

[42] Denk R, Volevich LR. Parameter-elliptic boundary value problems connected with the Newton polygon. Differential Integral Equations 2002, 15, No. 3, 289–326.

[43] Dines N, Harutyunyan G, Schulze B.-W. Mixed boundary value problems and parametrices in the edge calculas. Bull Sci Math 2007, No. 4, 325–60.

[44] Djakov PB, Mityagin BS. Instability zones of one-dimensional periodic Schrödinger and Dirac operators. Russian Math Surveys 2006, 61, No. 4, 663–766.

[45] Djakov PB, Mityagin BS. Spectral gaps of Schrödinger operators with periodic singular potentials. Dyn Partial Differ Equat 2009, 6, No. 2, 95–165.

[46] Donoghue WF. The interpolation of quadratic norms. Acta Math 1967, 118, No. 3–4, 251–70.

[47] Douglis A, Nirenberg L. Interior estimates for elliptic systems of partial differential equations. Comm Pure Appl Math 1955, 8, No. 4, 503–38.

[48] Edmunds DE, Gurka P, Opic B. On embeddings of logarithmic Bessel potential spaces. J Funct Anal 1997, 146, 116–50.

[49] Edmunds DE, Haroske DD. Spaces of Lipschitz type, embeddings and entropy numbers. Diss Math 1999, 380, 1–43.

[50] Edmunds DE, Triebel H. Spectral theory for isotropic fractal drums. C R Math Acad Sci Paris 1998, 326, 1269–74.

[51] Edmunds DE, Triebel H. Eigenfrequencies of isotropic fractal drums. Oper Theory Adv Appl 1999, 110, 81–102.

[52] Egorov YuV. Linear differential equations of principal type. New York, Consultants Bureau, 1986.

[53] Eidel'man SD, Zhitarashu NV. Parabolic boundary-value problems. Basel, Birkhäser, 1998 (Oper. Theory Adv. Appl., Vol. 101).

[54] Farkas W, Jacob N, Schilling RL. Feller semigroups, L_p-sub-Markovian semigroups, and applications to pseudo-differential operators with negative definite symbols. Forum Math 2001, 13, 59–90.

[55] Farkas W, Jacob N, Schilling RL. Function spaces related to continuous negative definite functions: ψ-Bessel potential spaces. Diss Math 2001, 393, 1–62.

[56] Farkas W, Leopold H-G. Characterizations of function spaces of generalized smoothness. Ann Mat Pura Appl 2006, 185, No. 1, 1–62.

[57] Foiaş C, Lions J-L. Sur certains théorèmes d'interpolation. Acta Scient Math Szeged 1961, 22, No. 3–4, 269–82.

[58] Franke J. Besov–Triebel–Lizorkin spaces and boundary value problems. Seminar analysis 1984/85, Berlin, Akad Wiss DDR, 1985, 89–104.

[59] Gagliardo E. Caratterizzazione construttiva di tutti gli spazi di interpolazione tra spazi di Banach. Symposia Math. Vol. II, INDAM, Rome, 1968, London, Academic Press, 1969, 95–106.

[60] Gel'fand IM, Shilov GE. Generalized functions. Vol. 1. Properties and operators. New-York, Academic Press, 1964.

[61] Gel'fand IM, Shilov GE. Generalized functions. Vol. 2. Spaces of fundamental and generalized functions. New-York, Academic Press, 1968.

[62] Geluk JL, Haan de L. Regular variation, extensions and Tauberian theorems. Amsterdam, Stichting Mathematisch Centrum, 1987.

[63] Geymonat G. Sui problemi ai limiti per i sistemi lineari ellittici. Ann Mat Pura Appl Ser 4, 1965, 69, 207–84.

[64] Gilbarg D, Trudinger NS. Elliptic partial differential equations of second order [revised 3rd edn]. Berlin, Springer, 1998.

[65] Gohberg ITs, Krein MG. The basic propositions on defect numbers, root vectors, and indices of linear operators. Amer Math Soc Transl Ser II 1960, 13, 185–264.

[66] Grisvard P. Caractérisation de quelques espaces d'interpolation. Arch Rat Mech Anal 1967, 25, No. 1, 40–63.

[67] Grubb G. A characterization of non-local boundary value problems associated with an elliptic operator. Ann Scuola Norm Sup Pisa (3) 1968, 22, No. 3, 425–513.

[68] Grubb G. On coerciveness and semiboundedness of general boundary problems. Isr J Math 1971, 10, 32–95.

[69] Grubb G. Functional Calculus of Pseudo-Differential Boundary Problems [2-nd ed]. Boston, Birkhäser, 1996.

[70] Gurka P, Opic B. Sharp embeddins of Besov-type spaces. J Comp Appl Math 2007, 208, No. 1, 235–69.

[71] Gustavsson J. A function parameter in connection with interpolation of Banach spaces. Math Scand 1978, 42, No. 2, 289–305.

[72] Haan de L. On regular variation and its application to the weak convergence of sample extremes. Amsterdam, Mathematisch Centrum, 1970.

[73] Haroske DD. Envelops and sharp embeddings of function spaces. Boca Raton FL, Chapman and Hall/CRC, 2007.

[74] Haroske DD, Moura SD. Continuity envelopes of spaces of generalised smoothness, entropy and approximation numbers. J Approxim Theory 2004, 128, No. 2, 151–74.

[75] Haroske DD, Moura SD. Continuity envelopes and sharp embeddings in spaces of generalised smoothness. J Funct Anal 2008, 254, No. 8, 1487–521.

[76] Harutyunyan G, Schulze B.-W. Parametrices of mixed elliptic problems. Math Nachr 2004, 277, 56–82.

[77] Harutyunyan G, Schulze B.-W. Elliptic mixed, transmission and singular crack problems. Zürich, European Math Soc, 2008.

[78] Hegland M. Variable Hilbert scales and their interpolation inequalities with applications to Tikhonov regularization. Appl Anal 1995, 59, No. 1–4, 207–23.

[79] Hegland M. Error bounds for spectral enhancement which are based on variable Hilbert scale inequalities. J Integral Equat Appl 2010, 22, No. 2, 285–312.

[80] Hörmander L. On the theory of general partial differential operators. Acta Math 1955, 94, No. 1, 161–248.

[81] Hörmander L. Linear partial differential operators. Berlin, Springer, 1963.

[82] Hörmander L. Pseudo-differential operators. Comm Pure Appl Math 1965, 18, No. 3, 501–17.

[83] Hörmander L. Pseudo-differential operators and non-elliptic boundary problems. Ann Math 1966, 83, No. 1, 129–209.

[84] Hörmander L. Pseudo-differential operators and hypoelliptic equations. Singular Integrals (Proc Sympos Pure Math, Vol X, Chicago, 1966), Providence, RI, Amer Math Soc, 1967, 138–83.

[85] Hörmander L. The analysis of linear partial differential operators. II: Differential operators with constant coefficients. Berlin, Springer, 1983.

[86] Hörmander L. The analysis of linear partial differential operators. III: Pseudo-differential operators. Berlin, Springer, 1985.

[87] Jacob N. Pseudodifferential operators and Markov processes (in 3 volumes). London, Imperial College Press, 2001, 2002, 2005.

[88] Janson S. Minimal and maximal methods of interpolation. J Funct Anal 1981, 44, No. 1, 50–73.

[89] Kalugina TF. Interpolation of Banach spaces with a functional parameter. The reiteration theorem. Moscow Univ Math Bull 1975, 30, No. 6, 108–16.

[90] Kalyabin GA, Lizorkin PI. Spaces of functions of generalized smoothness. Math Nachr 1987, 133, No. 1, 7–32.

[91] Karamata J. Sur certains "Tauberian theorems" de M. M. Hardy et Littlewood. Mathematica (Cluj) 1930, 3, 33–48.

[92] Karamata J. Sur un mode de croissance régulière des fonctions. Mathematica (Cluj) 1930, 4, 38–53.

[93] Karamata J. Sur un mode de croissance régulière. Théorêms foundamentaux. Bull Soc Math France 1933, 61, 55–62.

[94] Kashin BS, Saakyan AA. Orthogonal series. Providence, RI, Amer Math Soc, 1989.

[95] Khashanah K. A Douglis–Nirenberg elliptic operator in Biot theory. Appl Anal 1996, 61, No. 1–2, 87–96.

[96] Kohn JJ, Nirenberg L. An algebra of pseudo-differential operators. Commun Pure Appl Math 1965, 18, No. 1–2, 269–305.

[97] Kostarchuk YuV, Roitberg YaA. Isomorphism theorems for elliptic boundary value problems with boundary conditions that are not normal. Ukrainian Math J 1973, 25, No. 2, 222–6.

[98] Kovalenko IA. Theorems on isomorphisms for elliptic systems with boundary conditions which are not normal. Ukr Mat Zh 1973, 25, No. 3, 303–7.

[99] Kozhevnikov AN. Spectral problems for pseudodifferential systems elliptic in the Douglis-Nirenberg sense and their applications. Math Sb 1973, 21, No. 1, 63–90.

[100] Kozhevnikov A. Asymptotics of the spectrum of the Douglis-Nirenberg elliptic operators on a compact manifold. Math Nachr 1996, 182, No. 1, 261–93.

[101] Kozhevnikov A. Parameter-ellipticity for mixed-order elliptic boundary problems. CR Math Acad Sci Paris 1997, 324, No. 12, 1361–66.

[102] Kozhevnikov A. Complete scale of isomorphisms for elliptic pseudodifferential boundary-value problems. J London Math Soc (2) 2001, 64, No 2, 409–22.

[103] Kozhevnikov A. Parameter-ellipticity for mixed order systems in the sense of Petrovskii. Commun Appl Anal 2001, 5, No. 2, 277–91.

[104] Kozlov VA, Maz'ya VG, Rossmann J. Elliptic boundary value problems in domains with point singularities. Providence, RI, American Mathematical Society, 1997.

[105] Krein SG. On an interpolation theorem in operator theory. Soviet Math Dokl 1960, 1, 61–4.

[106] Krein SG. On the concept of a normal scale of spaces. Soviet Math Dokl 1960, 1, 586–9.

[107] Krein SG., ed. Functional analysis. Moscow, Nauka, 1972 [in Russian].

[108] Krein SG, Petunin YuI. Scales of Banach spaces. Russian Math Surveys 1966, 21, 85–159.

[109] Krein SG, Petunin YuI, Semenov EM. Interpolation of linear operators. Providence, RI, Amer Math Soc, 1982.

[110] Krugljak NYa. On the reiteration property of $\overrightarrow{X}_{\varphi,q}$ spaces. Math Scand 1993, 73, No. 1, 65–80.

[111] Ladyzhenskaya OA, Ural'tseva NN. Linear and quasilinear elliptic equations. New-York, Academic Press, 1968.

[112] Leopold H-G. Embeddings and entropy numbers in Besov spaces of generalized smoothness. Lecture Notes on Pure and Applied Mathematics 2000, 213, 323–36.

[113] Lewy H. An example of a smooth linear partial differential equation without solution. Ann Math (2) 1957, 66, 155–8.

[114] Lions J-L. Espaces intermédiaires entre espaces hilbertiens et applications. Bull Math Soc Sci Math Phys R P Roumanie 1958, 50, No. 4, 419–32.

[115] Lions J-L. Théorèmes de trace et d'interpolation. I. Ann Scuola Norm Sup Pisa (3) 1959, 13, 389–403.

[116] Lions J-L. Une construction d'espaces d'interpolation C R Math Acad Sci Paris 1961, 251, 1853–5.

[117] Lions J-L. Quelques méthodes de résolution des problèmes aux limites non linéaires. Paris, Dunod, 1969.

[118] Lions J-L. Optimal control of systems governed by partial differential equations. New-York, Springer, 1971.

[119] Lions J-L, Magenes E. Problémes aux limites non homogénes, V. Ann Scuola Norm Sup Pisa (3) 1962, 16, 1–44.

[120] Lions J-L, Magenes E. Problémes aux limites non homogénes, VI. J Anal Math 1963, 11, 165–88.

[121] Lions J-L, Magenes E. Non-homogeneous boundary-value problems and applications, Vol. I. Berlin, Springer, 1972.

[122] Lions J-L, Peetre J. Propriétés d'espaces d'interpolation. C R Math Acad Sci Paris 1961, 253, 1747–9.

[123] Lopatinskii YaB. A method of reduction of boundary-value problems for systems of differential equations of elliptic type to a system of regular integral equations. Ukr Mat Zh 1953, 5, No. 2, 123–51. [English transl in: Amer Math Soc Transl Ser II 1970, 85, 149–83.]

[124] Lopatinskii YaB. Theory of general boundary-value problems. Selected Works. Kiev, Naukova Dumka, 1984 [in Russian].

[125] Löfström J. Interpolation of boundary value problems of Neumann type on smooth domains. J. London Math. Soc (2) 1992, 46, No. 3, 499–516.

[126] Magenes E. Spazi di interpolazione ed equazioni a derivate parziali. in: Atti del Settimo Congresso dell'Unione Matematica Italiana (Genoa, 1963). Rome, Edizioni Cremonese, 1965, 134–97.

[127] Malgrange B. Sur une classe d'opératuers différentiels hypoelliptiques. Bull Soc Math France 1957, 85, 283–306.

[128] Maric V. Regular variation and differential equations. New York, Springer, 2000.

[129] Mathé P, Tautenhahn U. Interpolation in variable Hilbert scales with application to inverse problems. Inverse Problems 2006, 22, No. 6, 2271–97.

[130] Matuszewska W. On a generalization of regularly increasing functions. Studia Math 1964, 24, 271–9.

[131] Maz'ya VG, Shaposhnikova TO. Theory of multipliers in spaces of differentiable functions. Boston, Pitman, 1985.

[132] Maz'ya VG, Shaposhnikova TO. Theory of Sobolev multipliers. With applications to differential and integral operators. Berlin, Springer, 2009.

[133] Meaney C. On almost-everywhere convergent eigenfunction expansions of the Laplace–Beltrami operator. Math Proc Camb Phil Soc 1982, 92, 129–31.

[134] Menschoff D. [Men'shov D. E.] Sur les series de fonctions orthogonales I. Fund Math 1923, 4, 82–105.

[135] Merucci C. Interpolation réelle avec fonction paramètre: réitération et applications aux espaces $\Lambda^{\varrho}(\varphi)$. C R Math Acad Sci Paris 1982, 295, No. 6, 427–30.

[136] Merucci C. Application of interpolation with a function parameter to Lorentz, Sobolev and Besov spaces. Lecture Notes in Mathematics 1984, 1070, 183–201.

[137] Mikhailets VA. Asymptotics behaviour of the spectrum of elliptic operators and boundary conditions. Dokl Math 1982, 26, No. 2, 464–8.

[138] Mikhailets VA. A sharp estimate for the remainder in the spectral asymptotics of general elliptic boundary-value problems. Funct Anal Appl 1989, 23, No. 2, 137–9.

[139] Mikhailets VA. On the theory of general boundary-value problems for elliptic equations. in: Nonlinear boundary-value problems. Vol. 2, Kiev, Naukova Dumka, 1990, 66–70 [in Russian].

[140] Mikhailets VA, Molyboga V. Spectral gaps of the one-dimensional Schrödinger operators with singular periodic potentials. Meth Funct Anal Topol 2009, 15, No. 1, 31–40.

[141] Mikhailets VA, Molyboga V. Hill's potentials in Hörmander spaces and their spectral gaps. Meth Funct Anal Topol 2011, 17, No. 3, 235–43.

[142] Mikhailets VA, Molyboga V. Smoothness of Hill's potential and lengths of spectral gaps. Oper Theory Adv Appl 2012, 221, 467–78.

[143] Mikhailets VA, Murach AA. Elliptic operators in a refined scale of functional spaces. Ukr Math J 2005, 57, No. 5, 817–25.

[144] Mikhailets VA, Murach AA. Improved scales of spaces and elliptic boundary-value problems. I. Ukr Math J 2006, 58, No. 2, 244–62.

[145] Mikhailets VA, Murach AA. Improved scales of spaces and elliptic boundary-value problems. II. Ukr Math J 2006, 58, No. 3, 398–417.

[146] Mikhailets VA, Murach AA. Interpolation of spaces with functional parameter and spaces of differentiable functions. Dop Nats Akad Nauk Ukr Ser Mat Pryrodoznav Tekhn Nauk 2006, No. 6, 13–8 [in Russian].

[147] Mikhailets VA, Murach AA. Elliptic operator in the improved scale of spaces on a closed manifold. Dop Nats Akad Nauk Ukr Mat Pryrodoznav Tekhn Nauk 2006, No. 10, 27–33 [in Russian].

[148] Mikhailets VA, Murach AA. Regular elliptic boundary-value problem for a homogeneous equation in a two-sided improved scale of spaces. Ukr Math J 2006, 58, No. 11, 1748–67.

[149] Mikhailets VA, Murach AA. An elliptic operator with homogeneous regular boundary conditions in two-sided refined scale of spaces. Ukr Math Bull 2006, 3, No. 4, 529–60.

[150] Mikhailets VA, Murach AA. Refined scales of spaces and elliptic boundary-value problems. III. Ukr Math J 2007, 59, No. 5, 744–65.

[151] Mikhailets VA, Murach AA. Elliptic boundary-value problem in a two-sided refined scale of spaces. Ukr Math J 2008, 60, No. 4, 574–97.

[152] Mikhailets VA, Murach AA. Interpolation Hörmander spaces and elliptic operators. in: Collection of Works of the Institute of Mathematics, Ukrainian National Academy of Sciences 2008, 5, No. 1, 205–26 [in Russian].

[153] Mikhailets VA, Murach AA. Interpolation with a function parameter and refined scale of spaces. Meth Funct Anal Topol 2008, 14, No. 1, 81–100.

[154] Mikhailets VA, Murach AA. Elliptic systems of pseudodifferential equations in a refined scale on a closed manifold. Bull Pol Acad Sci Math 2008, 56, No. 3–4, 213–24.

[155] Mikhailets VA, Murach AA. Elliptic problems and Hörmander spaces. Oper Theory Adv Appl 2009, 191, 447–70.

[156] Mikhailets VA, Murach AA. On elliptic operators on a closed manifold. Dop Nats Akad Nauk Ukr Mat Pryrodoznav Tekhn Nauki 2009, No. 3, 29–35 [in Russian].

[157] Mikhailets VA, Murach AA. General forms of the Menshov–Rademacher, Orlicz, and Tandori theorems on orthogonal series. Meth Funct Anal Top 2011, 17, No. 4, 330–40.

[158] Mikhailets VA, Murach AA. On the unconditional almost-everywhere convergence of general orthonormal series. Ukr Math J 2012, 63, No. 10, 1543–50.

[159] Mikhailets VA, Murach AA. The refined Sobolev scale, interpolation, and elliptic problems. Banach J Math Anal 2012, 6, No. 2, 211–81.

[160] Mikhailets VA, Murach AA. Extended Sobolev scale and elliptic operators. Ukr Mat Zh 2013, 58, No. 3, 392–404.

[161] Mikhlin SG. Mathematical physics. An advanced course. Amsterdam, North-Holland, 1970.

[162] Moura S. D. Function spaces of generalized smoothness. Diss Math 2001, 398, 1–87.

[163] Murach AA. Elliptic boundary-value problems in complete scales of spaces of the Lizorkin–Triebel type. Dop/Dokl Akad Nauk Ukraïni 1994, No. 12, 36–9 [in Russian].

[164] Murach AA. Elliptic boundary-value problems in complete scales of Nikol'skii-type spaces. Ukr Math J 1994, 46, No. 12, 1827–35.

[165] Murach AA. Elliptic pseudodifferential operators in a refined scale of spaces on a closed manifold. Ukr Math J 2007, 59, No. 6, 874–93.

[166] Murach AA. Elliptic boundary-value problems in multiply connected domains in a refined scale of spaces. Dop Nats Akad Nauk Ukr Mat Pryrodoznav Tekhn Nauki 2007, No. 4, 29–35 [in Russian].

[167] Murach AA. The systems of differential equations elliptic in a sense of Petrovskii in a refined scale of spaces on a close manifold. Dop Nats Akad Nauk Ukr Mat Pryrodoznav Tekhn Nauki 2007, No. 5, 29–35 [in Russian].

[168] Murach AA. The boundary-value problem for a system of differential equations elliptic in the sense of Petrovskii in a refined scale of spaces. Dop Nats Akad Nauk Ukr Mat Pryrodoznav Tekhn Nauki 2007, No. 6, 24–31 [in Ukrainian].

[169] Murach AA. Systems elliptic in the Douglis–Nirenberg sense in spaces of generalized smoothness. Ukr Math Bull 2008, 5, No. 3, 345–59.

[170] Murach AA. Douglis–Nirenberg elliptic systems in the refined scale of spaces on a closed manifold. Meth Funct Anal Topol 2008, 14, No. 2, 142–58.

[171] Murach AA. Extension of some Lions–Magenes theorems. Meth Funct Anal Top 2009, 15, No. 2, 152–67.

[172] Murach AA. Elliptic systems in a two-sided scale of Hörmander spaces. in: Collection of Works of the Institute of Mathematics, Ukrainian National Academy of Sciences 2009, 6, No. 1, 126–41 [in Russian].

[173] Murach AA. On elliptic systems in Hörmander spaces. Ukr Math J 2009, 61, No. 3, 467–77.

[174] Murach AA, Zinchenko T. Parameter-elliptic operators on the extended Sobolev scale. Meth Funct Anal Topol 2013, 19, No. 1, 29–39.

[175] Nicola F., Rodino L. Global Pseudodifferential Calculus on Euclidean spaces. Basel, Birkhäser, 2010.

[176] Nikol'skii SM. Approximation of functions of several variables and imbedding theorems. Berlin, Springer, 1975.

[177] Opic B, Trebels W. Bessel potentials with logarithmic components and Sobolev-type embeddings. Anal Math 2000, 26, 299–319.

[178] Orlicz W. Zur Theorie der Orthogonalreihen. Bull Int Acad Sci Polon Cracovie 1927, 81–115.

[179] Ovchinnikov VI. The methods of orbits in interpolation theory. Mathematical Reports 1984, 1, No. 2, 349–515.

[180] Ovchinnikov VI. Interpolation orbits in couples of Lebesgue spaces. Funct Anal Appl 2005, 39, No. 1, 46–56.

[181] Paneah B. The oblique derivative problem. The Poincaré problem. Berlin, Wiley, 2000.

[182] Panich OI. Elliptic boundary problems with a parameter only in boundary conditions. Soviet Math Dokl 1967, 7, 1326–9.

[183] Panich OI. On boundary-value problems with a parameter in boundary conditions. in: Mathematical Physics, Issue 14, Kiev, Naukova Dumka, 1973, 140–6 [in Russian].

[184] Panich OI. Introduction to the general theory of elliptic boundary-value problems. Kiev, Vyshcha Shkola, 1986 [in Russian].

[185] Paszkiewicz A. A complete characterization of coefficients of a.e. convergent ortogonal series and majorizing measures. Invent Math 2010, 180, No. 1, 55–110.

[186] Peetre J. Mixed problems for higher order elliptic equations in two variables. I. Ann Scuola Norm Sup Pisa (3) 1961, 15, 337–53.

[187] Peetre J. Sur le nombre de paramètres dans la définition de certain espaces d'interpolation. Ric Mat 1963, 15, 248–61.

[188] Peetre J. On interpolation functions. Acta Sci Math 1966, 29, 167–71.

[189] Peetre J. On interpolation functions. II. Acta Sci Math 1968, 29, No. 1, 91–2.

[190] Persson L-E. Interpolation with a function parameter. Math Scand 1986, 59, No. 2, 199–222.

[191] Petrovskii IG. Sur l'analyticité des solutions des systèmes d'équations différentielles. Mat Sb 1939, 5, No. 1, 3–68.

[192] Petrovskii IG. Systems of equations with partial derivatives. Algebraic geometry. Selected Works. Moscow, Nauka, 1986 [in Russian].

[193] Pliś A. The problem of uniqueness for the solution of a system of partial differential equations. Bull Pol Acad Sci Math 1954, 2, 55–7.

[194] Pöschel J. Hill's potentials in weighted Sobolev spaces and their spectral gaps. in: Hamiltonian dynamical systems and applications. Dordrecht, Springer, 2008, 421–30.

[195] Pustyl'nik EI. On permutation-interpolation Hilbert spaces. Russian Math. (Izv VUZov) 1982, 26, No. 5, 52–7.

[196] Rabier P. Fredholm and regularity theory of Douglis–Nirenberg elliptic systems on $\mathbf{R}^n$. Math Z 2012, 270, No. 1–2, 369–93.

[197] Rademacher H. Einige Sätze über Reihen von allgemeinen Orthogonalfunctionen. Math Annalen 1922, 87, 111–38.

[198] Rempel S, Schulze B-W. Index theory of elliptic boundary problems. Berlin, Akademie, 1982.

[199] Reshnick SI. Extreme values, regular variation and point processes. New York, Springer, 1987.

[200] Roitberg IYa, Roitberg YaA. Green's formula and theorems on isomorphisms for general elliptic problems for Douglis–Nirenberg elliptic systems. Oper Theory Adv Appl 2000, 117, 281–99.

[201] Roitberg YaA. Local increase in the smoothness of solutions of elliptic equations up to the boundary. Ukr Mat Zh 1963, 15, No. 4, 444–8 [in Russian].

[202] Roitberg YaA. Elliptic problems with non-homogeneous boundary conditions and local increase of smoothness of generalized solutions up to the boundary. Soviet Math Dokl 1964, 5, 1034–8.

[203] Roitberg YaA. A theorem on homeomorphisms induced in L_p by elliptic operators and the local smoothness of generalized solutions. Ukr Mat Zh 1965, 17, No. 5, 122–9 [in Russian].

[204] Roitberg YaA. Theorems on homeomorphisms which can be realized by elliptic operators. Soviet Math Dokl 1968, 9, 656–60.

[205] Roitberg YaA. Green's formula and the homeomorphism theorem for general elliptic boundary value problems with boundary conditions that are not normal. Ukr Math J 1969, 21, No. 3, 343–9.

[206] Roitberg YaA. Homeomorphism theorems and Green's formula for general elliptic boundary value problems with boundary conditions that are not normal. Math Sb (USSR) 1970, 12, No. 2, 177–212.

[207] Roitberg YaA. The values on the boundary of the domain of generalized solutions of elliptic equations. Math Sb (USSR) 1971, 15, No. 2, 241–62.

[208] Roitberg YaA. A theorem about the complete set of isomorphisms for systems elliptic in the sense of Douglis and Nirenberg. Ukr Math J 1975, 27, No. 4, 447–50.

[209] Roitberg YaA. Elliptic boundary value problems in the spaces of distributions. Dordrecht, Kluwer, 1996.

[210] Roitberg YaA. Elliptic boundary value problems in the spaces of distributions. Dordrecht, Kluwer, 1999.

[211] Roitberg YaA, Sheftel ZG. A homeomorphism theorem for elliptic systems and its applications. Math Sb (USSR) 1969, 7, No. 3, 439–65.

[212] Roitberg YaA, Sheftel ZG. Green's formula and the theorem on homeomorphisms for elliptic systems with boundary conditions that are not normal. Mat Issled Kishinev 1972, 7, No. 2, 143–57 [in Russian].

[213] Saks RS. Boundary-value problems for weakly elliptic systems of differential equations. Soviet Math Dokl 1977, 18, No. 5, 1364–68.

[214] Saks RS. Weakly elliptic systems of pseudodifferential equations on a manifold without boundary. in: Proceedings of the S. L. Sobolev seminar "Partial differential equations." Novosibirsk, Institute of Mathematics, Siberian Branch of the Academy of Sciences of the USSR 1978, No. 2, 103–26 [in Russian].

[215] Saks RS. Weakly elliptic systems of pseudodifferential equations on a manifold without boundary. Soviet Math Dokl 1978, 19, No. 3, 690–4.

[216] Saks RS. Weakly elliptic systems of differential equations and their properties. in: Proceedings of the S. L. Sobolev seminar "Theory of cubature formulas and applications of functional analysis." Novosibirsk, Institute of Mathematics, Siberian Branch of the Academy of Sciences of the USSR 1979, No. 1, 91–118 [in Russian].

[217] Saks RS. Weakly elliptic boundary-value problems. in: Proceedings of the S. L. Sobolev seminar "Partial differential equations." Novosibirsk, Institute of Mathematics, Siberian Branch of the Academy of Sciences of the USSR 1980, No. 2, 57–78 [in Russian].

[218] Saks RS. Boundary-value problems for generalized elliptic systems of differential equations. in: Proceedings of the S. L. Sobolev seminar "Partial differential equations." Novosibirsk, Institute of Mathematics, Siberian Branch of the Academy of Sciences of the USSR 1981, No. 2, 86–108 [in Russian].

[219] Saks RS. Generalized elliptic pseudodifferential operators on a closed manifold. Dokl Ross Akad Sci, Math 1994, 48, No. 1, 214–9.

[220] Saks RS. On the properies of generalized elliptic pseudodifferential operators on a closed manifold. J Math Sci (New York) 2000, 99, No. 1, 936–68.

[221] Saks RS. On the local structure of generalized elliptic pseudodifferential operators and the Gauss method. J Math Sci (New York) 2002, 109, No. 5, 1965–83.

[222] Schechter M. General boundary value problems for elliptic partial differential equations. Comm Pure Appl Math 1959, 12, No. 3, 457–86.

[223] Schechter M. Mixed boundary value problems for general elliptic equations. Comm Pure Appl Math 1960, 13, No. 2, 183–201.

[224] Schechter M. A local regularity theorem. J Math Mech 1961, 10, No. 2, 279–88.

[225] Schechter M. On L_p estimates and regularity, I. Amer J Math 1963, 85, No. 1, 1–13.

[226] Schechter M. Complex interpolation. Compos Math 1967, 18, No. 1, 2, 117–47.

[227] Schechter M. Modern methods in partial differential equations. New York, McGraw-Hill, 1977.

[228] Schwartz L. Théorie des distributions. Vol. I. Paris, Herman, 1950.

[229] Schwartz L. Théorie des distributions. Vol. II. Paris, Herman, 1951.

[230] Shapiro ZYa. On general boundary-value problems for elliptic equations. Izv Akad Nauk SSSR Ser Mat 1953, 17, 539–62 [in Russian].

[231] Shlenzak G. Elliptic problems in a refined scale of spaces. Moscow Univ Math Bull 1974, 29, No. 3–4, 80–8.

[232] Shubin MA. Pseudodifferential operators and spectral theory [2nd ed]. Berlin, Springer, 2001.

[233] Seeley RT. Singular integrals and boundary value problems. Amer J Math 1966, 88, No. 4, 781–809.

[234] Seeley R. Interpolation in L_p with boundary conditions. Studia Math 1972, 44, 47–60.

[235] Seneta E. Regularly varying functions. Berlin, Springer, 1976 (Lecture Notes in Math., Vol. 508).

[236] Simanca SR. Mixed elliptic boundary value problems. Commun Partial Differential Equat 1987, 12, No. 2, 123–200.

[237] Skrypnik IV. Nonlinear higher-order elliptic equations. Kiev, Naukova Dumka, 1973 [in Russian].

[238] Slobodetskii LN. Estimates for the solutions of elliptic and parabolic systems. Dokl Akad Nauk SSSR 1958, 120, No. 3, 468–71 [in Russian].

[239] Slobodetskii LN. Estimates for the solutions of elliptic systems in L_p. Dokl Akad Nauk SSSR 1958, 123, No. 4, 616–9 [in Russian].

[240] Slobodetskii LN. Generalized Sobolev spaces and their applications to boundary-value problems for partial differential equations. Uchen Zap Leningrad Gos Ped Inst 1958, 187, 54–112. [English transl: Amer Math Soc Transl Ser II 1966, 57, 207–75.]

[241] Slobodetskii LN. Estimates for the solutions of linear elliptic and parabolic systems in L_2. I. Vestnik Leningrad Gos Univ, Ser Mat Mekh Astr 1960, 15, No. 7, 28–47.

[242] Sobolev SL. Applications of functional analysis in mathematical physics. Providence, RI, Amer Math Soc, 1963.

[243] Sobolev SL. Introduction to the theory of cubature formulas, Moscow, Nauka, 1974 [in Russian].

[244] Solonnikov VA. Bounds for the solutions of general boundary-value problems for elliptic systems. Soviet Math Dokl 1963, 4, 1089–91.

[245] Solonnikov VA. General boundary-value problems for systems elliptic in the sense of A. Douglis and L. Nirenberg. I. Amer Math Soc Transl, Ser II 1964, 56, 193–232.

[246] Solonnikov VA. General boundary-value problems for systems elliptic in the sense of A. Douglis and L. Nirenberg. II. Proc Steklov Inst Math 1968, 92, 3–32.

[247] Solonnikov VA. Estimates in L_p of solutions of elliptic and parabolic systems. Proc Steklov Inst Math 1970, 102, 157–185.

[248] Stepanets AI. Classification and approximation of periodic functions. Dordrecht, Kluwer, 1995.

[249] Stepanets AI. Methods of Approximation Theory. Utrecht, VSP, 2005.

[250] Strichartz RS. Multipliers on fractional Sobolev spaces. J Math Mech 1967, 16, No. 9, 1031–60.

[251] Tandori K. Über die orthogonalen Functionen X (unbedingte Kovergenz). Acta Sci Math 1962, 23, No. 3–4, 185–221.

[252] Tartar L. An introduction to Sobolev spaces and interpolation spaces. Berlin, Springer, 2007.

[253] Taylor M. Pseudodifferential operators. Princeton, Princeton University Press, 1981.

[254] Treves F. Introduction to pseudodifferential and Fourier integral operators. Vol. 1. Pseudodifferential operators. New York, Plenum Press, 1982.

[255] Treves F. Introduction to pseudodifferential and Fourier integral operators. Vol. 2. Fourier integral operators. New York, Plenum Press, 1982.

[256] Triebel H. Theory of function spaces. Basel, Birkhäuser, 1983.

[257] Triebel H. Theory of function spaces. II. Basel, Birkhäser, 1992.

[258] Triebel H. Interpolation theory, function spaces, differential operators [2nd ed]. Heidelberg, Johann Ambrosius Barth, 1995.

[259] Triebel H. The structure of functions. Basel, Birkhäser, 2001.

[260] Triebel H. Theory of function spaces. III. Basel, Birkhäser, 2006.

[261] Ul'yanov PL. Weyl factors for unconditional convergence. Mat Sb 1963, 60, No. 1, 39–62 [in Russian].

[262] Ul'yanov PL. Solved and unsolved problems in the theory of trigonometric and orthogonal series. Russ Math Surveys 1964, 19, No. 1, 1–62.

[263] Vainberg BR, Grushin VV. Uniformly nonelliptic problems, I. Math Sb (USSR) 1967, 1, 543–68.

[264] Vainberg BR, Grushin VV. Uniformly nonelliptic problems, II. Math Sb (USSR) 1967, 2, 111–33.

[265] Vishik MI, Éskin GI. Mixed boundary-value problems for elliptic systems of differential equations. Proc Tbilisi Inst Prikl Mat 1969, 2, 31–48 [in Russian].

[266] Vladimirov VS. Generalized functions in mathematical physics. Moscow, Nauka, 1979 [in Russian].

[267] Volevich LR. On the theory of boundary-value problems for general elliptic systems. Dokl Akad Nauk SSSR 1963, 148, No. 3, 489–92 [in Russian].

[268] Volevich LR. Solvability of boundary-value problems for general elliptic systems. Amer Math Soc Transl, Ser II 1968, 67, 182–225.

[269] Volevich LR, Paneah BP. Certain spaces of generalized functions and embedding theorems. Russian Math Surveys 1965, 20, 1–73.

[270] Wloka JT, Rowley B, Lawruk B. Boundary value problems for elliptic systems. Cambridge, Cambridge Univ Press, 1995.

[271] Zinchenko TN. Elliptic systems in the extended Sobolev scale. Dop Nats Akad Nauk Ukr Mat Pryrodoznav Tekhn Nauki 2013, No. 3, 14–20 [in Russian].

[272] Zinchenko TN, Murach AA. Douglis–Nirenberg elliptic systems in Hörmander spaces. Ukr Math J 2013, 64, No. 11, 1672–87.

Index

Bei Fragen zur Produktsicherheit wenden Sie sich bitte an:
If you have any questions regarding product safety,
please contact:

Walter de Gruyter GmbH
Genthiner Straße 13
10785 Berlin
productsafety@degruyterbrill.com